FUNDAMENTALS
OF THE
THEORY OF
MOVEMENT PERCEPTION

BY

DR. ERNST MACH

Prof. Ernst Mach late in his career in his office in Vienna. His hand rests on a volume of Enriques' Problems of science, in Italian (Enriques 1906). Mach's library covered a very wide range of topics and included many of the writings by and about Charles Darwin. Books in English and French were well represented along with his holdings in German.

FUNDAMENTALS

OF THE
THEORY OF

MOVEMENT PERCEPTION

BY
DR. ERNST MACH

Translated and Annotated

by

Laurence R. Young

Massachusetts Institute of Technology
Cambridge, Massachusetts

Volker Henn

University of Zürich
Zürich, Switzerland

Hansjörg Scherberger

California Institute of Technology
Pasadena, California

PACIFIC UNIVERSITY LIBRARY
FOREST GROVE, OREGON

Kluwer Academic / Plenum Publishers
New York, Boston, Dordrecht, London, Moscow

This volume includes a CD-ROM, the contents of which are noted on page 129.

ISBN 0-306-46711-9

© 2001 Kluwer Academic / Plenum Publishers, New York
233 Spring Street, New York, N.Y. 10013

http://www.wkap.nl/

10 9 8 7 6 5 4 3 2 1

A C.I.P. record for this book is available from the Library of Congress

Printed in the United States of America

Foreword to the Translated Edition

The value of this publication of Ernst Mach's work of 1875 on movement perception lies, first of all, in its scientific content and its useful applications to this day, even for space research, which would have delighted Mach. But the book also illuminates the kind of scientist-philosopher Mach was at that stage, and was to become in his later years.

During his doctoral work at the university in Vienna in the late 1850s, Mach had studied physics, mathematics, and philosophy, and then published on a wide range of primarily experimental subjects, from capillary phenomena to changes in musical pitch in coordinate systems in relative motion. His search for a specific field into which to throw his enormous energy led him to problems that exhibited a combination of physics, physiology, psychology of sensations, and psychophysics. It was a first indicator of his famous search for an Einheitswissenschaft, which in the words of the philosopher Max Schlick (1926), "arose from the wish to find a principal point of view to which he could cling in any research, one which he would not have to change when going from the field of physics to that of physiology or psychology. Such a firm point of view he reached by going back to what is given before all scientific research, namely, the world of sensations."

By 1860 he had become attracted to Gustav Fechner's pioneering ideas in *Elements of Psychophysics*, and for a while was also influenced by the work of the father of German physiology, Johannes Müller. Although Mach's innate skepticism and his own imaginative drive eventually caused him to question these authorities, he had become what Mach's biographer, John T. Blackmore, perceptively calls "an ontological phenomenalist who identified the external world with sensory impressions," and for that reason was more alert than most scientists to sensory novelties or peculiarities to which others pay only passing attention. Thus Mach suggested later in his *Popular Scientific Lectures* (1895, p. 272) that an important incentive to pursue his subsequent work on the effect of motion on the human body was the experience of accidentally observing a striking apparent inclination of the houses and trees as seen from a speeding railway carriage going around a curve.

Mach's search for an area in which he could excel in research on sense perception was, in his own opinion, also determined to some degree by the paucity of good experimental equipment available to him at the time. Thrown back largely on his own devices, he concentrated on his extraordinary mastery of observation. As he put it later, "Here, where I could observe my sensations, and against their environmental circumstances, I attained, as I believe, a natural Weltanschauung, freed from speculative,

metaphysical ingredients" (Scientia 1910, p. 234). An example of his power of perception was one of his first observations to become famous, the discovery of the so-called Mach bands, a change in the perceived brightness under circumstances for which to this day there is still discussion concerning its physical basis—but which could well have been observed by others long before, if they had been alert and sensitive enough.

Soon after moving to his professorship at the University of Prague, he launched on a whole spectrum of research activities on topics directly accessible to the human senses—retinal stimuli, stereoscopy, auditory perception, optical experiments on interference and spectra, wave motion, and the use of photographic devices to study the propagation of sound waves by imaging the change of density of the medium through which they move, from which arose the famous "Mach number" and "Mach angle."

It is in this context of Mach's sensation-based research that we should read the first paragraph in his introduction to the work at hand, beginning with "There are unmistakable characteristic sensations which accompany active or passive movements of our body...." Indeed, the whole work is a hymn to the impression the human physiology registers under various, even extreme, changes of position or motion. While Mach builds on the individual contributions of such predecessors as Purkinje, Flourens, Goltz, and others, he aims, as he says in his Foreword, to produce "a complete overview of a chapter of physiology"—although also attempting to give a physical basis as far as possible (for example, the importance of angular acceleration rather than angular velocity in the sensation during the experience of rotation). Superficially the book may seem not organized along traditional lines, but one must keep in mind that he saw the aim of doing science itself as giving a "compendious representation of the actual." Thus, we encounter constantly new, simple, original devices and procedures throughout, and even the effects on animals are not forgotten. Reports on direct sensations are foremost, although most are qualitative rather than quantitative. Mathematics (which appears only briefly and is of the simple kind) and hypotheses are secondary. Thus one of the important conclusions in this book (p. 69) is stated quite straightforwardly: "It has been asserted and also disputed that a special muscle sense exists.... We therefore accept the muscle sense as a fact; something that can be observed, without bothering to explain it."

In many respects, these characteristics of his 1875 book point forward to his masterwork of 1883, known in its English translation as *The Science of Mechanics: A Critical and Historical Account of Its Development.* That book was on the reading list of practically every alert scientist for decades, including Albert Einstein, who noted in his autobiographical essay of 1949 that it was this book of Mach's which, as a youth, shook him out of the "dogmatic faith" of the previous approach to physics (Einstein, in Schilpp 1949, p. 21). Indeed, in one of his letters to Ernst Mach (17 August 1909),

Einstein signed off with the phrase that he was Mach's "verehrender Schüler" (a student who venerates or reveres Mach).

Mach became world-renowned more through that book than through anything he did before or after. As his vast correspondence and the references to him in the literature of scientists prove, he amassed disciples for his science and his philosophy (even though he disavowed being a philosopher) throughout the world. These included such important figures in varied fields as Jacques Loeb and William James during his own lifetime, and Philipp Frank, P. W. Bridgman, and B. F. Skinner afterward. After the founding of the Nobel Prizes, Ernst Mach was repeatedly nominated in letters and petitions from distinguished scientists such as H. A. Lorentz, Ferdinand Braun, and Wilhelm Ostwald These nominations, even if unsuccessful in the end, do indicate Ernst Mach's standing among scientists at the time.

And yet, after the explosion of the new physics around the turn of the century, with one spectacular discovery or theory after another, Mach was left in the position of an Aussenseiter (to use the word in the title of the book edited by J. T. Blackmore and Klaus Hentschel, *Ernst Mach als Aussenseiter* [Vienna, 1985]), a person at the margin. His career, grounded in the splendid exploitation of sensations and the derogation of "hypotheses," together with his self-confessed poor understanding of higher mathematics, at that point did put him then at a position outside the center of the new science. His greatest personal assets as an experimenter had become internalized by the community, and in a sense were now taken for granted.

To me, there is a certain poetic injustice about this reversal of fortunes, which the publication of this book may rectify to some degree. For what this work may lack in the kind of sophisticated mathematical-theoretical prowess which we associate with many of the giants of twentieth-century physics, it makes up by the careful accounts of the sophisticated experimentation involving the sensations of the human body. My own view is that this fact points to a particular source of "scientific intelligence" of which Ernst Mach was a primary exemplar, one which in fact is shared with other scientists before and since, but which curiously has practically been kept secret.

To explain what this secret is, I turn to a historical example. In 1902–4, the French mathematician Jacques Hadamard published a lengthy questionnaire (reprinted as Appendix I of his book *The Psychology of Invention in the Mathematical Field* [Princeton University Press, 1945]), under the title "In Inquiry into the Working Methods of Mathematicians." This he submitted to his fellow scientists with a view to eliciting a typology of different aptitudes or, as we might now call them, intelligences, that helped them in their work. He included even such questions as whether they have ever worked in their sleep, or have found the answers to problems in

dreams, or whether artistic and literary occupations hindered or helped in mathematical invention.

At the very end, the thirtieth question ran as follows: "It would be very helpful for the purpose of psychological investigation to know what internal or mental images, what kind of 'internal word' mathematicians make use of; whether they are motor, auditory, visual, or mixed, depending on the subject which they are studying." Eventually, about 1944, Hadamard submitted this particular question also to Mach's erstwhile "student," Albert Einstein, and the resulting "Testimonial from Professor Einstein" is published as Appendix II in Hadamard's book.

The two-page "testimonial" is well known among historians of science. In essence, the reply emphasizes that "the words or the language, as they are written or spoken, do not seem to play any role in my mechanism of thought. The psychical entities which seem to serve as elements in thought are certain signs and more or less clear images which can be 'voluntarily' reproduced and combined." He continues to explain that before he can arrive at the scientifically valuable "logically connected concepts" he engages in a "rather vague play with the above-mentioned elements" and then adds what is to us the most revealing point: *"The above-mentioned elements are, in my case, of visual and some of muscular type"* [emphasis added]. He also answered to the question about his "logical type" as being "visual and motor." One part of this remarkable insight is fortified by a report from the psychologist and long-term friend of Einstein, Max Wertheimer, who reported that from 1916 on, he had questioned Einstein "in great detail about the concrete events in his thoughts." Einstein replied: "These thoughts did not come in any verbal formulation. I very rarely think in words at all. A thought comes, and I might try to express it in words afterwards." And Einstein added, "I have it in a kind of survey, in a way visually" (Wertheimer 1945, p. 184). It seems therefore that in addition to his other skills and intelligences, Einstein could, as it were, "feel" his way into the problem situation—not only, in the significant metaphor he often used, with a Fingerspitzengefühl—that helps to know whether one is on the right track in a complex research project--but in some more actually tangible way. I have occasionally heard from scientists about that same ability to feel in their bodies, in their muscles, the trajectory of an object under research. But of course this skill is never mentioned in textbooks or research papers, and is perhaps so rare among researchers as to remain slightly embarrassing to speak about. That ability seems to me also connected with Michael Polanyi's concept of the scientist's use of "tacit knowing," as explained in his book *The Tacit Dimension* (1967).

Having read and heard of these evidences of the participation of some internal motor skills or kinesthetic intelligence during science research, and not having any sense of it myself, I have often wondered how such a concept

could be demonstrated operationally. One insight came to me in the 1960s. I happened to be at a conference on the general topic of scientific creativity, where I encountered the psychologist and psychoanalyst George Klein. I asked him if he could throw any light on the matter, and his response was remarkable. He told me that, as it happened, he had persuaded Einstein in his late years to submit to a Rorschach test. As Dr. Klein described it, Einstein sat down before the opened book of Rorschach test images, staring at one of them for a long time; then, instead of speaking about his perception, Einstein slowly and silently rose from his seat, stretched out his arms, and waved them, like a giant bird. This stunning report gave me an inkling of what might be meant by having kinesthetic perception and allowing it to participate in a scientific query.

But to indicate that Einstein's (and Mach's) kinesthetic sensibilities are by no means unique, I draw on the clear description of the same guiding ability, provided by the great physicist and science statesman, I. I. Rabi, a Nobelist whose research on molecular beams and the magnetic resonance method was instrumental in bringing physics in the United States up to world-class level during the 1930s and 1940s. As his biographer John S. Rigden put it (in his book, *Rabi: Scientist and Citizen*, New York: Basic Books, Inc., 1987):

"Rabi relied strongly on his intuition, which allowed him, in a manner of speaking, to put himself into the beam and, along with the other beam particles, experience the sudden jolts and subtle nudges as he streamed through the apparatus. As Polykarp Kusch has said, "He [Rabi] appears to ride around on the electrons within an atom or asks the question, 'If I were an electron, what would I do?' Possibly, through sheer force of character he gets the electron to do precisely that."

"So Rabi rode the sodium atom, first by clinging to an electron, then by sitting on its nucleus. He could feel the beam split decisively into two beamlets by the force of a strong magnetic field acting directly on the large magnetic moment of the sodium atom's outermost electron." (p. 86)

Rigden then quotes Rabi directly from an interview concerning his crucial idea, in 1936, how to determine experimentally the sign of the magnetic moment of nuclei:

"One day I was walking up the hill on Claremont Avenue and I was thinking about it [the sign of the nuclear magnetic moment] as kinesthetically with my body. Now, yes, I was thinking about this as follows: Here's the moment and it's wobbling around in the direction of the field and [to find] the sign was to find out in which sense it was wobbling. To do this, I have to add another field which goes with it or against it. This is the idea, just concretely. The whole resonance method goes back to this." (p. 94)

To return to Ernst Mach's work at hand, on the subject of movement perception, I believe I can now understand a little better the kind of uncommon sensitivity to the phenomena Ernst Mach was investigating, often with his own body. So as we read in this work, let us keep his figure before our eyes, as he is standing on the rotating platform or sitting in the tilted chair mounted on a rotating frame, in every way participating in the events he describes as both subject and object, experiencing within himself kinesthetically this remarkable series of experiments.

by Gerald Holton
Jefferson Physical Laboratory
Harvard University
Cambridge, MA 02138

Schlick, M. (1926) Ernst Mach. Neue Freie Presse (Suppl.), Vienna. June 12. pp. 10–13.

Mach, E. (1895) *Popular Scientific Lectures*. Translated by: McCormick, T. J. Chicago: The Open Court Publishing Co.

Mach, E. (1910) Die Leitgedanken meiner naturwissenschaftlichen Erkenntnislehre und ihre Aufnahme durch die Zeitgenossen. *Scientia* 7:225-240.

Mach, E. (1883) *Die Mechanik in ihrer Entwicklung historisch-kritisch dargestellt*. Leipzig: Brockhaus. [First English translation: *The Science of Mechanics. A critical and historical exposition of its principals*. Translated by: McCormick, T. J. Chicago: The Open Court Publishing Co. (1893)].

Einstein, A. (1949) Autobiographical Notes. In: Schilpp PA (ed.) *Albert Einstein: Philosopher-Scientist*. Evanston: Library of Living Philosophers.

Einstein, A. (1916) Ernst Mach. Physikalische Zeitschrift 17:101-104.

Blackmore, J. T, Hentschel, K. (1985) *Ernst Mach als Aussenseiter*. Vienna: Wilhelm Braunmüller.

Hadamard, J. (1945) *The Psychology of Invention in the Mathematical Field*. Princeton: Princeton University Press.

Wertheimer, M. (1945) *Productive Thinking*. New York: Harper.

Polanyi, M. (1967) *The Tacit Dimension*. Garden City, NY: Doubleday.

Rigden, J. S. (1987) *Rabi: Scientist and Citizen*. New York: Basic Books, Inc.

Preface to the Translated Edition

At the dawn of the 21st century, why would anyone translate a 19th century work on motion sensation by a physicist best known for his research on fluid mechanics? The answer is simply that Ernst Mach's experiments and insight into sensory mechanics and spatial orientation, though critical and amazingly clever, are largely ignored or unknown by the community developing models of multi-sensory interaction. They are as relevant now as they were in 1875, and they serve to remind physiologists of the importance of dynamics and of the need for clear unambiguous experiments in the elucidation of the possible sources of motion sensation. The work also holds an important place in the history of science, as it represents the introduction of quantitative models and theory into the descriptive realm of 19th-century experimental psychology.

Our own introduction to this material took place in 1972, when I was a visiting professor in Günter Baumgartner's neurology department in Zurich, helping Volker Henn to set up the monkey laboratory. Henn had just returned from New York where he worked with Bernard Cohen on single-unit recordings underlying eye movements in alert monkeys. We were interested in combining animal and human experiments relating visual and vestibular stimuli to eye movements and motion perception. When I showed Henn the draft of my chapter for Mountcastle's physiology text, it contained the usual historical reference to the erroneous 19th-century notion that the semicircular canals were stimulated by continuous flow of endolymph, and I blamed Mach, Breuer, and Crum Brown collectively for the misconception. Henn assured me, as usual with a twinkle in his eye, that Mach could not possibly have made such an error. Indeed, a careful reading of Mach's contributions to the Vienna Academy of Science showed that he indeed had it right. Mach showed that for a spinning subject the friction forces in the canal would bring the endolymph to a stop shortly after the canals reached constant angular velocity. Therefore, the decaying sensation of turning could not result from endolymph flow, but must instead reflect a detection of pressure within the organ.

Mach's 1875 book was never published in English translation. Henn was convinced of the importance of this work for contemporary researchers. Together we wrote a short summary of some of Mach's contributions in 1975 and Henn elaborated this work in 1984. Henn later persuaded me to work with him on an annotated translation of the entire book including hypertext capabilities. Volker Henn[1] died unexpectedly in December 1997, after most of this translation and its notes were completed. Fortunately, Hansjörg Scherberger, one of Henn's last postdoctoral students, stepped in to complete this work with me.

Although earlier research on vestibular physiology had established some of the important facts concerning the function of the labyrinth in spatial orientation, no general theory relating motion to its sensation existed at the time of Mach's writings. There was not even agreement about the location of the principal motion sensors. Flourens had described the results of lesions of the labyrinth on the head movements of pigeons, but provided no theory to explain the results. Goltz had surmised that the ampullae in the semicircular canal were responsible for detecting motion. Purkinje, meanwhile, had made important observations concerning vertigo and the dynamic response of motion sensation during and following prolonged turning. He correctly concluded that the organ of motion sensation was in the head, but erroneously assigned this function to a movement of part of the brain. In an allied field, Helmholtz had published his momentous work on physiological optics, which Mach greatly admired. But nothing existed in the way of physical theories for motion sensation when Mach's interest was stimulated. His research began with the fortuitous observation that, while riding around a curve on a train, the houses and trees appeared to be tilted. He brought the traditional rigorous approach of physics to the empirical and descriptive field of sensory physiology.

The application of his reductionist approach to science is clear in this work. He begins by identifying the functional requirements that sensors must serve. In this case, Mach requires sensors that detect angular and linear acceleration, rather than velocity, for all directions in space. He next argues why these sensors must be located in the head, since the axes of motion sensation move with the head. He further assembles the evidence in favor of the semicircular canals and the otolith organs serving this function, and shows how these sensors can meet the physical requirements. Finally, he presents a testable general theory of motion perception. He does not claim that the theory is definitive, but expresses the hope "that the results of my work will remain valuable, even if one does not make the connection to the hypothesis which I developed in the last chapter about the organ of motion perception."

One of the delights of this book is the description of the simple but elegant experiments that Mach used to eliminate alternative sources of motion sensation, such as pressure on the feet, redistribution of blood, or force on the head. He sought simple, parsimonious explanations for complex phenomena. In his pursuit of a theory that would encompass all of the various observations, he eschewed any explanation that required the introduction of non-observable entities. In a larger sense, this view is consistent with Mach's controversial philosophy of science, relying on observed facts rather than hypothetical "laws."

In this book Mach uses a somewhat rambling and narrative style that would be quite unacceptable according to current scientific standards. He

combines his thought process with the weight of evidence to support his conclusions about the role of the semicircular canals and the otoliths in spatial orientation and motion sensation. In places his presentation seems almost a stream of consciousness, as he skips from one topic to another. Nevertheless he constantly seeks to remind the reader of the main point – the development of a theory of motion sensation based only upon observed input-output relationships and consistent with the laws of mechanics. For that purpose he also tries to teach some of the basic principles of mechanics to the physiologists of his day. In doing so he includes a number of illustrative examples to make his point, from Foucault pendulums to steam engines. Additionally, he offers a number of gratuitous comments about proper scientific methods and the shortcomings of some of the contemporary physiologists.

Mach was remarkable for the contributions he made in so many fields. In fluid mechanics his use of Schlieren photography to study shock waves and supersonic motion was recognized by attaching his name to the unit for the speed of sound in air. His work on visual psychophysics included the discovery of contrast enhancement at boundaries, now referred to as "Mach bands." Consistent with his positivism, Mach strenuously objected to the assumption of atoms as a necessary part of modern physics and disagreed with the theory of relativity. Albert Einstein, in his obituary in 1916, acknowledged Mach's great influence.

This translation is presented in a dual language format. The German text is preserved in order to provide direct access to the original font and layout. Brackets always delineate words or corrections added by the translators. This approach allowed us to deviate from the literal translation in favor of clarity by expressing Mach's ideas as he might have written them in English. Our numerous footnotes are intended both to explain Mach's text, including references to concepts and devices no longer known, and to put his ideas in the context of contemporary motion sensation research. Longer articles and related material are included only in the accompanying CD-ROM[2], which also contains the full text and links to all footnotes.

In the preparation of this translation we were assisted greatly by the members of the "Zurich Lab" that Henn headed at the University of Zurich, especially Caroline Saruhan, Markus Kotterer, Bernhard Hess, Domink Straumann, and Klaus Hess, Head of the Neurology Department. We are also grateful to Bruce Bridgeman, Lawrence Stark, Bernard Cohen, and Klaus Hepp for additional comments on the manuscript. The project was supported by the Betty and David Koetser Foundation for Brain Research, Zurich, the Apollo Program Chair of Astronautics at the Massachusetts Institute of Technology, and the National Space Biomedical Research Institute. Our historical research and access to Mach's papers and notebooks was made possible by the Ernst Mach Archive in Freiburg and the archive of

the Deutsches Museum in Munich—in particular through the assistance of Margrit Prussat and Wilhelm Füssl—and the Burndy Library of the Dibner Institute for the History of Science and Technology at MIT. Marsha Warren assisted with copyediting and indexing the manuscript. The entire project was made possible by the guidance and encouragement of Michael Hennelly, formerly of Kluwer Academic/Plenum Publishers. Kathleen P. Lyons and Beth Kuhne of the publisher made valuable editing contributions.

Gerald Holton's learned foreword to this translation brings the story up to modern times, especially regarding Mach's influence on the next generation of physicists and his use of body introspection.

Laurence R.Young   Cambridge, Massachusetts

Henn, V., Young, L. R. (1975). *Ernst Mach on the vestibular organ 100 years ago.* Otolaryngology. 37:138–148.
Henn, V. (1984). *E. Mach on the analysis of motion.* Human Neurobiol 3:145–148.

Contents of the Translated Edition

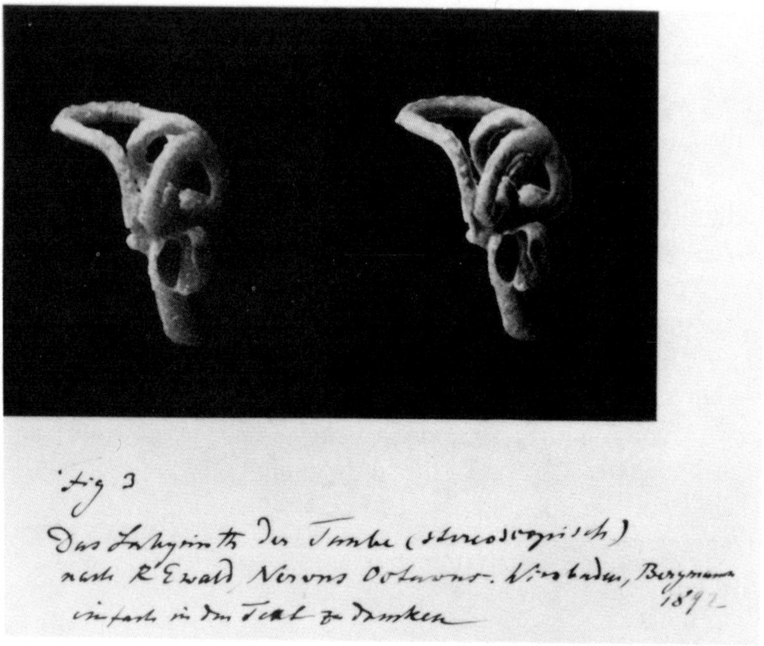

Figure A-2: Ewald's stereoscopic image of the labyrinth, owned by Mach (Figure 3 of Ewald 1892). "The stereoscopic labyrinth of the pigeon by R. Ewald, Eighth Nerve. Wiesbaden, Bergmann1892. Simply for insertion in the text". (photograph courtesy of E. Mach Archive, Deutsches Museum, Munich)

Table of Additional Figures

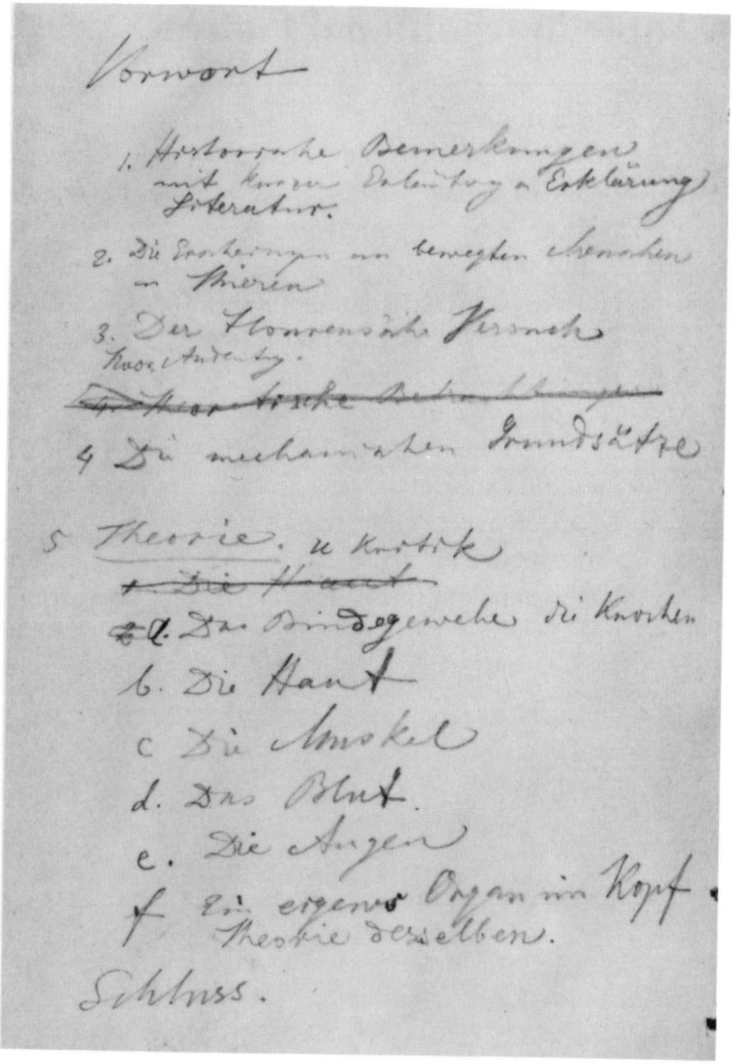

Figure A-3: Outline of the foreword including list of possible alternative gravity sensors. "Foreword. 1. Historical remarks with short introduction and explanation of the literature. 2. Discussion of moving men and animals. 3. Flourens' experiment. Theory, Application. 4. The mechanical bases. 5. <u>Theory</u> and discussion. 1.The skin 2a. The connective tissue, the bones. b. The skin. c. The muscles. d. The blood. e. The eyes. f. A specific organ in the head. Theory of it. Conclusions." (from Mach's notebook of 1874; photograph courtesy of E. Mach Archive, Deutsches Museum, Munich)

Chapter Summaries

Introduction

(Pages 1–5) Mach briefly reviews the modest scientific literature on motion sensation. He states that, since motion can be clearly defined in physical terms, any treatment of it must start by defining physical parameters. He laments the lack of any coherent theory. Even Flourens, whose surgical destruction of single semicircular canals regularly led to violent motion disturbances in the pigeon, refrained from interpreting these results. Mach recalls how he first became interested in movement perception when he went around a curve in a railway car. He finally states that the principal facts on which he based his theory were known since 1824 (the experiments by Purkinje [Purkyně] and Flourens).

He attributes the lack of theory to an unwillingness to consider the motion sensation functional requirements independent of the possible mechanisms for its biological realization. He was aware of the fragmentary status of anatomical and physiological knowledge, and concludes that the physical treatment, i.e., the functional or logical description (in modern terms a model), will be unaffected by possible novel interpretations of the biological phenomena.

The Mechanical Foundations

(Pages 6–22) The basic principles of Newtonian mechanics are summarized. Mach introduces the principle of the conservation of areas, which is an antiquated way to explain the conservation of angular momentum. He goes on to discuss several examples of engineering apparatus to show that any movement in one direction will evoke a movement in an opposite direction, so that the center of gravity remains stationary. Spinning tops, centrifuges, electric motors, the Foucault pendulum, falling weights, and sailors running on a ship are all presented to illustrate the conservation of momentum. The chapter closes with speculation about the site of motion sensation.

The Phenomena Observed in Moving Men and Animals

(Pages 22–40) Mach explains that during motion only accelerations are sensed. He recounts the observation that houses and trees seem to lean when viewed from a train rounding a narrow curve. To conduct his research, he

constructed a hand-driven three-axis human turntable. In a first series of experiments confined to angular motion, psychophysical responses are described for nine different studies including visual-vestibular interaction (pp. 25–28). A second series demonstrates sensations associated with the Coriolis effect (pp. 29–30). A third series shows the effects of linear acceleration (pp. 31–36), and measures thresholds. He discusses the relationship between vertical motion sensing and motion sickness. Further experiments combine angular and linear motion. He then turns to related animal experiments, rotating pigeons and rabbits, to observe postural disturbances. A short discussion follows that compares active to passive motion.

Flourens' Experiment

(Page 41–50) Flourens is extensively quoted concerning the experiments in which he lesioned single semicircular canals in animals and observed intense head instability in the planes of the affected canals. Mach wonders about the lack of theory. He refers to Goltz's interpretation of these experiments involving endolymph movement and pressure variation. Criticism of Flourens' experiments and of Goltz's interpretation, in particular those of Böttcher, are reviewed.

Phenomena Reminiscent of Flourens' Experiment

(Pages 51–54) The few human cases in the clinical literature reminiscent of Flourens' experiments are reviewed. Galvanic stimulation in human subjects and related electrical stimulation of fish are described.

Comparison of Movement Sensations with Other Sensations

(Pages 54–65) Mach describes experiments by Plateau to demonstrate that sensory organs seem to obey two common principles: (1) Aftereffects appear that are analogous to positive and negative after-images in the visual system and (2) adaptation occurs during continuing constant stimulation. Concerning visual experiments, several examples of optical illusions are presented to show that neither eye movements nor pupil variations are responsible for these phenomena.

Further Investigations of the Phenomena

(Pages 65–96) A series of further experiments is described to show that the sensory organ for motion detection most likely resides in the head. Mach systematically eliminates other sources of movement detection including connective tissue and bone, skin, muscles, blood, eyes, and the brain. An apparatus is constructed to investigate whether or not pressure under the feet plays a role. To investigate muscle load, a system of levers is fixed to the head with suspended buckets of water, which can be emptied quickly. To eliminate the influence of the spatial distribution of pressure, Mach placed subjects on a board above a kind of bathtub that can be raised so that the subject becomes buoyant in water without ever being moved. Visual vertigo and Breuer's extensions of Purkyně's experiments are reviewed. Mach describes visual-vestibular interaction including vection. He discusses the role that induced eye movements might play and concludes that nystagmus is a reflex elicited by acceleration. To prove the point he devises experiments in which eye movements can be dissociated from motion sensation. As all the experiments point toward the head as the seat for the organ to detect motion, Mach reviews the literature concerning possible explanations for these sensations. While in agreement with Purkyně that motion sensation takes place in the head, Mach disputes his view that the mechanical movement of the brain is the source of this sensation. Finally, he describes rotation experiments with the head partly or completely stationary in space, which support his claim that a sensory organ for motion sensation resides in the head.

Theory of the Phenomena

(Pages 97–124) The communications of Breuer and of Crum Brown, published at the same time and based on independent observations, are quoted extensively. Mach takes their findings and views them as support for his own theory. He concludes that the most likely organ that could detect angular and linear motion is the labyrinth, and discusses possible mechanisms. He performed experiments using an analog model of the canals to study the movement induced by angular acceleration, and demonstrated that the flow of endolymph was inadequate to explain the time course of motion sensation. He concludes that the adequate stimulus most likely is a pressure gradient across the cupula, which minimizes actual endolymph flow. Comparisons with other sensory organs show that the energy delivered by such a pressure gradient should be sufficient to induce a sensation for the accelerations associated with normal head movements. Mach then explains

how the combined activity from the afferent nerves leads to a common perception of motion rather then a mosaic of separate sensations from each canal.

Conclusions

(Pages 124–126) Mach summarizes his theoretical and experimental results in eleven statements. The first ten relate to the dynamic nature of the sensation, the characteristics of the adequate stimulus, the location of the sense organ in the head, and the elimination of other possible origins of movement sensation. The eleventh statement postulates that the six oppositely paired semicircular canals constitute the organ of movement sensation. He emphasizes that the first ten statements are independent of the last, and will remain valid, whether or not later generations of physiologists can support his hypothesis that the labyrinth is the sensory organ of motion detection. Finally, to support his eleventh statement, he suggests further electrical stimulation and lesion experiments, which he wishes to leave to others with more appropriate knowledge and skills.

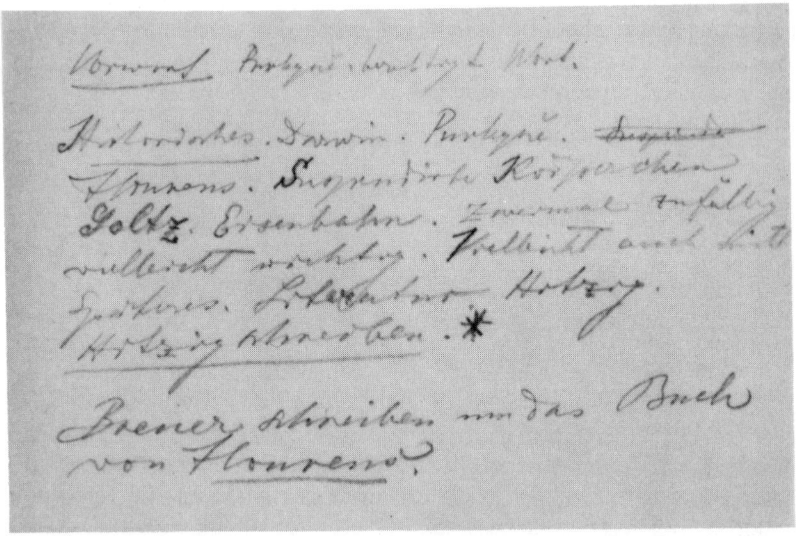

Figure A-4: Notes on references from Mach's Notebook of 1874: "Foreword – Purkyně. Historical: Darwin, Purkyně, Flourens, suspended particles, Goltz, Railroad, Doubly random might be important. Maybe with ... write to Hitzig*. Write to Breuer, asking for Flourens' book." (photograph courtesy of E. Mach Archive, Deutsches Museum, Munich)

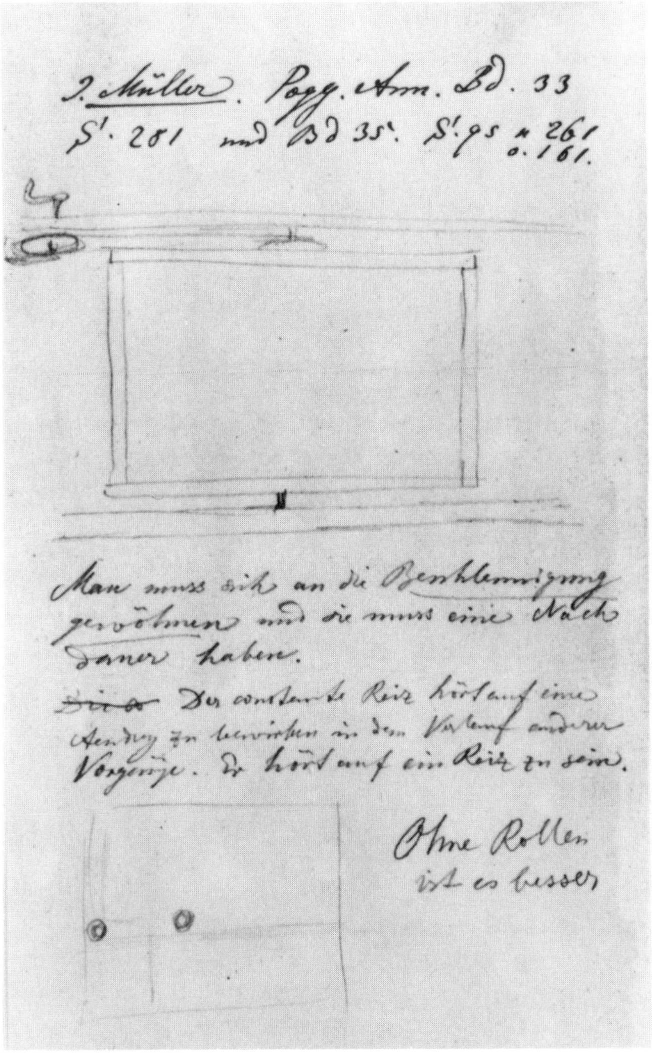

Figure A-5: Early sketch of the rotating frame, showing the cranking mechanism, not necessarily included in the final design (see Fig. 7, p.24). "J. Müller. Pogg. Ann. Vol. 33 p. 281 and Vol. 35, p. 95 and 261or 161. One must adapt to the <u>acceleration</u> and it must have an after-effect. The constant stimulus ceases to cause a change in the course of other processes. It ceases to be a stimulus. Better without rollers." (from Mach's notebook of 1874; photograph courtesy of E. Mach Archive, Deutsches Museum, Munich)

FUNDAMENTALS

OF THE
THEORY OF

MOVEMENT PERCEPTION

BY

DR. E. MACH

FUNDAMENTALS

OF THE

THEORY OF

MOVEMENT PERCEPTION

BY

D[r.] E. MACH.[0.1]

PROFESSOR OF PHYSICS AT PRAGUE UNIVERSITY.

WITH 18 WOODCUTS.

LEIPZIG

Verlag von Wilhelm Engelmann

1875

GRUNDLINIEN

DER

LEHRE VON DEN

BEWEGUNGSEMPFINDUNGEN

VON

Dr. E. MACH.

PROFESSOR DER PHYSIK AN DER UNIVERSITAT PRAG.

MIT 18 HOLZSCHNITTEN.

LEIPZIG

Verlag von Wilhelm Engelmann

1875

Contents.

Inhalt.

Foreword.

This publication presents for the first time a complete overview of a chapter of physiology, built upon the groundwork laid by Purkinje[0.2] (Purkyně), Flourens[0.3] and Goltz.[0.4] Both in my experiments as well as in my writing I especially stressed the physics approach, which can easily be justified. As the most developed of the natural sciences, it also should be the first aspect to be dealt with. Not only is this aspect the one which is nearest to my own specialty, but it is also the view which has been least investigated thus far.

That an investigation concerning the mechanical processes which lead to the sensation of motion could not be ignored much longer, is shown by the fact that soon after my first communication on this subject, Breuer[0.5] and Brown[0.6] published their results, not to ignore countless other papers with similar contents.

I hope that the results of my work will remain valuable, even if one does not make the connection to the hypothesis which I

Vorwort.

Vorliegende Schrift versucht zum erstenmal die vollständigere Darstellung eines Capitels der Physiologie, zu dem unstreitig Purkinje (Purkyně), Flourens und Goltz den ersten Grund gelegt haben. Sowohl bei meinen Versuchen wie bei meiner Darstellung habe ich ein besonderes Gewicht auf die physikalische Seite der Sache gelegt, was sich leicht rechtfertigen lässt. Nicht nur hat diese Seite als der meist entwickelten Naturwissenschaft angehörig Anspruch darauf, zuerst erledigt zu werden, nicht nur lag sie mir fachlich näher wie jede andere, sondern sie war auch die bisher am wenigsten bearbeitete.

Dass die Untersuchung der mechanischen Vorgänge, durch welche die Bewegungsempfindungen erregt werden, nicht mehr lange hätte ausbleiben können, zeigen die bald nach meiner ersten Mittheilung über diesen Gegenstand erschienenen Arbeiten von Breuer und Brown, der andern zahlreichen Abhandlungen verwandten Inhaltes gar nicht zu gedenken.

Man wird, wie ich hoffe, finden, dass die Ergebnisse meiner Arbeit ihren Werth behalten, auch wenn man sich

develop in the last chapter about the organ of motion perception. In considering the many facts which all clearly lead to this hypothesis, and despite the arguments presented by <u>Boettcher</u>[0.7], I believed that I had to present my theory.

Prague, August 1874.

The Author

nicht der am Schlusse entwickelten Hypothese über das Organ der Bewegungsempfindungen anschliesst. Angesichts der vielen Thatsachen aber, welche alle zu dieser Hypothese drängen, habe ich, ungeachtet der von Boettcher vorgebrachten Einwendungen, geglaubt, dieselbe aufrecht halten zu müssen.

Prag, August 1874.

D.V.

Introduction.

1.

Unmistakable characteristic sensations accompany active or passive movements of our body. We use them to detect the direction and extent of such motions. Inside a ship's cabin, without help from vision, we detect the sway of the ship, its rotations as well as the beginning of linear translation. We know such motions almost as well as we could detect our own voluntary motions, although in this case we are moved passively. Even though these sensations are easily observable, only very few and unsystematic experiments have been done to elucidate the associated facts and their laws. The goal of this publication is to investigate the sources of such motion sensation and to study their dependence on the movements of the body.

2.

The elder Darwin[1.1] and Purkyně[1.2] studied the surprising subjective motion sensation phenomena that occur when one turns around several times and then abruptly stops and remains still. They also established some rules that apply to these phenomena of rotational vertigo. Of course, their attention also was first drawn to the unusual situation that produced these illusions. Within this context, Purkyně had the opportunity to repeat an experiment done by Ritter[1.3], in which a similar spinning sensation was created with the help of galvanic current.

Einleitung.

1.

Die characteristischen Empfindungen, welche die activen oder passiven Bewegungen unseres Körpers begleiten und die wir verwenden um den Sinn und das Ausmaass dieser Bewegungen zu erkennen, entgehen auch der oberflächlichen Beobachtung nicht. Wir wissen in der Cajüte des Schiffes eingeschlossen ohne Hilfe unseres Gesichtssinnes die Schwankungen des Schiffes, die Drehungen, so wie auch die beginnenden Progressivbewegungen fast so gut zu beurtheilen als ob es unsere eigenen willkührlichen Bewegungen wären, die wir hier gezwungen, passiv ausführen. So leicht nun auch die erwähnten Empfindungen der Beobachtung zugänglich sind, beziehn sich doch nur sehr wenige vereinzelte und unvollständige Versuche auf die Ermittlung der hieher gehörigen Thatsachen und ihrer Gesetze. Die Quellen dieser Bewegungsempfindungen aufzusuchen und ihre Abhängigkeit von den Bewegungen des Körpers zu finden, ist die Aufgabe dieser Schrift.

2.

Der ältere Darwin und Purkyně haben die sonderbaren subjectiven Bewegungserscheinungen studirt, welche auftreten, wenn man sich einigemal rasch umdreht und dann plötzlich stehn bleibt. Sie haben auch einige Regeln dieser Drehschwindelphänomene festgestellt. Natürlich haben auch hier die ungewöhnlichen, zu Täuschungen Anlass gebenden Erscheinungen zuerst die Aufmerksamkeit auf sich gezogen. Purkyně hat bei dieser Gelegenheit auch, einen Versuch Ritters wiederholend, ähnliche Schwindelerscheinungen mit Hilfe des galvanischen Stroms hervorgebracht.

Flourens[1.4] performed experiments of an apparently totally different sort, but which were probably closely related to those just mentioned. These are the famous experiments in which he severed the semicircular canals of the labyrinth of the ear in animals. Following surgery the animals displayed truly unusual rotational movements.

3.

I personally was led to consider motion sensation on the occasion of a series of fluid flow experiments with suspended particles, which one of my pupils performed on my suggestion *). One observes that the load on a scale depends on the state of acceleration of the body placed upon it. These experiments led me to consider that similar conditions might prevail for our body in the sense that each body part can detect the changing load of others depending on its motion. According to my earlier view it was the whole body that contributed to motion sensation. The idea of a special organ for the sensation of motion was very far from me at that time.

Goltz[1.5] provided a change in the interpretation of the Flourens' experiments after they had been repeated by several different scientists. Although Flourens always spoke of vertigo and disequilibrium when describing the phenomena, he never considered the semicircular canals[1.6] as organs of equilibrium. For him the labyrinth[1.7] is simply an organ for hearing and the consequence of his surgical intervention is a painful hearing disturbance.[1.8] According to Goltz, however, the semicircular canals constitute an organ of equilibrium for balancing the head.

A chance occurrence[1.9] led me back to the study of motion sensation. I observed how buildings and trees appeared tilted when riding in a train going around a curve. This could easily be explained if one postulated a direct sensation of the resultant mass acceleration. Although the

---

*) C. Bondy, On buoyancy in fluids in which particles of heavier or lighter weight specific weight are suspended. (Ger.) Proc. Vienna Acad. vol 51. – Mach, On fluids containing suspended particles. Pogg. Ann. vol. 126. p. 327.

2

Versuche von scheinbar ganz anderer Art, die aber wahrscheinlich mit den erwähnten im engsten Zusammenhange stehn, sind von Flourens angestellt worden. Es sind dies die berühmten Versuche mit der Durchschneidung der Bogengänge des Ohrlabyrinthes an Thieren, welche höchst eigenthümliche Drehbewegungen der operirten Thiere zur Folge haben.

3.

Ich selbst bin zum Nachdenken über die Bewegungsempfindungen bei Gelegenheit einer Versuchsreihe über Flüssigkeiten, welche suspendirte Körperchen enthalten, angeregt worden, die einer meiner Schüler auf meinen Vorschlag ausgeführt hat *). Die von dem Beschleunigungszustand der aufgelegten Körper abhängige Belastung der Wage, die bei diesen Versuchen auftritt, legt den Gedanken nahe, dass ähnliche Verhältnisse bei Bewegung unseres Körpers bestehn und dass jeder Körpertheil den andern als veränderliche von der Bewegung abhängige Last empfinden kann. Nach meiner damaligen Anschauung liefert der ganze Körper Beiträge zu den Bewegungsempfindungen. Die Annahme eines besondern Organs der Bewegungsempfindungen blieb mir damals fern.

Die Auffassung der Flourens'schen Versuche hat, nachdem dieselben von sehr verschiedenen Forschern wiederholt worden waren, durch Goltz eine neue Wendung genommen. Obgleich Flourens bei Beschreibung der Erscheinungen immer von Schwindel und Gleichgewichtsstörung spricht, hat er die Halbcirkelcanäle nie als Gleichgewichtsorgan betrachtet. Das Labyrinth ist ihm durchaus Gehörorgan und die Folge des Eingriffes eine schmerzhafte Gehörsstörung. Nach Goltz aber sind die Halbcirkelcanäle ein Gleichgewichtsorgan für die Kopfstellung.

Ein Zufall führte mich auf das Studium der Bewegungsempfindungen zurück. Ich beobachtete die Schiefstellung der Häuser und Bäume beim Durchfahren einer Eisenbahncurve. Sie liess sich leicht erklären, wenn man eine directe Empfindung der resultirenden

---

*) C. Bondy, über den Auftrieb in Flüssigkeiten, welche specifisch schwerere oder leichtere Körperchen suspendirt enthalten. Sitzgsber. d. Wien. Akademie. Bd. 51. – Mach, über Flüssigkeiten, welche suspendirte Körperchen enthalten. Pogg. Ann. Bd. 126. S. 327.

physiological side of the subject was quite foreign to me, even when I was confronted with it again, and furthermore I knew little of the works or even the names of F l o u r e n s and G o l t z, this fermenting clue proved sufficient to guide my thoughts in the direction which they actually followed. Through physiological-optical studies meanwhile I had gained a very uncomfortable feeling regarding "unconscious conclusions" and consequently the psychological assembly of a motion perception from very different elements. The balance organ was therefore greeted as a welcome piece of information.

4.

The viewpoint that I took, which seems to me to join all the currently known facts in a natural unifying theory, is simply that certain nerves of the labyrinth respond to every stimulus with a motion sensation according to their <u>specific energy</u>[1.10], which explains F l o u r e n s' phenomena. As a rule, this stimulus is transduced by the contents of the labyrinth itself, which responds to the animal's movement in accordance with the principles of the conservation of the center of gravity and the conservation of areas. In this way, animals achieve motion sensation, whether they move actively, or are passively moved.

That such a view is welcome and can also attract others is demonstrated by the fact that immediately after my first publication this theory in slightly different form was developed independently by two others, first by B r e u e r and later by B r o w n.

5.

The facts which lead to my view and which provide its foundation, were all known since 1824, half a century ago. If such a view had not been discussed so far, then this demonstrates that to a physiologist even simple physical considerations might not always be at hand, and therefore I, as a physicist, claim the right to participate in this matter. Even if physiologists might not agree with all parts of my work, the elaboration of the physical

Massenbeschleunigung annahm. Obwohl mir die physiologische Seite des Gegenstandes, auch als ich wieder auf denselben verfiel, noch ganz fremd war und ich die Arbeiten von F l o u r e n s und G o l t z kaum dem Namen nach kannte, so war diese Spur Ferment doch genügend, um meine Gedanken in der Richtung anzuregen, welche sie wirklich eingeschlagen haben. Durch physiologisch – optische Studien waren mir einstweilen die »unbewussten Schlüsse« und damit die psychologische Zusammensetzung einer Bewegungswahrnehmung aus sehr verschiedenen Elementen sehr unangenehm geworden und das Gleichgewichtsorgan wurde als ein willkommenes Auskunftsmittel begrüsst.

4.

Die Anschauung die ich gewonnen habe, und welche mir die sämmtlichen bisher bekannten Thatsachen in natürlichen übersichtlichen Zusammenhang zu bringen scheint, ist nun die, dass gewisse Labyrinthnerven vermöge ihrer specifischen Energie jeden Reiz mit einer Bewegungsempfindung beantworten, wodurch sich die F l o u r e n s 'schen Erscheinungen erklären. Dieser Reiz wird aber in der Regel durch den Labyrinthinhalt selbst gesetzt, welcher bei Bewegungen der Thiere das Schwerpunkt- und Flächenprincip zu erfüllen strebt. Auf diese Weise erhalten die Thiere Bewegungsempfindungen, gleichgültig ob sie sich activ bewegen, oder passiv bewegt werden.

Dass diese Anschauung einladend ist und auch Andere anziehen kann, beweist der Umstand, dass sie in wenig abweichender Form unmittelbar nach meiner ersten Publication noch zweimal selbstständig von Andern entwickelt worden ist, einmal von B r e u e r und später von B r o w n.

5.

Die Thatsachen, welche zu meiner Anschauung führen und zur Begründung derselben ausreichen, sind sämmtlich seit 1824, also seit einem halben Jahrhundert bekannt. Wenn diese Anschauung bisher noch nicht aufgetreten ist, so beweist dies, dass auch einfache physikalische Betrachtungen den Physiologen nicht immer nahe genug liegen, und hieraus schöpfte ich eben die Berechtigung als

side of the subject will not become totally worthless.

I did not want to treat my theory just as an idea and suppress it, because its consequences are too rich. It would have been very easy to restrict myself to a short note; however, this could have been interpreted as too much like selective extraction of the scientific jewels. Therefore I decided to perform specific experiments to be able to offer more than a simple commentary.

In the course of these investigations the existence of an organ of equilibrium became extremely likely. Whether my efforts increase the probability of development into accepted science, the reader may judge by himself from the following pages, which contain the foundations on which I base my view. I do not hide gaps nor cover up weaknesses. The wish to bring this subject to a definite conclusion is anyway only an illusory success. I am content if I have furthered this chapter of science.

It always seems to me desirable for experiments to be repeated. Therefore I mention that the cost of all experiments and apparatus (using simple construction) did not exceed 100 thaler.[1.11] One has to also consider that the repetition of experiments should always be less costly, because no modification of equipment is needed which is unavoidable when doing the first experiments.

6.

The following is a summary of the relevant literature[1.12] as far as it is known to me:

* Marcus Herz, About Vertigo. Berlin. 1791.
Darwin, Zoonomia (transl. by Girtanner).
Fr. u. P. Gruithuisen, Anthropology. Munich. 1810.
Ritter in Hufelands Journal for Pract. Medicine. Vol XVII.
* Purkyně, Contributions to a better knowledge of vertigo. Medical Yearbooks of the Austrian State. VI. Vol. II. St. P. 79. Vienna 1820.
—— Physiological experiments about vertigo. 10. Bulletin of the Natural Sciences Section of the Silesian Society. p. 35. Breslau 1825.

4

Physiker ein Wort in dieser Angelegenheit mitzureden. Mögen immerhin die Physiologen nicht mit allen Theilen meiner Arbeit einverstanden sein, die Bearbeitung der physikalischen Seite des Gegenstandes wird nicht ganz ihren Werth verlieren

Meine Theorie ganz als Einfall behandeln und sie unterdrücken wollte ich nicht, weil sie zu reich an Folgerungen war. Mich auf eine blosse Notiz beschränken, wäre sehr bequem gewesen, hätte jedoch zu freibeuterisch erscheinen können. So entschloss ich mich eine besondere Untersuchung anzustellen, um mehr wie ein blosses Aperçu bieten zu können.

Im Laufe dieser Untersuchung hat das Gleichgewichtsorgan für mich sehr an Wahrscheinlichkeit gewonnen. Ob diese Wahrscheinlichkeit durch meine Bemühungen auch für die Wissenschaft gewachsen ist, möge der Leser aus den folgenden Blättern entnehmen, welche die Begründung meiner Anschauung enthalten, in der ich keine Lücke verdecken, keine Schwäche verhüllen will. Der Wunsch, diesen Gegenstand rasch seinem definitiven Abschluss zuzuführen, dürfte ohnehin nur einen illusorischen Erfolg haben. Ich bin zufrieden, wenn ich das Capitel gefördert habe.

Immer ist es mir wünschenswerth, dass die Versuche wiederholt werden. Ich füge desshalb bei, dass die Kosten aller Versuche und Apparate (in einfacher Ausführung) 100 Thaler nicht überstiegen haben. Hiebei ist noch zu bedenken, dass die Wiederholung von Versuchen immer weniger kostspielig ist, weil alle Umänderungen der Apparate, welche bei der ersten Ausführung unvermeidlich sind, wegfallen.

6.

Es möge gleich hier eine Zusammenstellung der mir bekannten hieher gehörigen Literatur folgen:

* Marcus Herz, über den Schwindel. Berlin. 1791.
Darwin, Zoonomia (übers. v. Girtanner).
Fr. u. P. Gruithuisen, Anthropologie. München. 1810.
Ritter in Hufelands Journal für prakt. Heilkunde. Band XVII.
* Purkyně, Beiträge zur näheren Kenntniss des Schwindels. Medicin. Jahrbücher des österreich. Staates. VI. Bd. II. St. S. 79. Wien 1820.
—— Physiologische Versuche über den Schwindel. 10. Bulletin d. naturwiss. Section der schlesischen Gesellschaft. S. 35. Breslau 1825.

5

Purkyně, About the physiological significance of vertigo. 12. Bulletin. etc 1826. p. 1.

* Flourens, Experimental investigation about the properties and functions of the nervous system. 2nd edition. Paris 1842. p. 438.

Harless, Wagners Handbook of Physiology. vol. IV. p. 422.

Czermak, Comptes rendus. 1860. T. LI. S. 821.

Brown-Sequard, Course of lectures on the physiology and pathology of the central nervous system. Philadelphia 1860. S. 194.

Vulpian, Lectures about the general and comparitive physiology of the nervous system. Paris 1866. p. 800.

* Schiff, Textbook of physiology. Lahr 1858 – 1859. S. 899.

* Löwenberg, About the movement disorders after cutting the nerves of the circular canals of the ear labyrinth. Archives for eye and ear medicine by Knapp u. Moos. vol. III.

* Goltz, About the physiological functions of the circular canals of the ear labyrinth. Pflügers Archives. Vol III. p. 172.

* Mach, About fluids, which contain suspended particles. Pogg. Ann. Vol 126. p. 327 (1865).

Schklarewsky, Göttingen News. 1872. No. 15.

* Hitzig, Archives of Reichert and Dubois. 1871.

* Mach, Physical experiments about the sense of equilibrium in the human. Proc. Vienna Acad. Vol. 68. 6. Nov. 1873.

* Breuer, Bulletin of the k. and k. Society of Physicians. Vol. 7 (1873). Meeting of 14. Nov. 1873.

* —— Yearbook of the Society of Physicians. I. Issue. 1874.

* Mach, Experiments about the sense of equilibrium. Proc. Vienna Acad. 2nd. communication. Vol. 69. 1874.

* —— third communication. Vol. 69. 1874.

* Curschmann, About the relation of the semicircular canals of the ear labyrinth to the equilibrium of the body. German clinics No. 3. 1874.

* Wundt, Physiological Psychology. p. 208. 1873.

* Cyon, Pflügers Archives. VIII. Vol. Issue 6, 7. p. 306.

* Böttcher, About the cutting of the circular canals of the ear labyrinth. Archives for Ear Diseases by Tröltsch. Vol. 9. p. 1.

—— Dorpat Medical Journal 1872.

* Berthold, About the function of the circular canals of the ear labyrinth. Archives for Ear Diseases by Tröltsch. Vol. 9. p. 77.

* A Crum Brown, on the sense of rotation. Journal of Anatomy and Physiology. Vol. VIII.

* Sigm. Exner, Bulletin of the Vienna Acad. 1874. No. 20.

Publications marked with an asterik I have in the original, the others I am aquainted with from citations or excerpts.

5

Purkyně, Ueber die physiologische Bedeutung des Schwindels. 12. Bulletin etc. 1826. S. 1.

* Flourens, recherches experimentales sur les propriétés et les fonctions du système nerveux. 2me édition. Paris 1842. S. 438.

Harless, Wagners Handwörterbuch d. Physiologie. Bd. IV. S. 422.

Czermak, Comptes rendus. 1860. T. LI. S. 821.

Brown-Sequard, Course of lectures on the physiology and pathology of the central nervous system. Philadelphia 1860. S. 194.

Vulpian, leçons sur la physiologie géenérale et comparée du système nerveux. Paris 1866. S. 800.

* Schiff, Lehrbuch d. Physiologie. Lahr 1858 – 1859. S. 899.

* Löwenberg, über die nach Durchschneidung d. Bogengänge d. Ohrlabyrinthes auftretenden Bewegungsstörungen. Archiv fur Augen- und Ohrenheilkunde von Knapp u. Moos. Bd. III.

* Goltz, über die physiol. Bedeutung der Bogengänge des Ohrlabyrinthes. Pflügers Archiv. Bd. III. S. 172.

* Mach, über Flüssigkeiten, welche suspendirte Körperchen enthalten. Pogg. Ann. Bd, 126. S. 327 (1865).

Schklarewsky, Göttinger Nachrichten. 1872. No. 15.

* Hitzig, Archiv von Reichert und Dubois. 1871.

* Mach, Physikalische Versuche über den Gleichgewichtssinn d. Menschen. Sitzgsber. d. Wiener Akademie. Bd. 68. 6. Nov. 1873.

* Breuer, Anzeiger d. k. k. Gesellschaft d. Aerzte. No. 7 (1873). Sitzung v. 14. Nov. 1873.

* —— Jahrbuch d. Gesellschaft d. Aerzte. I. Heft. 1874.

* Mach, Versuche über den Gleichgewichtsinn. Sitzgsber. d. Wiener Akademie. 2. Mittheilung. Bd. 69. 1874.

* —— dritte Mittheilung. Bd. 69. 1874.

* Curschmann, über das Verhältniss der Halbcirkelcanäle des Ohrlabyrinthes zum Körpergleichgewicht. Deutsche Klinik No. 3. 1874.

* Wundt, physiologische Psychologie. S. 208. 1873.

* Cyon, Pflügers Archiv. VIII. Bd. Heft 6, 7. S. 306.

* Böttcher, über die Durchschneidung d. Bogengänge d. Gehörlabyrinthes. Archiv für Ohrenheilkunde von Tröltsch. Bd. 9. S. 1.

—— Dorpater medicinische Zeitschr. 1872.

* Berthold, über die Function d. Bogengänge des Ohrlabyrinthes. Archiv für Ohrenheilkunde von Tröltsch. Bd. 9. S. 77.

* A Crum Brown, on the sense of rotation. Journal of Anatomy and Physiology. Vol. VIII.

* Sigm. Exner, Anzeiger d. Wiener Akademie. 1874. No. 20.

Die mit Sternchen bezeichneten Abhandlungen liegen mir bei dieser Arbeit im Original vor, die übrigen kenne ich bloss aus Citaten und Auszügen.

6

The Mechanical Foundations.

1.

The movement of our body is always a <u>mechanical process</u>.[2.1] Any particular movement sensations must in any case also finally be based on mechanical processes. We must therefore next recall the mechanical foundations that must be applied when studying these phenomena. We shall explain them by several concrete examples, so that they will appear familiar even to those who are not accustomed to mathematical approaches.

Since the times of <u>Galileo</u>[2.2] and <u>Newton</u>[2.3] it has been understood that a change of velocity (acceleration) of a body always implies the change of velocity of another <u>body</u>.[2.4] By observation of such a case of reciprocal velocity change of two bodies —A, B— one can deduce a certain movement depending upon the proportion of masses, which determines all other relationships between these two bodies. If A is accelerated with φ over a certain time, and B with φ', then φ/φ', is the ratio of masses of B to A. It becomes $-\varphi/\varphi'$ if one takes into account that the bodies are accelerated in opposite directions along their connecting line. The mass relationship of a series of bodies to the body A is just the mass of each body if A is taken as the unit mass*). From this definition one sees that for the masses m, m', which undergo accelerations φ, φ', taking their signs into account, the following equation applies

$$m\,\varphi + m'\varphi' = 0.$$

The product $m\varphi$ cannot undergo any change without changing the other product, which is opposite in sign, by the same factor. Therefore, these products $m\varphi$, $m'\varphi'$ provide an effective means for the study of

---

*) Mach, The history and the root of the law of the conservation of energy. Prague, Calve 1872. p. 50.

6

Die mechanischen Grundsätze.

1.

Die Bewegung unseres Körpers ist stets ein mechanischer Vorgang. Gibt es besondere Bewegungsempfindungen, so liegt die letzte Wurzel derselben jedenfalls auch in mechanischen Vorgängen. Wir müssen uns also zunächst die mechanischen Grundsätze in Erinnerung bringen, welche wir beim Studium dieser Erscheinungen anzuwenden haben. Wir wollen uns dieselben durch mehrere anschauliche Beispiele erläutern, damit sie auch demjenigen, der an mathematische Betrachtungen nicht gewohnt ist, geläufig werden.

Die Galilei-Newton'sche Zeit hat die Einsicht gewonnen, dass die Geschwindigkeitsänderung (Beschleunigung) eines Körpers stets durch die Geschwindigkeitsänderung eines andern Körpers bedingt ist. Aus der Beobachtung eines solchen Falles gegenseitiger Geschwindigkeitsänderung zweier Körper A, B lässt sich ein bewegungsbestimmendes Merkmal, das Massenverhältniss ableiten, welches für alle andern Beziehungen dieser zwei Körper maassgebend ist. Erhält A in der Zeiteinheit die Beschleunigung φ, B aber φ', so ist φ/φ', das Massenverhältniss von B zu A, oder wenn man berücksichtigt, dass die Körper sich Beschleunigungen von entgegengesetztem Sinn nach der Richtung ihrer Verbindungslinie ertheilen $-\varphi/\varphi'$. Die Massenverhältnisse einer Reihe von Körpern zu demselben Körper A, sind, wenn A als Masseneinheit angenommen wird, die Massen dieser Körper*). Aus dieser Definition ersieht man, dass für die Massen m, m', welche sich die Beschleunigugnen φ, φ' ertheilen, mit Rücksicht auf die Zeichen, die Gleichung besteht

$$m\varphi + m'\varphi' = 0.$$

Das Product $m\varphi$ kann keine Aenderung erfahren, ohne dass das andere dem Zeichen nach entgegengesetzte Product dieselbe Aenderung erfährt. Diese Producte $m\varphi$, $m\varphi'$ haben also einen öcono-

---

*) M a c h, die Geschichte und die Wurzel des Satzes der Erhaltung der Arbeit. Prag, Calve 1872. S. 50.

mechanics. They are generally referred to as <u>forces</u>.[2.5] The forces which two bodies exert on one another are equal and oppositely directed. More generally, one can also state that the sum of all internal forces in a system of masses $= 0$.

If our own body interacts reciprocally with a mass m, then we sense a <u>pressure</u>[2.6], which depends on the value of the product $m\varphi$. One has therefore also used the expression pressure for $m\varphi$ itself. Hence one also speaks of the equivalence of pressure and counterpressure when considering the mutual interaction of two bodies. Whenever a mass m undergoes a change of velocity φ, a pressure $m\varphi$ must also necessarily have been exerted on another body as counterpressure.

2.

Under uniform <u>gravity</u>[2.7] conditions (e.g., on the same spot on the earth), the mass and weight of a body are proportional to one another. Therefore one can normally take the center of gravity and the center of mass of a body as identical. As discussed above, the center of gravity of a body follows an important law which had already been formulated by Newton: the law of the <u>conservation of the center of gravity</u>.[2.8]

The center of gravity of two masses lies along a line connecting the two masses, at a distance from each mass inversely proportional to each mass. If these masses experience mutual accelerations inversely proportional to each mass along the line connecting them, the position of the center of gravity will be unchanged by their mutual influence.

If a system consisting of several masses undergoes accelerations not only as a result of internal forces, but also by outside forces from bodies that do not belong to the system, one can easily show that the internal forces have no influence on the movement of the center of gravity. In fact, if any arbitrary mass m of the system is moved in a certain direction x, then the center of gravity of the system will be moved by $\frac{mx}{M}$ in the same direction, where M is the mass of the system. This shift of m has required,

mischen Werth für die mechanische Betrachtung. Sie heissen nach dem allgemeinen Gebrauch Kräfte. Die Kräfte, welche zwei Körper auf einander ausüben, sind gleich und entgegengesetzt. Man kann auch allgemeiner sagen, die Summe aller Kräfte in einem System von Massen ist = 0.

Ist unser eigener Körper in Wechselwirkung mit einer Masse m, so empfinden wir einen von der Grösse des Productes $m\varphi$ abhängigen Druck. Man hat daher auch für $m\varphi$ selbst den Ausdruck Druck gebraucht. Man spricht demnach von Gleichheit des Druckes und Gegendruckes bei der Wechselwirkung zweier Körper. Ueberall, wo einer Masse m eine Geschwindigkeitsveränderung φ ertheilt wird, muss ein Druck $m\varphi$ vorhanden gewesen sein, den demnach auch noch ein anderer Körper als Gegendruck erlitten haben muss.

2.

Unter gleichen Gravitationsumständen (z. B. an demselben Orte der Erde) sind Massen und Gewichte der Körper einander proportional. Desshalb kann man für gewöhnlich den Schwerpunkt und den Mittelpunkt der Masse eines Körpersystems für identisch halten. Für den Schwerpunkt eines Körpersystems ergibt sich aber dem Obigen gemäss ein wichtiger Satz, welcher schon von Newton aufgestellt worden ist, der Satz der Erhaltung des Schwerpunktes.

Der Schwerpunkt zweier Massen theilt die Verbindungslinie derselben in Abschnitte, welche sich umgekehrt wie die anliegenden Massen verhalten. Ertheilen sich nun die Massen Beschleunigungen nach der Richtung ihrer Verbindungslinie, welche sich verkehrt wie sie selbst verhalten, so vermögen sie durch ihre gegenseitige Einwirkung diesen Schwerpunkt nicht zu verschieben.

Besteht ein System aus mehreren Massen, welche nicht nur (durch innere Kräfte) sich Beschleunigungen ertheilen, sondern auch von nicht zum System gehörigen Körpern (von äusseren Kräften) afficirt werden, so lässt sich leicht zeigen, dass die inneren Kräfte auf die Bewegung des Schwerpunktes des Systems keinen Einfluss haben. In der That, wenn irgend eine Masse m des Systems durch eine innere Kraft eine Verschiebung x nach einer bestimmten Rich-

however, that another mass of the system m' had been shifted by $-x'$ thereby shifting the center of gravity by $-\dfrac{m'x'}{M}$. According to the above relationship, $mx - m'x'/M = 0$, the resulting shift of the center of gravity is $\dfrac{mx}{M} - \dfrac{m'x'}{M} = 0.$ This applies for any two masses of the system and for any direction. The movement of the center of gravity is determined only by means of external forces. This is in essence the law of the conservation of the center of gravity.

Another general law that applies for any free system of masses is the law of the <u>conservation of areas</u>[2.9] introduced by <u>D'Arcy</u>[2.10] and <u>Euler</u>.[2.11] Let us consider two masses m, m' which mutually influence each other. Because of their mutual influences they move along the colinear connecting lines AB and CD, respectively (<u>Fig. 1</u>).[2.12] Taking into account the sign of these displacements, according to the above discussion,

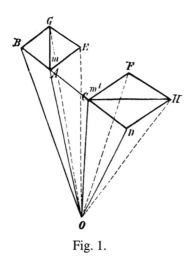

Fig. 1.

$$m\ AB + m'\ CD = 0$$

If one draws a radial vector from any arbitrary point 0 to the moving masses, and considers the surfaces which are traced out by these radii and the mass movements [AB and CD, respectively], taking into account the opposite signs for opposite movements, one can state the relationship $m\ OAB + m'\ OCD = 0$.

If two masses act upon each other and one draws radial vectors to them from some arbitrary point, the sum of the traversed surfaces multiplied by their corresponding masses = 0.

tung erfährt, so wird der Schwerpunkt des Systems um $\frac{mx}{M}$, wobei M die Masse des Systems, nach derselben Richtung verschoben. Diese Verschiebung von m hat aber zur Voraussetzung, dass eine andere Masse des Systems m' um $-x'$ sich verschoben hat, was also dem Schwerpunkt die Verschiebung $-\frac{m'x'}{M}$ ertheilt. Weil nun nach dem Obigen $mx - m'x' = 0$, so ist auch die totale Schwerpunktverschiebung $\frac{mx}{M} - \frac{m'x'}{M} = 0.$ Dies gilt für je zwei Massen des Systems und für jede Richtung. Die Bewegung des Schwerpunktes ist also bloss durch die äussern Kräfte bestimmt. Darin besteht eben der Satz der Erhaltung des Schwerpunktes.

Ein anderer allgemeiner Satz, der für ein freies Massensystem gilt, ist der Satz der Erhaltung der Flächen. Derselbe ist von D'Arcy und Euler gefunden worden. Betrachten wir zwei in Wechselwirkung stehende Massen m, m'. Dieselben legen vermöge ihrer Wechselwirkung allein die Wege AB, CD (Fig. 1) nach der Richtung der Verbindungslinie zurück. Nimmt man auf das Zeichen dieser Bewegungen Rücksicht, so ist dem Obigen gemäss

$$m\,AB + m'\,CD = 0$$

Zieht man von irgend einem Punkte 0 aus zu den bewegten Massen Radienvectoren

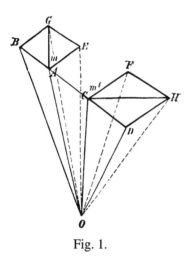

Fig. 1.

und betrachtet die in entgegengesetztem Sinne von denselben durchstrichenen Flächenräume als von entgegengesetztem Zeichen, so ist auch $m\,OAB + m'\,OCD = 0$.

Wenn zwei Massen in Wechselwirkung stehn und man zieht von irgend einem Punkte aus zu denselben Radienvectoren, so ist die Summe der von denselben durchstrichenen Flächenräume multiplicirt mit den zugehörigen Massen $= 0$.

If the masses had been acted upon by external forces, which would cause them to trace out the areas *OAE* and *OCF*, then the combined action of both forces (during a very small time) would trace out the surfaces *OAG* and *OCH*. Now, with the help of <u>Varignon's</u>[2.13] law of the parallelogram of forces, one can easily show that

$$m \cdot OAG \ + \ m' \cdot OCH \ = \ m \cdot OAE \ + \ m' \cdot OCF \ + \ m \cdot OAB \ + \ m' \cdot OCD$$
$$= \ m \cdot OAE + m' \cdot OCF$$

i.e., the sum of the traversed surfaces is not altered by internal forces.

If several masses are present, then one can state that the projection of the entire movement process onto a given plane is the same as for any two masses.

If one draws radial vectors in the direction of the masses from one point, and projects the traced surfaces onto a given plane, then the sum of these traversed areas, multiplied by their respective masses, is determined only by external forces. This is the principle of the conservation of areas.

3.

For completeness the analytical elaboration of both laws, which have been in common use since <u>Lagrange's</u>[2.14] time, is also discussed. A free system, consisting of masses *m, m', m"* .... acted upon by forces *X, Y, Z, X', Y', Z'*.... obeys <u>D'Alembert's</u>[2.15] Principle

$$\sum \left[\left(X - m\frac{d^2x}{dt^2} \right) \delta x + \left(Y - m\frac{d^2y}{dt^2} \right) \delta y + \left(Z - m\frac{d^2z}{dt^2} \right) \delta z \right] = 0$$

i.e., the forces $\left(X - m\dfrac{d^2x}{dt^2} \right), \left(Y - m\dfrac{d^2y}{dt^2} \right), \left(z - m\dfrac{d^2z}{dt^2} \right)$... keep the system in equilibrium. As <u>Stevin</u>[2.16] had already noted, equilibrium is not disturbed if one introduces new fixed constraints in the system. The equilbrium will therefore be maintained if we think of the whole free system as rigid. It follows that the forces described above must fulfill the equilibrium conditions of a rigid body. Therefore we have

Wären die Massen von äussern Kräften afficirt und würden vermöge dieser die Flächenräume *OAE* und *OCF* beschrieben, so gibt die Zusammenwirkung beider Kräfte (während einer sehr kleinen Zeit) die Flächenräume *OAG* und *OCH;* Nun lässt sich aber mit Hilfe des Varignon'schen Satzes vom Kräfteparallelogramm leicht nachweisen, dass

$$m{\cdot}OAG \; + \; m'{\cdot}OCH \; = \; m{\cdot}OAE \; + \; m'{\cdot}OCF \; + \; m{\cdot}OAB \; + \; m'{\cdot}OCD$$
$$= \; m{\cdot}OAE + m'{\cdot}OCF$$

d. h. die Summe der durchstrichenen Flächenräume wird durch die innern Kräfte nicht geändert.

Sind mehrere Massen vorhanden, so kann man von der Projection des ganzen Bewegungsvorganges auf eine gegebene Ebene für je zwei Massen dasselbe behaupten.

Zieht man von einem Punkte aus nach den Massen eines Systems Radienvectoren, und projicirt die durchstrichenen Flächenräume auf eine gegebene Ebene, so ist die Summe dieser mit den zugehörigen Massen multiplicirten Flächenräume bloss durch die äusseren Kräfte bestimmt. Dies ist das Princip der Erhaltung der Flächen.

3.

Der Vollständigkeit wegen mag hier auch die analytische Entwicklung beider Sätze Platz finden, wie sie seit Lagrange gebräuchlich ist. Ein freies System, bestehend aus den Massen *m*, *m'*, *m"*...., auf welche die Kräfte *X, Y, Z, X', Y', Z'*.... wirken, befolgt das D'Alembert'sche Princip

$$\sum\left[\left(X - m\frac{d^2x}{dt^2}\right)\delta x + \left(Y - m\frac{d^2y}{dt^2}\right)\delta y + \left(Z - m\frac{d^2z}{dt^2}\right)\delta z\right] = 0 \cdot$$

d. h. die Kräfte $\left(X - m\dfrac{d^2x}{dt^2}\right), \left(Y - m\dfrac{d^2y}{dt^2}\right), \left(z - m\dfrac{d^2z}{dt^2}\right)...$ halten

sich an diesem System das Gleichgewicht. Wie nun Stevin zuerst bemerkt hat, wird ein Gleichgewicht nicht gestört, wenn man in dem System neue feste Verbindungen anbringt. Das Gleichgewicht wird also fortbestehn, wenn wir uns das ganze freie System fest denken. Es werden also die obigen Kräfte die Gleichgewichtsbedingungen eines festen Körpers erfüllen müssen. Demnach haben wir

$$\sum\left(X - m\frac{d^2x}{dt^2}\right) = 0$$

$$\sum\left(Y - m\frac{d^2y}{dt^2}\right) = 0$$

$$\sum\left(Z - m\frac{d^2z}{dt^2}\right) = 0$$

$$\sum\left(X - m\frac{d^2x}{dt^2}\right)y - \left(Y - m\frac{d^2y}{dt^2}\right)x = 0$$

$$\sum\left(Y - m\frac{d^2y}{dt^2}\right)z - \left(Z - m\frac{d^2z}{dt^2}\right)y = 0$$

$$\sum\left(Z - m\frac{d^2z}{dt^2}\right)x - \left(X - m\frac{d^2x}{dt^2}\right)z = 0$$

Of these six equations the first three express the principle of the center of gravity, and the other three the principle of conservation of areas. If one defines M as the total mass of the system, and if $\dfrac{\sum mx}{M} = \xi$, $\dfrac{\sum my}{M} = \eta$, $\dfrac{\sum mz}{M} = \zeta$ signifies the coordinates of the center of gravity, then the first three equations readily take the form

$$\sum X = M\frac{d^2\xi}{dt^2}$$

$$\sum Y = M\frac{d^2\eta}{dt^2}$$

$$\sum Z = M\frac{d^2\zeta}{dt^2}$$

In this form one can see that the center of gravity moves just as though all the masses and forces were concentrated at that point. Because the internal forces consist of equal and opposite pairs, they exert no influence on the movement of the center of gravity.

The last three equations can be treated in the following manner. The first of the equations yields

$$\sum(Xy - Yx) = \sum m\left(y\frac{d^2x}{dt^2} - x\frac{d^2y}{dt^2}\right) = \frac{d}{dt}\sum m\left(y\frac{dx}{dt} - x\frac{dy}{dt}\right)$$

Let us draw radial vectors from the origin of the coordinate system to the masses, project the resulting surfaces onto the XY plane, and label them α. Then the expression $\sum m\left(y\dfrac{dx}{dt} - x\dfrac{dy}{dt}\right)$ may be given as the sum $2\sum m\dfrac{d\alpha}{dt}$, and accordingly the whole expression on the right side becomes the sum $2\sum m\dfrac{d^2\alpha}{dt^2}$. The term $\sum(Xy - Yx)$ is the torque[2.17] about the X axis [read Z axis] that we will call $\mathfrak{A}$.[2.18] The three equations, which can all be treated in the same manner, may now be put in the form

$$\sum \left(X - m\frac{d^2 x}{dt^2} \right) = 0$$

$$\sum \left(Y - m\frac{d^2 y}{dt^2} \right) = 0$$

$$\sum \left(Z - m\frac{d^2 z}{dt^2} \right) = 0$$

$$\sum \left(X - m\frac{d^2 x}{dt^2} \right) y - \left(Y - m\frac{d^2 y}{dt^2} \right) x = 0$$

$$\sum \left(Y - m\frac{d^2 y}{dt^2} \right) z - \left(Z - m\frac{d^2 z}{dt^2} \right) y = 0$$

$$\sum \left(Z - m\frac{d^2 z}{dt^2} \right) x - \left(X - m\frac{d^2 x}{dt^2} \right) z = 0$$

Von diesen sechs Gleichungen liefern die drei ersten das Schwerpunkts-, die drei andern das Flächenprincip. Nennt man M die Gesammtmasse des Systems und bedenkt, dass $\dfrac{\sum mx}{M} = \xi$, $\dfrac{\sum my}{M} = \eta$, $\dfrac{\sum mz}{M} = \zeta$ die Coordinaten des Schwerpunktes vorstellen, so bringt man die drei ersten Gleichungen leicht auf die Form

$$\sum X = M\frac{d^2 \xi}{dt^2}$$

$$\sum Y = M\frac{d^2 \eta}{dt^2}$$

$$\sum Z = M\frac{d^2 \zeta}{dt^2}$$

Aus dieser Form ersieht man, dass die Bewegung des Schwerpunktes dieselbe ist, wie wenn alle Massen und Kräfte in demselben vereinigt wären. Da nun die innern Kräfte paarweise gleich und entgegengesetzt sind, haben sie auf diese Bewegung keinen Einfluss.

Die drei letzten Gleichungen lassen sich in folgender Weise behandeln. Die erste dieser Gleichungen gibt

$$\sum (Xy - Yx) = \sum m\left(y\frac{d^2 x}{dt^2} - x\frac{d^2 y}{dt^2} \right) = \frac{d}{dt}\sum m\left(y\frac{dx}{dt} - x\frac{dy}{dt} \right)$$

Denken wir uns vom Anfangspunkte der Coordinaten zu den Massen Radienvectoren gezogen, projiciren die von denselben beschriebenen Flächenräume auf die XY Ebene und bezeichen sie mit α, so stellt der Ausdruck $\sum m\left(y\frac{dx}{dt} - x\frac{dy}{dt} \right)$ die Summe $2\sum m\frac{d\alpha}{dt}$, demnach der ganze Ausdruck rechter Hand die Summe $2\sum m\frac{d^2\alpha}{dt^2}$ vor. Der Ausdruck $\sum (Xy - Yx)$ ist das Drehungsmoment des ganzen Systems in Bezug auf die X Axe, welches wir $\mathfrak{A}$ nennen wollen. Die

$$2\sum m \frac{d^2\alpha}{dt^2} = \mathfrak{A}$$

$$2\sum m \frac{d^2\beta}{dt^2} = \mathfrak{B}$$

$$2\sum m \frac{d^2\gamma}{dt^2} = \mathfrak{C}$$

These equations show that the sum of the traversed surfaces depends only on external forces acting on the system, since the internal forces, as equal and opposite pairs, do not appear in these equations. If we integrate the equations over a sufficiently small movement element, such that the external forces (and rotational moments) can be considered as constant, we obtain

$$2\sum m\alpha = \frac{\mathfrak{A}t^2}{2} + At + a$$

$$2\sum m\beta = \frac{\mathfrak{B}t^2}{2} + Bt + b$$

$$2\sum m\gamma = \frac{\mathfrak{C}t^2}{2} + Ct + c$$

For a system which is not acted upon by external forces, so that $\mathfrak{A} = \mathfrak{B} = \mathfrak{C} = 0$, the traversed surfaces can only increase proportional to time (which is the case for constant velocity rotation). If within such a system the sum is $= 0$ at any instant, then it must remain $= 0$ for all time, if no external disturbance[2.19] is encountered.

Let us assume that our system of free masses is a rigid body that we rotate about an axis. We designate the rotation angle as φ. The equation $2\sum m \frac{d^2\alpha}{dt^2} = \mathfrak{A}$ then can be written as: $\left(\sum mr^2\right)\frac{d^2\varphi}{dt^2} = \mathfrak{A}$. In this equation $\sum mr^2$ is the moment of inertia and $\frac{d^2\varphi}{dt^2}$ the angular acceleration. The moment of inertia[2.20] is an extension of the notion of masses[2.21], and the moment of rotation[2.22] is an extension of the notion of force.[2.23]

In order for a mass to be accelerated, or for the center of gravity of a system of masses to be accelerated, a certain level of force is necessary, which must be located outside of this mass or this system of masses. If the

drei Gleichungen, welche sich alle in derselben Weise behandeln lassen, nehmen nun die Form an

$$2 \sum m \frac{d^2 \alpha}{dt^2} = \mathfrak{A}$$

$$2 \sum m \frac{d^2 \beta}{dt^2} = \mathfrak{B}$$

$$2 \sum m \frac{d^2 \gamma}{dt^2} = \mathfrak{C}$$

Aus diesen Gleichungen ersieht man, dass die Summe der durchstrichenen Flächenräume nur durch die äusseren Kräfte des Systems bestimmt ist, indem die inneren Kräfte als paarweise gleich und entgegengesetzt in diese Gleichungen gar nicht eingehn. Integriren wir die Gleichungen für ein so kleines Bewegungselement, dass für dasselbe die äusseren Kräfte (und Drehungsmomente) als constant angesehen werden können, so erhalten wir

$$2 \sum m \alpha = \frac{\mathfrak{A} t^2}{2} + At + a$$

$$2 \sum m \beta = \frac{\mathfrak{B} t^2}{2} + Bt + b$$

$$2 \sum m \gamma = \frac{\mathfrak{C} t^2}{2} + Ct + c$$

In einem Systeme, welches von äusseren Kräften gar nicht afficirt ist, in welchem also $\mathfrak{A} = \mathfrak{B} = \mathfrak{C} = 0$, können die durchstrichenen Flächenräume (wie bei gleichförmigen Drehungen) nur proportional der Zeit wachsen. Ist diese Summe in einem solchen System für irgend einen Zeittheil = 0, so muss sie für jeden Zeittheil = 0 bleiben, wenn nicht eine äussere Störung eintritt.

Nehmen wir an, unser freies Massensystem sei ein fester Körper, den wir in Drehung um eine Axe setzen. Wir bezeichnen den Drehungswinkel mit φ. Der Gleichung $2 \sum m \frac{d^2 \alpha}{dt^2} = \mathfrak{A}$ entspricht dann diese: $\left(\sum mr^2 \right) \frac{d^2 \varphi}{dt^2} = \mathfrak{A}$. In derselben bedeutet $\sum mr^2$ das Trägheitsmoment und $\frac{d^2 \varphi}{dt^2}$ die Winkelbeschleunigung. Das Trägheitsmoment ist eine Erweiterung des Massebegriffs, das Drehungsmoment eine Erweiterung des Kraftbegriffs.

Soll eine Masse eine Beschleunigung oder ein Massensystem eine Beschleunigung des Schwerpunktes erfahren, so ist hiezu ein bestimmbarer Kraftaufwand nöthig, der seine Quelle ausserhalb die-

sum of the traversed surfaces (calculated as mentioned above) is to be accelerated, then a moment of rotation is required, which must originate outside of the system. A body or a system of bodies cannot start to move in a linear or angular motion without an outside cause.

4.

We will now explain these theoretical remarks by some experiments and illustrative considerations. If a mass m subject to gravitational acceleration g lies on a table, then the table cancels this acceleration. Pressure [force] and counterpressure of the mass and the table are given by mg. If the table sinks with constant velocity, then the acceleration of the mass is still canceled and the pressure still is mg. However, if the table sinks with acceleration g', then the change of velocity for the mass m is $g-g'$ per unit time, and therefore the pressure is $m(g-g')$. This term becomes $m(g+g')$, if g' is directed against gravity. The mass would exert a smaller pressure than its weight on a downward accelerating table, a larger pressure on an upward accelerating table, and no pressure at all on a freely falling table.

<u>Poggendorff's falling machine</u>[2.24] (Fig. 2) demonstrates these relations very nicely. A balance carries a pulley instead of the scale. A cord running over this pulley carries the weights P on one end and $P + p$ on the other. The larger weight is fixed to the axis of the pulley by a string. Once the scale has been balanced, one burns the string. The weights on the string start falling with the acceleration $pg/(2P+p)$. Since of the two weights P [on either side] one rises with the same acceleration as the other falls, this movement has no influence on the system, and it behaves as if p alone were falling with an acceleration of $pg/(2P+p)$. This causes a decrease of

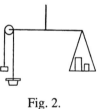

Fig. 2.

ser Masse oder dieses Massensystems haben muss. Soll die Summe der durchstrichenen Flächenräume (in obiger Weise gezählt) eine Beschleunigung erfahren, so ist hiezu ein bestimmbares Drehungsmoment nöthig, welches seine Quelle ausserhalb des Systems haben muss. Ein Körper oder ein Körpersystem kann ohne äussere Ursache weder in fortschreitende noch in drehende Bewegung gerathen.

4.

Wir wollen uns nun diese theoretischen Bemerkungen durch einige Experimente und anschaulichere Betrachtungen erläutern. Wenn eine Masse m mit der Schwerebeschleunigung g auf einem Tisch liegt, so hat der Tisch diese Beschleunigung vernichtet. Druck und Gegendruck der Masse und des Tisches sind also durch mg bestimmt. Sinkt der Tisch mit constanter Geschwindigkeit, so bleibt die Beschleunigung der Masse noch immer vernichtet, der Druck noch immer mg. Sinkt aber der Tisch mit der Beschleunigung g', so ist die an der Masse m hervorgebrachte Geschwindigkeitsänderung in der Zeiteinheit $g - g'$ und demnach der Druck $m(g-g')$. Dieser Ausdruck übergeht in $m(g+g')$, wenn g' der Schwere entgegen gerichtet ist. Auf den mit Beschleunigung sinkenden Tisch würde die Masse einen kleineren Druck als ihr Gewicht, auf den mit Beschleunigung steigenden Tisch einen grösseren Druck und auf den frei fallenden Tisch gar keinen Druck ausüben.

Die Poggendorff'sche Fallmaschine (Fig. 2) demonstrirt diese Verhältnisse in sehr hübscher Weise. Eine Wage trägt statt der einen Wagschale eine Rolle. Ueber diese Rolle geht eine Schnur, die an beiden Enden die Gewichte P und $P + p$ trägt, deren grösseres durch einen Faden an der Axe der Rolle festgebunden ist. Nachdem die Wage aequilibrirt ist, brennt man diesen Faden ab. Die Gewichte an der Schnur beginnen eine Fallbewegung mit der Beschleunigung, $pg/(2P+p)$. Da von den beiden Gewichten P das eine mit derselben Beschleunigung steigt als das

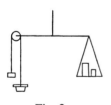

Fig. 2.

pressure $pp/(2P+p)$ [2.25] on the pulley, which therefore becomes lighter by this amount, and the balance shows this by deflecting. Such deflection disappears if the weight $pp/(2P+p)$ is removed from the other side of the balance.

We can observe similar relationships on our own body. If we stand on a stationary support, it carries our entire weight and every body part similarly carries all the weight of the parts above it. If the support and our body upon it starts to accelerate downward, then the pressure on the support is immediately reduced, and every body part rests lighter on the parts below it. During free fall of both the support and the body, all pressure and counterpressure between the support and our body is eliminated, as well as those between the parts of our body. The extremities no longer pull on the joints, the blood loses its weight, and is no longer carried by the walls of the blood vessels, the whole body behaves as if weightless. Therein lies the reason for the sensations that we encounter when falling or even accelerating downward. One can predict similar sensations, if one were suddenly transposed to a heavenly body of smaller mass and therefore weaker gravity, e.g., the moon.[2.26] There one experiences a sensation of a continuous sinking movement. On a body of larger mass a sensation of continuously rising is anticipated. These rapid and important changes of gravitational force might also be responsible for sea sickness.[2.27]

In the course of Hirn's[2.28] work on testing laws of the mechanical theory of heat, he conducted an experiment inside a constant speed rotating treadmill, whereby a subject had first to descend on the rising side, and then ascend on the descending side. In both cases, therefore, he remained absolutely stationary in space. If one now asks how much work that man produced in

andere fällt, so hat diese Bewegung keinen Einfluss und es verhält sich so, als ob p allein mit der Beschleunigung $pg/(2P+p)$ fallen würde. Es wird also dabei die Druckverminderung $pp/(2P+p)$ eintreten, die Rolle wird während der Fallbewegung um diesen Betrag leichter und die Wage zeigt dies durch einen Ausschlag an. Der Ausschlag verschwindet, wenn von der andern Seite der Wage das Gewicht $pp/(2P+p)$ weggenommen wird.

Aehnliche Verhältnisse können wir an unserem eigenen Körper beobachten. Stehen wir auf einer ruhenden Unterlage, so trägt diese unser ganzes Gewicht und jeder Körpertheil trägt das ganze Gewicht des darüber befindlichen. Beginnt die Unterlage und damit unser Körper eine beschleunigte Bewegung abwärts, so vermindert sich sofort der Druck auf die Unterlage und jeder Körpertheil wird für die darunter befindlichen Körpertheile leichter. Beim freien Fall der Unterlage und des Körpers hört jeder Druck und Gegendruck vermöge der Schwere sowohl zwischen der Unterlage und unserem Körper, als auch zwischen den Theilen unseres Körpers auf. Die Extremitäten lasten nicht mehr auf den Gelenken, das Blut verliert sein Gewicht und wird durch die Gefässwände nicht mehr getragen, der ganze Körper verhält sich wie schwerlos. Dadurch sind jedenfalls die sonderbaren Empfindungen bedingt, die man beim Fallen und schon beim beschleunigten Sinken erhält. Ganz ähnliche Empfindungen müsste man haben, wenn man plötzlich auf einen Weltkörper von kleinerer Masse, also auch geringerer Schwerebeschleunigung, z. B. auf den Mond versetzt würde. Man müsste dort das Gefühl eines fortwährenden Versinkens, auf einem Weltkörper von grösserer Masse aber das Gefühl des fortwährenden Steigens haben. Diese rapiden und bedeutenden Veränderungen des Schweredrucks mögen es auch mit sein, welche die Seekrankheit hervorbringen.

Hirn hat zum Zwecke der Prüfung einiger Sätze der mechanischen Wärmetheorie in einem gleichförmig gedrehten Tretrade einen Mann einmal an der aufsteigenden Seite herabsteigen, dann an der herabsteigenden Seite heraufsteigen lassen und zwar so, dass er in beiden Fällen in absoluter Ruhe verblieb. Frägt man nun

the second case, then our earlier considerations teach us immediately that it depends on the relative acceleration between the body of the man and the steps of the wheel, because that is what must be overcome. For a constant velocity rotation[2.29] of the wheel, however, this is normal gravity. The work is therefore the same as if the man were to climb a mountain with the same vertical velocity. One is easily misled here if one simply notes that the man does not actually move, and apparently needs not to overcome the force of gravity.

Let us consider a bird[2.30], which hovers at a constant height. What work must it produce per unit time to maintain its altitude? If its mass is m' [read m], and its gravitational acceleration g, then an upward vertical force mg must act on it. Within every unit of time it must provide the [surrounding] mass m' with the incremental velocity g'. Therefore $mg = m'g'$. The vital force [kinetic energy] or work per unit time is $m'(g')^2/2$. Since $g' = mg/m'$, this work is $m^2g^2/2m'$. One sees from this example that the work for the bird is reduced with an increase of the mass m' to be moved in unit time, that is, with larger wings and slower wing movements. The work becomes $= 0$, if m' were infinite. This is the case when the bird rests on ground.

5.

Let us now move to the illustration of the principle of the center of gravity. An animal free in space[2.31] would not be able to move itself away from its starting position. It is bound to its own center of gravity, and since the latter is unmovable, it must always stay in exactly the same place, if its body parts return to the same relative position.[2.32] We can't say what happens to the movement of the parts of a bomb that explodes in midflight. But according to the principle of the center of gravity it is clear that except for air resistance, and as long as no bomb part is impeded, the common center of gravity continues to follow the same trajectory as if no explosion had occurred.

hier nach der Arbeit, welche der Mann etwa in dem zweiten Falle leistet, so lehren uns die obigen Bemerkungen sofort, dass es auf die relative Beschleunigung zwischen Körper des Mannes und den Stufen des Rades ankommt, indem diese zu überwinden ist. Diese ist aber bei gleichförmiger Drehung des Rades die gewöhnliche Schwerebeschleunigung. Die Arbeit ist aber dieselbe, als ob der Mann mit derselben Verticalgeschwindigkeit einen Berg ersteigen würde, worüber man sich täuschen könnte, wenn man bloss beachten würde, dass der Mann gar nicht von der Stelle kommt, also die Schwere scheinbar gar nicht überwindet.

Denken wir uns einen Vogel, welcher in constanter Höhe schwebt. Welche Arbeit muss er in der Zeiteinheit produciren, um sich zu erhalten? Ist seine Masse m', seine Schwerebeschleunigung g, so muss auf ihn vertical aufwärts die Kraft mg ausgeübt werden, er muss also in jeder Zeiteinheit der Masse m' die Geschwindigkeit g' ertheilen. Hiebei ist $mg = m'g'$. Die in der Zeiteinheit producirte lebendige Kraft oder Arbeit ist $m'(g')^2/2$ oder weil $g' = mg/m'$, so ist diese Arbeit $m^2g^2/2m'$. Man sieht hieraus, dass die Arbeit desto geringer ausfällt, je grösser die in der Zeiteinheit bewegte Masse m' ist, also mit je grösseren Flügeln und je langsamerem Flügelschlag der Vogel arbeitet. Die Arbeit wird = 0, wenn m' unendlich wird. Dieser Fall tritt ein, wenn der Vogel am Boden ruht.

5.

Uebergehn wir nun zur Veranschaulichung des Schwerpunktsprincipes. Ein Thier frei im Weltraume wäre nicht im Stande, sich von der Stelle zu bewegen. Es ist an seinen eigenen Schwerpunkt gebunden, und da letzterer unbeweglich ist, so muss er immer genau in derselben Lage sein, wenn seine Theile wieder in die gleiche relative Stellung kommen. Von der Bewegung der Theile einer Bombe, welche im Wurf geplatzt ist, wissen wir nichts zu sagen. Es ist aber nach dem Schwerpunktsprincip klar, dass von dem Luftwiderstande abgesehn und so lange kein Bombenstück einen Widerstand trifft, der gemeinsame Schwerpunkt fortfährt die Wurfbahn in derselben Weise zu beschreiben, als ob kein Zerspringen eingetreten wäre.

If a machine that is free in space is set in motion, such that parts of it move relative to each other, then it must follow the principle of the <u>center of gravity</u>.[2.33] The rule states that if one part moves, then the body of the machine also has to move, in order for the center of gravity to remain at the same position. For example, we observe that a steam engine that is not heavy enough or not well enough secured to the floor starts making hopping movements when put into <u>action</u>.[2.34]

Let us consider a <u>Page</u>[2.35]-type electric motor with a horizontal coil, as illustrated in Fig. 3. Each time the iron core inside the coil AB moves towards the right because of the internal forces between the coil and the core, the body of the motor is forced to the left. If it is movable on small wheels $rrrr$, this force causes the body to move leftward, in such a way that the overall center of gravity remains stationary. One therefore sees how the motor rapidly moves to and fro when in action. This jerking can be compensated if one adds a suitable weight which can move along the

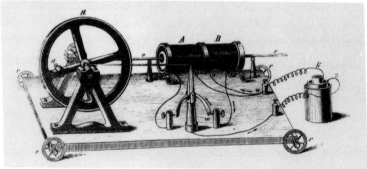

Fig. 3.

spoke of the flywheel such that it always moves opposite to the motion of the iron core, and thereby neutralizes the induced shift of the center of gravity*). Then it becomes unnecessary to compensate it by motion of the motor body, which now remains stationary. In locomotives, where such kicking and jerking must be avoided, we see such compensation on the drive wheels. If the crew on a ship were to run towards one end, the ship would be jerked in the opposite direction. However, this movement would

---

*) M a c h, Carls Repetitorium of Experimental Physics. Vol. IV. p. 350.

Wenn eine im Raume freie Maschine in Bewegung geräth, indem sich Theile derselben gegen einander verschieben, erfüllt sie das Schwerpunktsprincip. In der Regel muss bei Bewegung eines Theils auch der Körper der Maschine sich bewegen, damit der Schwerpunkt an Ort und Stelle bleiben kann. Desshalb sehen wir z. B. ein Dampfmaschinenmodell, welches nicht genug schwer oder nicht hinreichend befestigt ist, beim Gange in hüpfende Bewegung gerathen.

Betrachten wir einen Electromotor nach Page (Fig. 3) mit horizontaler Spule. Jedesmal wenn der Eisenkern in der Spule AB durch die innern Kräfte zwischen Spule und Kern nach rechts geht, erfährt der Körper des Motors einen Druck nach links, dem er auch folgt, wenn er etwa auf Rädchen $rrrr$ beweglich ist, so zwar, dass der Gesammtschwerpunkt an Ort und Stelle bleibt. Man sieht desshalb den Motor im Gange rasch hin- und herrücken. Dieses Rucken kann aufgehoben werden, wenn man auf eine

Fig. 3.

Speiche des Schwungrades R ein passendes Laufgewicht so aufschraubt, dass sich dasselbe dem Eisenkern stets entgegenbewegt und die durch denselben hervorgebrachte Schwerpunktsverschiebung gerade compensirt*). Dann wird es also unnöthig, dass diese Compensation durch den Motorkörper ausgeführt wird und derselbe bleibt in Ruhe. Bei den Locomotiven, bei welchen ein solches Rücken und Stossen vermieden werden muss, sehen wir eine derartige Com-

---

*) Mach, Carls Repertorium d. Experimentalphysik. Bd. IV. S. 350.

be cancelled when the crew moves back.

If the crew of a ship forms a circle and starts moving clockwise, the ship will move in an opposite direction. This is based upon the principle of the conservation of areas. A similar phenomenon can be demonstrated with any electric motor*). Let us position Gruel's motor (Fig. 4) with its oscillating core vertically, so that the flywheel is horizontal, and purposely replace the normal wooden flywheel with a more massive one made out of metal. The whole motor is now movable about this vertical axis and the current is fed in

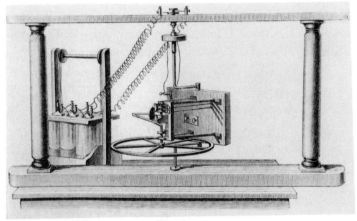

Fig. 4.

via two mercury filled circular ducts $\underline{k}^{2.36}$ so that this movement is not restrained. We now use a string to fasten the body of the motor to the frame supporting the vertical axis and switch on the current. As soon as the flywheel begins to move in a clockwise direction (as seen from above), the motor tends to move in the opposite direction. This rotation also appears rapidly, if the string is burned. For this experiment it is important that the axis is truly vertical.

---

*) M a c h, as quoted

pensation an den Triebrädern angebracht. Wenn die Mannschaft auf einem Schiffe nach einem Ende hinlaufen würde, so erhielte dadurch das Schiff einen Ruck in entgegengesetzter Richtung. Diese Bewegung würde aber mit der Rückbewegung der Mannschaft wieder vernichtet.

Stellt sich die Mannschaft eines Schiffes in einen Kreis und beginnt eine Drehung etwa im Sinne des Uhrzeigers, so erhält das Schiff dadurch die verkehrte Drehung. Dies beruht auf dem Flächenprincip. Eine ähnliche Erscheinung können wir an jedem Electromotor demonstriren*). Bringen wir den Motor von Grüel (Fig. 4) mit oscillirendem Anker auf einer verticalen Axe so an, dass das Schwungrad horizontal liegt, wobei zweckmässig das gewöhn-

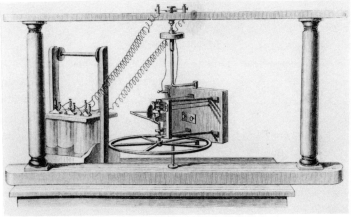

Fig. 4.

lich hölzerne Schwungrad durch ein massiveres von Metall ersetzt wird. Der ganze Motor ist nun um diese Verticalaxe beweglich und der Strom wird durch zwei kreisförmige Quecksilberrinnen k an der Axe so eingeleitet, dass diese Bewegung nicht behindert ist. Wir binden nun den Motorkörper mit einem Faden an das Stativ der Axe fest und leiten den Strom ein. Sobald das Schwungrad (von oben gesehn) sich im Sinne des Uhrzeigers zu drehen beginnt, zeigt der Motorkörper das Bestreben sich verkehrt zu drehen. Diese

---

*) M a c h , a. a. O.

The motor is a freely moving system with respect to the axis of rotation. If it is at rest, the sum of the areas swept out in unit time = 0. If it starts moving as a result of the internal (electric) forces between the coil and the iron core, then the flywheel also immediately rotates through an angle, in unit time, determined by the well-known rule of conservation of areas. The overall sum has to remain = 0, however, since only internal forces are involved. Therefore, the traversed area [angular momentum] of the flywheel must be compensated by the oppositely oriented area of the motor body.

An unusual phenomenon occurs when the current is interrupted. Wheel and motor initially continue to rotate in opposite directions. Soon friction becomes noticeable and, little by little, the parts of the motor are brought to rest relative to one another. One sees now how the movement of the motor body slows down and finally reverses. The whole motor then rotates the way the flywheel alone turned initially. The explanation for this occurrence is simple. The system is not completely free, but constrained by the resistance and friction of the axis. In the case of a totally free system, the sum of the traversed areas would always have to be = 0, because only inner forces were at play. As soon as the parts became motionless relative to each other, the motor would also have had to come to rest. Not so here. The friction of the flywheel axis reduces the area swept out [angular momentum] by the wheel as well as by the body. The friction of the main axis, however, only reduces the area swept out by the body, whereas the wheel retains an excess [of momentum], which becomes apparent after switching off the current. The whole result following switching off the current is a nice illustration of the process that occurred with the moon, according to astronomers. The tidal motions excited on the earth reduced the rotation of the moon, such that the moon-day[2.37] grew to the duration of a month. The flywheel corresponds to the water masses being moved by the tides.

Drehung wird auch rapid ausgeführt, wenn man den Faden abbrennt. Wichtig ist bei dem Versuch, dass die Axe gut vertical steht.

Der Motor ist in Bezug auf die Axendrehung ein freies System. Wenn er in Ruhe, ist die in der Zeiteinheit durchstrichene Flächenraumsumme = 0. Kommt er durch die innern (electrischen) Kräfte zwischen Anker und Eisenkern in Bewegung, so liefert das Schwungrad für die Zeiteinheit sofort eine nach der bekannten Regel bestimmbare Flächenraumsumme. Die Gesammtsumme muss aber, weil bloss innere Kräfte spielen, = 0 bleiben. Demnach müssen die Flächenräume des Schwungrades durch entgegengesetzt beschriebene des Motorkörpers compensirt werden.

Eine eigenthümliche Erscheinung tritt ein, wenn man den Strom unterbricht. Rad und Motor fahren zunächst fort entgegengesetzt zu rotiren. Bald aber wird die Reibung merklich und bringt nach und nach die Theile des Motors in relative Ruhe gegen einander. Hiebei sieht man nun, wie die Bewegung des Motorkörpers immer langsamer wird und schliesslich sich dem Sinne nach umkehrt. Der ganze Motor rotirt also dann so, wie anfänglich das Schwungrad allein sich bewegt hat. Die Erklärung des Vorganges ist einfach. Das System ist kein vollkommen freies, es ist durch den Widerstand der Axe und durch die Axenreibung behindert. Bei einem vollkommen freien System hätte die Summe der durchstrichenen Flächenräume, da bloss innere Kräfte wirkten, immer = 0 sein müssen. Sobald also die Theile in relative Ruhe gegeneinander gekommen waren, hätte der Motor wieder in Ruhe sein müssen. Nicht so hier. Die Reibung an der Schwungradaxe vermindert die Flächenraumsumme sowohl für das Rad wie für den Körper. Die Reibung an der Hauptaxe aber wirkt bloss vermindernd auf die Flächenraumsumme des Körpers, jene des Rades bleibt also im Ueberschuss und dies äussert sich bei Unterbrechung des Stromes. Der ganze Vorgang nach Unterbrechung des Stromes ist ein schönes Bild desjenigen, welcher nach Voraussetzung der Astronomen am Monde eingetreten ist. Die von der Erde erregte Fluthwelle hat die Rotation des Mondes derart vermindert, dass der Mondtag zur Dauer eines Monats angewachsen ist. Das Schwungrad stellt die durch die Fluth bewegte Wassermasse vor.

The last result can also be demonstrated in another way. A bar *HH* (Fig. 5) that can easily rotate on a base *S* has two small indentations *bb*, into each of which one inserts the axis of a well-known <u>Schmidt gyro</u>[2.38] *kk*. As is

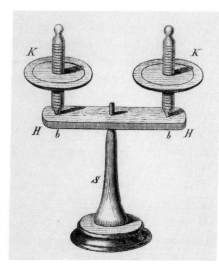

commonly known it can be started by hand. If both tops rotate in the same sense, then the whole bar *HH* soon achieves a rapid rotation in correspondence with the gyros.

Even without allowing current to flow, one can observe that the flywheel of the motor will at first not turn if one rapidly rotates the motor body. On the other hand, if one suddenly stops the motor body after the motor had been turning for a long time, the flywheel will maintain its rotation

Fig. 5.

at first. This phenomenon, which incidentally can easily be seen on every <u>pocket watch</u>[2.39] and its balance, demonstrates that every part of the body follows the principle of the conservation of areas to the extent possible.

6.

We will perform the experiment in still another way, which permits the necessary variations. We take a gyro, the axis of which rotates very easily in a ring [gimbal] and mount it with the axis vertical on the disk of a centrifuge. Previously we had attached colored paper in sectors on the gyro and on the centrifuge to recognize their rotations more easily. When we start rotating the centrifuge the disk of the gyro initially remains stationary and only slowly is brought into rotation. Upon stopping the machine [centrifuge] after prolonged rotation the gyro, in contrast, continues to rotate.

Letzteren Vorgang kann man noch auf andere Weise demonstriren. Ein an dem Stativ *S* leicht drehbarer Hebel *HH* (Fig. 5) ist mit zwei Bohrungen *bb* versehen, in welche man die Axen der

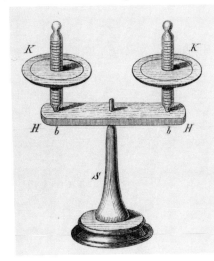

bekannten **Schmidt'schen** Kreisel *kk* steckt, die man bekanntlich in der Hand kann anlaufen lassen. Laufen beide Kreisel in demselben Sinn, so kommt bald der ganze Hebel *HH* in rapide Rotation übereinstimmend mit den Kreiseln.

Auch ohne den Strom wirken zu lassen, kann man beobachten, dass das Schwungrad des Motors die Drehung zunächst nicht mitmacht, wenn man den Körper rasch in Rotation versetzt. Umgekehrt, hat der Motor längere Zeit rotirt und man hält den Körper plötzlich an, so behält das

Fig. 5.

Rad zunächst seine Rotation bei. Diese Beobachtung, die sich übrigens an jeder Taschenuhr und deren Unruhe hinreichend gut anstellen lässt, zeigt, dass jeder Körpertheil, so viel er kann, das Flächenprincip befolgt.

6.

Wir werden das Experiment noch in einer andern Art anstellen, welche die nöthigen Variationen erlaubt. Wir nehmen eine Kreiselscheibe, deren Axe in einem Ring sehr leicht läuft, und befestigen sie mit verticaler Axe an der Scheibe einer Centrifugalmaschine, nachdem wir zuvor, um die Drehungen leicht zu erkennen, an der Kreiselscheibe und an der Centrifugalscheibe Papiere mit Sectoren bemalt angebracht haben. Beim Beginn der Drehung der Centrifugalmaschine bleibt die Kreiselscheibe in Ruhe und wird erst allmälig mitgenommen. Beim Anhalten der Maschine nach längerer Rotation dreht sich im Gegentheil die Kreiselscheibe weiter.

If a part of the body cannot completely fulfill the principle of the conservation of areas, partial adherence will take place, whenever possible.

We screw a wooden block of the form *AB* (Fig. 6) onto the disk of a centrifuge. We fasten the ring *R* of the disk *k* through a hinge *a* such that the axis *ab* can assume any arbitrary inclination relative to the vertical rotation axis of the centrifuge. We fix this inclination with a string *bn* that is connected through a hole in the wooden frame to a small hook *h* positioned rather accurately in line with the rotation axis. As soon as the angle between the axis of the gyro disk and the axis of the centrifuge differs from 90°one observes the above phenomenon with every movement of the apparatus. However, this becomes weaker as the angle approaches 90°, and it completely disappears at 90°. This is based on the alignment of rotations. If the sum of the traversed areas which the totally unrestrained disk would achieve in unit time is called *f*, then for an angle of inclination α between the two axes it is only *f*. cos α.

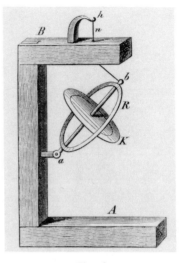

Fig. 6.

With the help of the string *bn* we now position the gyro axis horizontal, so that the angle α = 90°. The above phenomena will no longer occur. After the constant rotation has continued for some time, we burn the string at *n* with a long darting flame. The disk will fall so that *ab* is directed downwards. At the same time we see that the disk does not rotate, but remains motionless. It had therefore not taken up the rotation. It is clear that for this and the following experiment the piece $\underline{ac}^{2.40}$ must be considerably longer than indicated in the drawing, in order to save space.

Kann ein Körpertheil das Flächenprincip nicht vollständig erfüllen, so tritt, wo dies möglich ist, eine theilweise Erfüllung ein.
Wir schrauben ein Holzstück von
der Form *AB* (Fig. 6) auf die
Scheibe der Centrifugalmaschine.
An demselben befestigen wir den
Ring *R* der Kreiselscheibe *k* mit
einem Gelenk *a* derart, dass die
Axe *ab* eine beliebige Neigung
gegen die verticale Rotationsaxe
der Centrifugalscheibe annehmen
kann. Diese Neigung fixiren wir
durch einen Faden *bn*, der durch
ein Loch des Holzstückes an ein
Häkchen *h* geht, das ziemlich
genau in der Rotationsaxe liegt.
Sobald nun der Winkel der Krei-
selscheibenaxe mit der Axe der
Centrifugalmaschine von 90° ver-

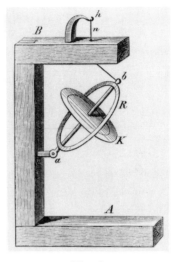

Fig. 6.

schieden ist, beobachtet man bei jeder Bewegung des Apparates die
eben beschriebenen Erscheinungen, welche allerdings desto schwächer sind, je mehr sich dieser Winkel 90° nähert, und welche bei
90° vollständig verschwinden. Es beruht dies auf der Zusammensetzung der Drehungen. Wäre die Flächenraumsumme, welche die
Kreiselscheibe bei vollkommener Freiheit in der Zeiteinheit geben
würde, *f*, so ist sie bei der Neigung der beiden Axen um den Winkel α nur *f*. cos α.

Wir stellen nun mit Hilfe des Fadens *bn* die Axe der Kreiselscheibe *ab* horizontal, wobei also der Winkel α = 90° wird. Die
erwähnten Erscheinungen treten nun nicht ein. Nachdem die
gleichförmige Rotation eine Zeit lang unterhalten worden ist,
brennen wir bei fortdauernder Rotation mit einer langen Stichflamme
bei *n* den Faden ab. Die Scheibe fällt nun herab, so dass *ab* vertical abwärts gerichtet ist. Zugleich sehn wir, dass die Kreiselscheibe sich nicht dreht, sondern ruhig ist. Sie hat also die Rotation gar nicht angenommen. Es versteht sich, dass für dieses und
das folgende Experiment das Stück *ac* bedeutend länger sein muss,

We now tie the gyro disk with the thread such that the axis ab is directed vertically upwards. After a prolonged constant velocity rotation, we burn the thread without interfering with the rotation. The disk will fall such that ab is directed vertically downwards, and the disk turns completely upside down. The disk, which had assumed the angular velocity of the centrifuge will, of course, immediately rotate in an opposite direction, while maintaining its own velocity but presenting its other side to an outside observer. This case allows us to present a nice application.

Let us continue with a physical observation. Consider a heavy, easily rotatable horizontal, seemingly motionless disk. It must take on the angular velocity resulting from the inclination of its axis relative to the earth's rotation axis. If we could suddenly turn this disk over in its own plane, then, because of its inertia, it would reach twice the angular velocity of a Foucault[2.41] pendulum at the same location.

The experiment is difficult to perform in this manner, but little difficulty should be encountered in the following version.[2.42] Consider a heavy vertical disk with its axis horizontal. The axis itself should be suspended at one point (or be floating), so that it can very easily be moved horizontally. The center of gravity of the disk lies almost directly above the axis and the disk is fixed with a string. If one burns this string, the disk[2.43] will perform nearly half a rotation. Its southern rim will, for example, exchange place with its northern rim, and the axis must now rotate with an angular velocity double that of the Foucault pendulum.

7.

An unrestrained system of masses A satisfies the principles of the center of gravity and of the conservation of area. If the system of masses A is not free, but rather fully or partially attached to a system of masses

als dies der Raumersparniss wegen in der Zeichnung angegeben ist.

Wir binden die Scheibe mit Hilfe des Fadens so fest, dass die
Axe ab vertical aufwärts gerichtet ist, und brennen den Faden nach
längerer gleichförmiger Rotation ohne diese zu unterbrechen ab.
Die Scheibe fällt so herab, dass ab vertical nach unten gerichtet
und die Scheibe nun ihre vorher obere Seite nach unten kehrt, dass
sie also vollständig umgeklappt wird. Hiebei rotirt nun die Scheibe,
welche die Winkelgeschwindigkeit der Centrifugalmaschine angenommen hat, selbstverständlich sofort im entgegengesetzten Sinne,
indem sie ihre Geschwindigkeit beibehält, dem Beobachter aber die
andere Seite präsentirt. Von diesem Fall werden wir eine hübsche
Anwendung machen.

Knüpfen wir hieran eine physikalische Bemerkung. Denken
wir uns eine schwere, sehr leicht bewegliche horizontale scheinbar
ruhende Scheibe. Diese muss die Winkelgeschwindigkeit annehmen,
welche ihr vermöge der Neigung gegen die Axe der Erdrotation zukommt. Könnten wir nun diese Scheibe rasch in ihre eigene Ebene
umklappen, so müsste sie vermöge ihrer Trägheit die doppelte Winkelgeschwindigkeit zeigen, welche dem Foucault'schen Pendel an
demselben Orte zukommt.

In dieser Form ist das Experiment schwer ausführbar, es dürfte
aber in der folgenden kaum auf Schwierigkeiten stossen. Wir denken uns eine schwere verticale Scheibe mit horizontaler Axe. Die
Axe selbst soll in einer Spitze aufgehängt sein (oder schwimmen),
so dass sie sich sehr leicht horizontal bewegen kann. Der Schwerpunkt der Scheibe liegt fast vertical über der Axe und die Scheibe
ist durch einen Faden fixirt. Brennt man letzteren Faden ab, so
macht die Scheibe fast eine halbe Drehung, ihr südlicher Rand tritt
beispielsweise an die Stelle des nördlichen und die Axe muss nun
die doppelte Winkelgeschwindigkeit des Foucault'schen Pendels
zeigen.

7.

Ein freies Massensystem *A* erfüllt das Schwerpunkts- und Flächenprincip. Wenn das Massensystem *A* nicht frei, sondern ganz
oder theilweise an ein Massensystem *B* gebunden ist, so wird das

B, then the principles of the center of gravity and conservation of areas will no longer apply for A. However, the deviations of A from both principles are then determined by the forces that B exerts on A, and the same forces must also be exerted by A upon B. With every movement producing a displacement by a force on A, there will be a push and pull on B as well.

We will now make the simplifying assumption, which is sufficient for most cases, that both A and B are simple rigid bodies. The following principles then follow directly from the above. If A, having mass m and acceleration φ, is to be brought by a body B to its acceleration φ', then B must exert a force $m(\varphi - \varphi')$ on A. The directions of both accelerations are taken into acount, of course. If A is to be brought to an acceleration ψ relative to B, it will require a force $m[\psi + (\varphi' - \varphi)]$ by B. This force must exert its action by B upon A and also by A upon B.

The same law applicable to linear movements is also valid for angular motion. If a body A has moment of inertia T and angular acceleration φ measured with respect to the rotation that is to take place, and if it is to be brought to the angular acceleration of B, then the required static moment is $T(\varphi' - \varphi)$. Similarly, if A is to be brought to the angular acceleration level ψ by B, a moment $T[(\psi + (\varphi' - \varphi)]$ is required. This rotational moment must be exerted by B upon A, and by A upon B.

8.

When we move actively, the mechanical output of muscle could be associated with a sensation that provides us with information about the actual movement. Furthermore, according to the above principles, many other sources for movement sensation are certainly available. The common causes affecting a movement or obstacles to a movement will normally not act on our entire body mass simultaneously, but only on a small part. If we sit in a carriage that starts moving, our skin will be the first to be prevented from fulfilling the principles of the center of gravity and of the conservation of area. This in turn will affect the muscles, bones, connective tissue, blood, and the other fluids of the body. Every part of the body will push and pull on every other part. To the extent that many of these pressures and strains can

Schwerpunkts- und Flächenprincip für *A* nicht mehr gelten, aber die Abweichungen des *A* von beiden Principien sind dann durch die Kräfte, die *B* auf *A* ausübt, hervorgebracht und dieselben Kräfte muss nun auch *A* auf *B* ausüben. Bei jeder Bewegung werden sich also die Abweichungen, zu welchen *A* gezwungen wird, als Züge und Pressungen auf *B* äussern.

Wir wollen nun die erleichternde Annahme machen, welche für die meisten Fälle auch ausreicht, dass sowohl *A* wie *B* einfach feste Körper sind. Dem Obigen gemäss sind dann folgende Grundsätze unmittelbar verständlich. Wenn *A* die Masse *m* und die Beschleunigung φ hat, welche letztere durch den Körper *B* von der Beschleunigung φ' relativ gegen diesen aufgehoben oder relativ gegen denselben in ψ umgewandelt werden soll, wobei alle Beschleunigungen in demselben Sinne positiv gezählt werden, so ist hiezu die Kraft $m(\varphi' - \varphi)$, beziehungsweise $m[\psi + (\varphi' - \varphi)]$ erforderlich. Diese muss aber von *B* auf *A* und auch von *A* auf *B* ausgeübt werden.

Das gleiche Gesetz wie für die Progressivbewegungen gilt auch für die drehenden Bewegungen. Hat der Körper *A* in Bezug auf die auszuführende Drehung das Trägheitsmoment *T* und die Winkelbeschleuuigung φ und wird diese durch den Körper *B* von der Winkelbeschleunigung φ' relativ gegen letzteren aufgehoben oder in ψ umgewandelt, so ist das hiezu erforderliche statische Moment $T(\varphi' - \varphi)$, beziehungsweise $T[\psi + (\varphi' - \varphi)]$. Dieses Drehungsmoment muss hiebei *B* auf *A* und *A* auf *B* ausüben.

8.

Wenn wir uns activ bewegen, so mögen schon die mechanschen Leistungen der Muskel mit Empfindungen verbunden sein, welche uns über die ausgefürte Bewegung belehren. Ausserdem werden sich nach den angegebenen Grundsätzen noch viele andere Quellen von Bewegungsempfindungen auffinden lassen. Die gewöhnlich wirkenden Bewegungsursachen und Bewegungshindernisse wirken nicht gleichzeitig auf die ganze Masse unseres Körpers, sondern stets nur auf einen kleinen Theil. Sitzen wir in einem Wagen, welcher in Bewegung geräth, so wird zunächst unsere Haut verhinhindert, das Schwerpunkts- und Flächenprincip zu erfüllen, diese afficirt, weiter Muskel, Knochen, Bindegewebe, das Blut und die übri-

be sensed, just as many means of motion sensation might occur. For active movements as well, one part of the body will be moved by another, and therefore the same considerations apply.

The sensation of motion could therefore arise from pressure and strain on the skin, by an excitation of the sensory nerves in the bones, muscles and connective tissue, or by the muscular sense, which could be experienced by maintaining or changing position of a part of the body during motion. The brain also could possibly sense its own weight (as assumed by Purkyně). If the vasomotor nerves are connected to sensory ones, a direct sensing of the weight of the blood and its distribution could be possible. Even the visual sense will not remain unaffected. We will see from further investigations that probably all these enumerated sensations, to the extent they exist at all, are of only secondary importance, and the most important aspects of motion sensation originate in a special organ. Even if this result should be modified by future investigations, the principles that have been elaborated in this chapter must always be applied in finding the <u>origins of motion sensation</u>.[2.44] In this sense the following experiments will always retain a certain value.

The Phenomena Observed in Moving Men and Animals.

1.

When one says that a motion can be sensed, it is not a sufficient and, strictly taken, not an appropriate term for expressing the facts. Very violent motions can be encountered, which are not sensed at all. One must specify the circumstances in which a motion is sensed. The mechanical interaction of masses consists in mutual accelerations. Therefore, it it is very likely that only accelerations are sensed.

gen Flüssigkeiten des Körpers. Jeder Körpertheil wird auf jeden andern Pressungen und Zerrungen ausüben müssen und so vielfach diese Pressungen und Zerrungen empfunden werden können, so vielfach werden die Bewegungsempfindungen sein. Bei der activen Bewegung wird ebenfalls ein Körpertheil durch den andern bewegt und es gelten also dieselben Betrachtungen.

Bewegungsempfindungen können also entstehn durch Pressungen und Zerrungen der Haut, durch Affection der sensiblen Nerven in den Knochen, Muskeln und im Bindegewebe, durch die Muskelgefühle, welche bei Erhaltung oder Aenderung der Stellung der Körpertheile während der Bewegung auftreten. Möglicher Weise könnte auch (wie Purkyně angenommen hat) das Hirn seinen eigenen Druck empfinden. Sind die vasomotorischen Nerven mit sensiblen verbunden, so wäre auch eine directe Empfindung des Gewichtes und der Vertheilung der Blutmasse möglich. Auch der Gesichtssinn bleibt nicht unafficirt. Wir werden nun durch die weitere Untersuchung sehn, dass wahrscheinlich alle bisher aufgezählten Empfindungen, so weit sie überhaupt existiren, nur nebensächlich sind und die wichtigsten Bewegungsempfindungen muthmaasslich von einem besondern Organ herrühren. Sollte aber auch dieses Resultat durch künftige Untersuchungen alterirt werden, so werden doch die in diesem Capitel aufgestellten Grundsätze wohl immer benutzt werden müssen, um die Quellen der Bewegungsempfindungen aufzufinden. Damit werden aber die folgenden Versuche immer einen gewissen Werth behalten.

Die Erscheinungen an bewegten Menschen und Thieren.

1.

Wenn man sagt, dass die Bewegung empfunden werde, so ist dies kein erschöpfender und genau genommen kein richtiger Ausdruck der Thatsachen. Es gibt sehr heftige Bewegungen, die gar nicht empfunden werden. Man muss sagen, welcher Umstand der Bewegung empfunden wird. Die mechanische Wechselwirkung der Massen besteht in gegenseitiger Beschleunigung. Es hat also

When travelling on a train moving at constant velocity, one feels only a little shaking, consisting of alternating accelerations and decelerations of our body. These balance each other because the average velocity remains constant. This shaking remains the same whether the train is moving forward or backward. As a matter of fact, with closed eyes one can easily imagine either direction equally well, and one can switch from one idea to the other without difficulty. This is no longer possible if the train starts moving or comes to a stop, that is, when the average acceleration is non-zero and has a certain direction.

If one travels around a narrow <u>curve in a train</u>[3.1], houses and trees seem to lean considerably from the vertical, so that the tops of trees seem to tilt away from the train towards the convex side of the curve. On the other hand one often notices the railway cars to be tilted and then the houses and trees appear to be vertical.

It is well known that the rail on the convex side of a curve is placed a little higher to compensate for the effects of centrifugal forces. The difference in height, however, is appropriate for only a single value of train velocity. These two seemingly contradictory facts can easily be explained, if it is assumed that one senses the direction of the vertical, and takes as the vertical the direction of mass acceleration resulting from the combination of weight and <u>centrifugal force</u>.[3.2]

If one travels with that velocity which just corresponds to the curvature and the height difference of the rails, one senses no tilt of the railway car. Then the houses and trees are tilted. In every other case the railway car seems to be tilted.

2.

After these observations were made the following experiments were performed in order to study the phenomena under more controlled

von vorn herein eine grosse Wahrscheinlichkeit, dass bloss Beschleunigungen empfunden werden.

Befindet man sieh auf einem mit constanter Geschwindigkeit bewegten Eisenbahnzuge, so fühlt man nur die kleinen Erschütterungen, welche in geringen abwechselnden Beschleunigungen und Verzögerungen unseres Körpers bestehn, die sich weil die mittlere Geschwindigkeit eben konstant bleibt die Wage halten. Diese Erschütterungen bleiben dieselben, ob der Zug vor- oder rückwärts fährt. In der That kann man sich bei geschlossenen Augen beides gleich leicht vorstellen und kann ohne Schwierigkeit von der einen Vorstellung zur andern übergehn. Dies ist nicht mehr möglich, wenn der Zug im Abfahren oder Anhalten begriffen ist, also de mittlere Beschleunigung von Null verschieden ist und einen bestimmten Sinn hat.

Fährt man auf der Eisenbahn durch eine starke Krümmung, so scheinen die Häuser und Bäume oft beträchtlich von der Verticalen abzuweichen und zwar scheint sich der Gipfel der Bäume auf der convexen Seite der Krümmung von der Bahn wegzuneigen. Anderseits bemerkt man sehr oft auch eine Schiefstellung des Wagens und hält nun die Häuser und Bäume für vertical.

Bekanntlich wird die Schiene auf der convexen Seite der Krümmung etwas höher gelegt um die Wirkung der Centrifugalkraft zu compensiren. Der Höhenunterschied kann aber nur einer einzigen Fahrgeschwindigkeit entsprechen. Die beiden erwähnten einander scheinbar widersprechenden Facta klären sich nun einfach auf, wenn man annimt, dass man die Richtung der Verticalen empfindet und stets die Richtung der aus Schwere und Centrifugalkraft resultirenden Massenbeschleunigung für die verticale hält.

Fährt man mit jener Geschwindigkeit, welche der Krümmung und dem Höhenunterschiede der Schienen entspricht, so weiss man nichts von der Schiefstellung der Wagens. Dann scheinen die Häuser und Bäume schief. In jedem andern Falle erscheint der Wagen schief.

2.

Die folgenden Versuche, welche zunächst durch die angeführten Beobachtungen und den Wunsch die Erscheinungen unter Umstän-

conditions. They were done following the same principle used earlier by Knight*)[3.3] to study the influence of gravity on the growth of plants. Purkyně[3.4] had previously performed some of my experiments, and I could have saved myself some effort, if I had known of Purkyně's publications from the beginning. But then I would probably not have thought of many new experiments at all.

I will now describe my equipment. Imagine a vertical wooden frame R (Fig. 7), 4^m long, 2^m high, which can easily be rotated about a vertical axis A passing through its center. Inside there is a smaller second vertical frame r, which can also be rotated about a vertical axis a, which can be positioned at any distance relative to the first axis. The second frame carries a chair, which can be tilted about a horizontal axis α. The observer who sits on this chair can be positioned aligned with the rotation axis A about which the whole device can be rotated, or he can be positioned at a considerable

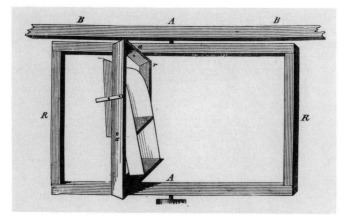

Fig. 7.

distance from it. In the latter case, because of the possibility of rotation about a, he could look towards A, or in a direction orthogonal to the plane determined by A and a. Finally, the subject can be brought into a more

---

*) Philosophical transactions 1806.

den zu studiren, die man mehr in der Hand hat, veranlasst sind, wurden nach demselben Princip angestellt, welches schon Knight*) angewandt hat, um den Einfluss der Schwere auf das Wachsthum der Pflanzen zu untersuchen. Die Versuche sind zum Theil schon von Purkyně ausgeführt und ich hätte mir, wenn mir gleich bei Beginn meiner Arbeit die Purkyně'sche Abhandlung bekannt gewesen wäre, manche Mühe ersparen können. Wahrscheinlich wäre ich aber dann auf viele neue Versuche gar nicht verfallen.

Ich will nun zunächst meinen Apparat beschreiben. Man denke sich einen verticalen Holzrahmen R (Fig. 7), 4^m lang, 2^m hoch, um eine durch seine Mitte gehende verticale Axe A leicht drehbar. In diesem befindet sich ein zweiter kleinerer verticaler Rahmen r, ebenfalls um eine vexticale Axe a drehbar, welche in beliebige Distanz von der ersten Axe gebracht werden kann. Letzterer Rahmen trägt einen Stuhl, welcher sich um eine horizontale Axe α neigen lässt. Der Beobachter, welcher auf diesem Stuhl Platz nimmt, kann in die

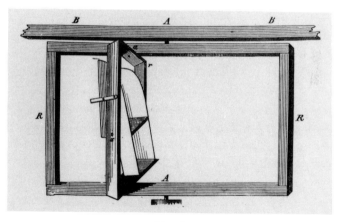

Fig. 7.

Rotationsaxe A, um die der ganze Apparat in Umschwung versetzt wird, oder in beträchtliche Entfernung von derselben gebracht werden. Er kann in letzterem Falle vermöge der Drehbarkeit um a

---

*) Philosophical transactions 1806.

horizontal or more vertical position through the possibility of rotation about α. To finally avoid visual vertigo, the whole subject can be enclosed in a kind of paper box. The apparatus was often modified in the course of experiments and new parts were added, which will be mentioned at the appropriate places in the following discussion.

3.

Next we will consider experiments involving rotation.

Experiment 1. The subject is enclosed in the paper box and tilted about the axis α, so that he slowly shifts from a vertical to a supine position. He indicates the direction that he considers as vertical, using a pointer[3.5] which projects from the box. This indication is initially quite exact. As he approaches the supine position, the subject indicates his tilt from the vertical[3.6] to be smaller than it actually is.

Experiment 2. The subject is placed on the rotation axis A in an upright sitting position within the paper box, and is rotated about the vertical axis. Each rotation is immediately detected, as to both direction and magnitude. When one continues the constant rotation for several seconds, however, the sensation of turning gradually diminishes. When one leaves the apparatus alone so that it decelerates, one senses a rotation in the opposite direction. This sensation of opposite rotation becomes very strong if one stops the apparatus suddenly, and decreases gradually over several seconds with duration[3.7] depending on the speed of the rotation. If the experiment is carried too far, sensations of head heaviness and disgusting nausea also occur. One therefore does not sense angular velocity, but rather angular acceleration.

An optical[3.8] phenomenon is also noticeable in this experiment. The interior of the enclosure contains a vertical line with a star that can be fixated. As soon as the apparatus is stopped, one gets the feeling of turning in the opposite direction, along with the box. When the box is opened, the entire visible room[3.9] seems to turn, along with all its contents. It seems as if

sowohl nach *A* hin sehen, als auch senkrecht auf die Ebene von *A* und *a* hinausblicken. Endlich kann dem Beobachter vermöge der Drehbarkeit um α eine mehr horizontale oder mehr verticale Lage gegeben werden. Um endlich den Gesichtsschwindel auszuschliessen, kann der ganze Beobachter in eine Art Papierkasten eingeschlossen werden. Der Apparat hat im Laufe der Versuche vielfache Veränderungen erfahren und neue Theile erhalten, welche an den betreffenden Stellen zur Sprache kommen.

3.

Wir betrachten zunächst die auf die Rotation bezüglichen Versuche.

Versuch 1. Der Beobachter wird in den Papierkasten eingeschlossen und um die Axe α geneigt, so dass er allmälig aus der Verticalstellung in die Rückenlage kommt. Hiebei gibt er mit einem Stabe, der aus dem Kasten hervorragt, die Richtung an, welche er für vertical hält. Diese Angabe ist anfänglich ziemlich genau. Bei starker Annäherung an die Rückenlage hält jedoch der Beobachter seine Abweichung von der Verticalen für kleiner als sie wirklich ist.

Versuch 2. Der Beobachter wird aufrecht sitzend in die Rotationsaxe *A* gebracht, in den Papierkasten eingeschlossen und in Rotation um die Verticalaxe versetzt. Jede Drehbewegung wird sofort dem Sinne nach und der beiläufigen Grösse nach erkannt. Erhält man aber einige Secunden lang die Rotation gleichförmig, so hört allmälig das Gefühl der Drehung ganz auf. Es tritt das Gefühl einer entgegengesetzten Drehung auf, wenn man den Apparat sich selbst überlässt, so dass er einen verzögerten Gang annimmt. Dies Gefühl der Gegendrehung wird äusserst heftig, wenn man den Apparat plötzlich anhält, und dauert je nach der Stärke der Rotation allmälig abnehmend einige Secunden. Treibt man das Experiment zu weit, so stellt sich auch Eingenommenheit des Kopfes und Ekel ein. Man empfindet also nicht die Winkelgeschwindigkeit, sondern die Winkelbeschleunigung.

Bei diesem Versuche bemerkt man auch eine optische Erscheinung. Im Innern des Papierkastens befindet sich ein verticaler Strich und darauf ein Stern, den man fixiren kann. Sobald der Apparat angehalten wird, hat man den Eindruck, als ob man sammt dem Kasten eine Gegendrehung ausführen würde. Oeffnet man

the visible room is itself being turned within a second room. This second room, though not visible, is assumed to be completely stationary. One might believe that a second room stands behind the visible room, and provides a constant frame of reference for the first one. This fact of fundamental importance has to be experienced for one's self. It cannot easily be described. Furthermore, this phenomenon had never been characterized the way I did it here, not even by Purkyně, the most careful observer.

Remarkably, following a sudden stop of the apparatus one notices a sensation as though the rotation in the opposite direction encountered a certain resistance. The feeling, especially for the head, is as though one was moving through a viscous mash.

The experiment works for any position of the body and the head relative to the rotation axis. As long as one remains still, the rotation axis after stopping will also be the axis of the illusory counter-rotation. The body position can be varied about the axis α. I also installed a horizontal board within the frame R, and lay on it in a comfortable position so that all parts of my body were well supported. I observed no change in the phenomena. It makes no difference whether one lies on the side or on the back.

Experiment 3. We stop the apparatus abruptly as in experiment 2, and after 1½ - 2 seconds we suddenly start it rotating again in the same direction. The sensation of counter-rotation, which occurred during the pause, disappeared as soon as the rotation resumed. The sensation of rotation induced by an angular acceleration persists for considerable time, and can be abolished by an angular rotation in the opposite direction.

Experiment 4. If one performs experiment 2 with the head pitched forward, and moves it back to the erect position after stopping the apparatus, one has the following impression. If the subjective rotation had been from the right to the front and left, the rotation sensation now is from right to above and left, so that one fears falling to the side.[3.10] Based on these

rasch den Kasten, so dreht sich der ganze sichtbare Raum mit seinem ganzen Inhalt. Es sieht so aus, als ob der sichtbare Raum sich in einem zweiten Raum drehen würde, den man für unverrückt fest hält, obgleich letzteren nicht das mindeste Sichtbare kennzeichnet. Man möchte glauben, dass hinter dem Sehraum ein zweiter Raum steht, auf welchen ersterer immer bezogen wird. Diese Thatsache von fundamentaler Wichtigkeit muss man selbst erfahren. Sie lässt sich nicht gut beschreiben. Die Erscheinung ist auch nie so characterisirt worden, wie ich es hier gethan habe, auch von Purkyně nicht, welcher am besten beobachtet hat.

Merkwürdig ist beim raschen Anhalten des Apparates noch das Gefühl, als ob die Gegendrehung mit einem gewissen Widerstande ausgeführt wurde. Man meint sich namentlich mit dem Kopfe wie in einem zähen Brei zu bewegen.

Der Versuch gelingt bei jeder Lage des Körpers und Kopfes gegen die Rotationsaxe. So lange man sich ruhig verhält, ist die Rotationsaxe nach dem Anhalten auch immer die Axe der scheinbaren Gegendrehung. Die Körperlage kann mit Hilfe der Axe α variirt werden. Auch habe ich in dem Rahmen R ein horizontales Brett angebracht und mich bequem darauf gelegt, so dass alle Körpertheile gut unterstützt waren, ohne eine Aenderung der Erscheinungen zu beobachten. Es ist hiebei einerlei, ob man auf dem Rücken oder auf der Seite liegt.

Versuch 3. Halten wir bei dem Versuch 2 den Apparat plötzlich an und setzen wir ihn nach $1^{1}/_{2}$ – 2 Secunden wieder plötzlich in demselben Sinne in Bewegung. Das Gefühl der Gegendrehung, welches in der Pause auftritt, verschwindet, so wie die Bewegung wieder eingeleitet ist. Die durch eine Winkelbeschleunigung erzeugte Drehempfindung hat also eine beträchtliche Nachdauer und kann durch eine entgegengesetzte Winkelbeschleunigung aufgehoben werden.

Versuch 4. Neigt man bei Versuch 2 den Kopf nach vorn und richtet ihn nach dem Anhalten des Apparates wieder auf, so meint man, wenn die subjective Drehung von rechts nach vorn und links stattgefunden hat, sich von rechts nach oben und links zu drehen, so dass man ein Umstürzen nach der Seite fürchtet. Man

experiments as well as the related earlier, less precise and convenient investigations of Purkyně, the head position is seen to be a decisive factor. By placing the head in an arbitrary position during the outlasting sensation of rotation, one can show that the apparent rotation axis (which was determined by the initial actual rotation) moves with the head. Its position remains fixed in the head.

Experiment 5. We now place the subject approximately upright, facing the rotation axis A at a distance of 1^m, and start him turning. Soon after the angular velocity of the subject enclosed in the paper box becomes constant, any sensation of rotation subsides. The subject then feels only that he is tilted further backwards than is actually the case. One senses the direction of the resultant gravito-inertial acceleration and takes it to be the vertical.

Experiment 6. We leave everything as in experiment 5, except that now by turning the frame r about axis α the subject is positioned such that he faces orthogonal to the plane containing the axes A and a. Furthermore, he remains enclosed in the paper box containing a vertical stripe on the inside. As soon as the angular velocity becomes constant, the subject feels that he and the enclosure are permanently tilted to the side. He feels his head tilted away from the rotation axis. Once the apparatus stops he feels as though he and the environment have returned to the erect orientation, as had also occurred in the preceding experiment. We deduce from this experiment, just as from experiment 5, that one senses the direction of the gravito-inertial acceleration as the vertical. Purkyně had previously observed this tilting to the side on the platform of a merry-go-round.

Experiment 7. If, in experiment 6, one hangs a pendulum with an attached protractor inside the paper enclosure, one sees a 10 - 20° offset of the pendulum during the highest attainable rotation speed, corresponding to a centripetal acceleration of about $\frac{1}{5}$ - $\frac{1}{3}$ gravity. This pendulum direction is taken as the vertical during rotation, whereas the enclosure and one's self are perceived as tilted. However,

sieht aus diesem Versuch, welchen mit vielen analogen in weniger reiner und bequemer Form schon Purkyně angestellt hat, dass die Stellung des Kopfes maassgebend ist. Man kann so zu sagen mit der nachdauernden Drehempfindung den Kopf in eine beliebige Lage bringen und die Axe der scheinbaren Drehung, welche durch die anfängliche wirkliche Drehung bestimmt ist, macht alle Bewegungen des Kopfes mit, ihre Lage im Kopfe ist unveränderlich.

Versuch 5. Wir bringen nun den Beobachter 1^m weit von der Rotationsaxe A in nahe verticale Stellung, lassen ihn gegen die Axe A hinsehn und bringen ihn in Rotation. Bald nachdem die Winkelgeschwindigkeit des in den Papierkasten eingeschlossenen Beobachters constant geworden ist, hört jedes Drehgefühl auf. Der Beobachter meint bloss mehr auf dem Rücken zu liegen, als dies wirklich der Fall ist. Man empfindet die Richtung der resultirenden Massenbeschleunigung und hält diese für die Verticale.

Versuch 6. Wir lassen Alles wie im Versuch 5, nur stellen wir den Beobachter durch Drehung des Rahmens r um die Axe α so, dass er senkrecht gegen die Ebene der Axen A und a hinausblickt. Uebrigens bleibt er in dem Papierkasten eingeschlossen, auf dessen Innenseite sich ein verticaler Strich befindet. Sobald die Rotationsgeschwinligkeit constant geworden ist, meint der Beobachter sammt dem Kasten constant seitwärts geneigt zu sein, und zwar glaubt er sich mit dem Kopfe von der Rotationsaxe weg zu neigen. Beim Anhalten des Apparates glaubt er sich sammt der Umgebung wieder aufzurichten, wie dies auch im vorigen Versuche geschieht. Wir folgern aus diesem Versuche wie aus 5, dass man die Richtung der Massenbeschleunigung als Verticale empfindet. Diese Seitwärtsneigung hat schon Purkyně auf der Scheibe eines Ringelspiels beobachtet.

Versuch 7. Hängt man bei Versuch 6 in den Papierkasten ein Pendel mit einem Gradbogen, so sieht man, dass bei den erreichbaren Rotationsgeschwindigkeiten dasselbe $10 - 20^o$ ausschlägt, was eine Centrifugalbeschleunigung angibt, die rund $1/5 - 1/3$ der Schwerebeschleunigung beträgt. Dieses Pendel hält man nun bei der Rotation für vertical, den Kasten und sich selbst

sometimes I had the impression that the vertical lay somewhere between the direction of the pendulum and my body axis.

Experiment 8. When we stop the apparatus suddenly in experiment 5, the subject believes for a moment that he and the enclosure are tilted sideways, such that the head is tipped in the direction of the rotation. Every momentary linear acceleration, whose direction is not aligned with the true vertical, immediately alters the subjective vertical.[3.11]

Experiment 9. Up to this point the second frame r was locked to the main frame R by a bolt. Now we loosen the bolt so that it can easily be rotated. If we set the apparatus in rotation by a rapid push, the subject in the enclosure (according to the principle of the conservation of areas) will maintain his orientation for a certain time. That is to say, he is swung in a circle without rotating. The direction of the centrifugal force then moves around him. Consequently, the subject thinks that he is moving around the surface of a cone[3.12] whose axis is vertical.

4.

The phenomena described above generally belong to the field of so-called rotational vertigo. At this point we will not enter into discussion of its theory. We will only mention that all these phenomena become easily understood if we assume the existence of a sense for angular acceleration. If angular acceleration is sensed, stopping of a rotation must be sensed as a rotation in the opposite direction, since it accelerates in the opposite direction, that is, it decelerates.

It is also clear that the sensation of acceleration continues[3.13] much longer than the acceleration itself. Soon after stopping the rotation, all inertial accelerations will have stopped, whereas one still feels in motion.

The experiments described above also make it clear that one senses the

aber für schief. Doch schien es mir zuweilen, als ob die Verticale zwischen der Richtung des Pendels und der Axe meines Körpers enthalten wäre.

Versuch 8. Wenn wir in Versuch 5 den Apparat plötzlich anhalten, glaubt sich Beobachter sammt dem Kasten für einen Moment seitwärts zu neigen und zwar mit dem Kopfe im Sinne der Rotation. Jede momentane Progressivbeschleunigung, deren Richtung nicht mit der wahren Verticalen zusammenfällt, ändert momentan die scheinbare Verticale.

Versuch 9. Bisher war der Rahmen r in R durch eine Schraube festgestellt. Nun machen wir ihn durch Lockern der Schraube leicht drehbar. Setzen wir den Apparat durch einen raschen Stoss in Rotation, so behält nun der im Kasten eingeschlossene Beobachter (nach dem Flächenprincip) eine Zeit lang seine Stellung bei, d. h. er wird im Kreise geschwungen, ohne sich zu drehen. Die Richtung der Centrifugalkraft geht dann um ihn herum. Dem entsprechend glaubt sich der Beobachter in einer Kegelfläche zu bewegen, deren Axe die Verticale ist.

4.

Die eben beschriebenen Erscheinungen gehören grösstentheils dem Gebiet des sogenannten Drehschwindels an. Auf eine Theorie des Drehschwindels wollen wir uns hier zunächst nicht einlassen. Wir bemerken nur, dass alle Erscheinungen sich von selbst verstehn, wenn wir annehmen, dass es eine Empfindung der Winkelbeschleunigung gibt. Wird die Winkelbeschleunigung empfunden, so muss die aufhörende Drehung als eine entgegengesetzte Drehung empfunden werden, weil sie entgegengesetzt beschleunigt, d. h. verzögert ist.

Es ist ferner klar, dass die Empfindung der Beschleunigung viel länger anhält, als die Beschleunigung selbst. Denn sehr bald nach Unterbrechung der Drehung werden alle Massenbeschleunigungen aufgehört haben, während man noch immer eine Bewegung empfindet.

Dass auch die Richtung von Progressivmassenbe-

direction of linear acceleration as well.

To this point everything could still be generally explained by sensations involving the whole body. However, in several of the experiments described above the enormous influence of head position on the phenomena is demonstrated. From this one can conclude, as will be explained in greater detail later, that a very important, if not the most important, contribution to the sensation of motion is evoked in the head.

5.

Let us now move to another series of rotation experiments.

Experiment 1. If the subject lies down horizontally on a board within the frame *R*, say on the right ear, and the rotation remains constant, the sensation[3.14] of it will soon cease. If the subject now rapidly moves from the right ear to the left ear down position, the sensation of rotation will occur with much greater strength than during the initial acceleration. Each time one turns one's self about a horizontal axis by 180° during a constant velocity rotation, a strong reappearance of the rotation sensation occurs. The physical analog for this experiment is easily demonstrated. One sits in the apparatus and brings along a smoothly running siren, whose disk is maintained horizontal. Once the rotation has become constant and the disk of the siren has stopped its initial rotation, each time the disk is turned over by 180° it will rotate with twice the velocity in a direction opposite to the motion of the frame *R*.

Two similar phenomena appear, one objective, in which the position of the siren is changed during constant velocity rotation, and the other subjective, in which the position of the body or only of the head is altered. In principle, the explanation for these phenomena is simple. If the frame is accelerated clockwise (as seen from above), the motion will be transferred to our body, particularly to the head, and will be sensed as a clockwise rotation. According to the principle of the conservation of areas or the law of inertia, when the body is turned over, it would rotate counterclockwise, except that it

schleunigungen empfunden wird, geht aus den beschriebenen Versuchen ebenfalls hervor. So weit liesse sich alles durch die Empfindlichkeit des ganzen Körpers noch leidlich erklären, allein in mehreren der beschriebenen Versuche äussert sich der enorme Einfluss der Kopfstellung auf die Erscheinungen und damit ergibt sich, was später noch genauer begründet werden soll, dass ein sehr wichtiger, wo nicht der wichtigste Beitrag zu den Bewegungsempfindungen im Kopfe ausgelöst wird.

5.

Uebergehn wir nun zu einer andern Reihe von Rotationsversuchen.

Versuch 1. Legt sich der Beobachter horizontal auf ein in dem Rahmen *R* angebrachtes Brett etwa auf's rechte Ohr, so verschwindet bei gleichförmiger Drehung bald die Empfndung derselben. Kehrt sich nun der Beobachter rasch vom rechten auf's linke Ohr um, so tritt die Empfindung der Rotation sehr viel heftiger auf wie beim ersten Anstosse. So oft man sich während der gleichförmigen Rotation um eine Horizontalaxe um 180° dreht, tritt eine starke Auffrischung des Drehgefühls auf. Das physikalische Analogon des Versuches ist leicht herzustellen. Man setzt sich in den Apparat und nimmt eine gut laufende Sirene mit, deren Scheibe man horizontal hält. Ist die Rotation gleichförmig geworden und hat die beim ersten Anstoss in Bewegung gerathene Sirenenscheibe aufgehört zu rotiren, so tritt die Rotation derselben immer der Bewegung des Rahmens *A* entgegen mit doppelter Geschwindigkeit auf, sobald man die Scheibe in ihre eigene Ebene von 180° umklappt.

Aehnliche Erscheinungen zeigen sich objectiv, wenn man die Stellung der Scheibe während der gleichförmigen Rotation und subjectiv, wenn man die Lage des Körpers oder nur jene des Kopfes ändert. Die Erklärung der Erscheinungen im Princip ist einfach. Erfährt der Rahmen eine Winkelbeschleunigung (von oben gesehn) im Sinne des Uhrzeigers, so wird diese den Massentheilen unseres Körpers, speciell des Kopfes, aufgedrungen und als Drehung im Sinne des Uhrzeigers empfunden. Der umgedrehte Körper würde

is prevented from doing so by the apparatus, which forces it to rotate clockwise, that is, with an acceleration in the previous direction. Thereby a sensation of clockwise rotation is recreated.

Experiment 2. One is seated inside the frame, approximately on the axis of rotation. With the head kept stationary, the sensation of rotation disappears during constant velocity rotation. Every rotation of the head about an axis not parallel to that of the frame, however, will instantly produce a <u>strange sensation of turning</u>.[3.15]

Let us define an axis A, parallel to the rotation axis of the apparatus, an axis B about which the head is rotated, and another one C, about which the illusory rotation occurs. We denote the axes by arrows in the sense that the subject's head is at the arrowhead and his feet at its tail such that his feet appear to rotate in a clockwise direction. The rule governing all such phenomena is the following: If we place the tails of the arrows A, B, C at the same point, C must lie orthogonal to the plane AB, directed so that A will turn into B by a clockwise rotation of less than 180°, as seen by a subject observing from the arrowhead of C.

The axis of the illusory rotation is therefore always oriented orthogonal to the plane that contains the axis of the apparatus and the rotation axis of the head. The illusory rotation that is elicited by a simple head rotation lasts several seconds if the head movement is strong enough. It can be abolished immediately by a quick rotation of the head in the opposite direction.

If the apparatus rotates clockwise as seen from above (usually one full rotation in 4 seconds) and one nods the head forward, every visible plane then appears to rotate clockwise. With alternating up and down movement of the head, the paper box, in which the subject is placed in all these experiments, seems to resemble the rolling of a ship at sea.

As one can immediately see, the described subjective sensations

nun das Flächenprincip oder Trägheitsgesetz erfüllend gleichförmig dem Uhrzeiger entgegen rotiren, wird aber durch den Apparat verhindert, der ihm die Uhrzeigerbewegung mittheilt, also Beschleunigungen von demselben Sinn wie zuvor aufdringt. Damit entsteht also von Neuem die Empfindung der Uhrzeigerdrehung.

Versuch 2. Man setzt sich in den Rahmen, etwa in die Rotationsaxe. Hält man den Kopf ruhig, so verschwindet bei gleichförmiger Drehung das Drehungsgefühl. Jede Drehung des Kopfes um eine der Rotationsaxe des Rahmens nicht parallele Axe bringt aber sofort ein eigenthümliches Drehungsgefühl hervor.

Nennen wir eine der Rotationsaxe des Apparates parallele Axe *A, B* die Axe, um welche der Kopf gedreht wird, und *C* diejenige, um welche die scheinbare Drehung geschieht. Die Axen wollen wir durch Pfeile bezeichnet denken und zwar so, dass für einen Beobachter mit dem Kopf in der Spitze, dem Fuss im Schwanze des Pfeiles, die Rotation unter den Füssen im Sinne des Uhrzeigers vor sich geht. Die Regel aller hieher gehörigen Erscheinungen ist nun diese: Legen wir die Schwänze der Pfeile *A, B, C* in denselben Punkt, so haben wir *C* senkrecht gegen die Ebene *AB* zu stellen, jedoch so, dass für den Beobachter mit dem Kopfe in der Spitze von *C, B* durch Drehung um weniger als 180° im Sinne des Uhrzeigers aus *A* hervorgeht.

Die Axe der scheinbaren Drehung steht also immer senkrecht auf der Ebene, welche die Axe des Apparates und die Drehungsaxe des Kopfes enthält. Die scheinbare Drehung, welche durch eine einfache Kopfwendung erregt wird, dauert bei gehöriger Stärke einige Secunden nach. Sie kann sofort durch eine rasch folgende entgegengesetzte Kopfwendung aufgehoben werden.

Rotirt der Apparat von oben gesehn wie der Uhrzeiger (gewöhnlich etwa mit einem Umlauf in 4 Secunden) und man nickt bejahend, mit dem Kopfe, so dreht sich jede Ebene, die man ansieht, in sich selbst wie der Uhrzeiger. Bei abwechselndem Heben und Senken des Kopfes scheint der Papierkasten, in welchem sich der Beobachter bei allen diesen Versuchen befindet, das Rollen eines Seeschiffes nachzuahmen.

Wie man sofort erkennt, sind die beschriebenen sub-

are analogous to those which one can observe objectively in a Fessel or Bohnenberger type of flywheel and which had been used to demonstrate the precession of the equinox. A detailed explanation can easily be derived from Poinsot's theory of rotations. We will return to this subject later.

This series of experiments also demonstrates the sensation of acceleration and its aftereffects, and shows that the seat of motion sensation is in the head.

6.

The sensations of turning are much more striking than the sensations depending upon translational motion and lead more easily to illusions by way of rotational vertigo. However, one may assume that such sensations do occur. This even follows in part from the experiments previously described. New experiments bearing directly on the investigation of such sensations can easily be imagined.

The sensations connected with falling and downward acceleration were already mentioned. They can be demonstrated with reduced intensity on a seesaw, and we will therefore first discuss such an experiment.

Experiment 1. Let us imagine a large balance[3.16] with equal arms. Each balance beam is 2^m long. The subject is placed on one of the weight dishes, and is balanced by counterweights on the other arm. If the scale is placed in motion, the subject will experience almost pure sinusoidal vertical oscillation. In this way the experiment is simpler and purer than one could perform with a conventional seesaw. If we denote the vertical excursion from the equilibrium point as x, its maximum a, the period of oscillation of the scale T, and time t, then we can write

$$x = a \sin \frac{2\pi t}{T}$$

and for the acceleration

$$\frac{d^2 x}{dt^2} = -a \left(\frac{2\pi}{T} \right)^2 \sin \frac{2\pi t}{T}$$

jectiven Erscheinungen analog denjenigen, welche man objectiv an der Fessel'schen oder Bohnenberger'schen Schwungmaschine wahrnehmen kann und die man zur Demonstration der Präcession der Nachtgleichen verwendet. Die detaillirte Erklärung ergibt sich sehr einfach mit Hilfe der Poinsot'schen Drehungstheorie. Wir kommen auf dieselbe noch zurück.

Auch diese Versuchsreihe demonstrirt die Empfindung der Beschleunigung, die Nachdauer derselben und weist den Sitz der Bewegungsempfindungen im Kopfe nach.

6.

Die Empfindungen der Drehung sind viel auffallender und führen in Form des Drehschwindels leichter zu Täuschungen, als diejenigen Empfindungen, die mit der progressiven Bewegung zusammenhängen. Es lässt sich aber vermuthen, dass solche Empfindungen existiren. Es geht dies schon zum Theil aus den vorher beschriebenen Experimenten hervor. Neue Experimente, welche direct auf die Untersuchung solcher Empfindungen ausgehn, lassen sich leicht auffinden.

Die Empfindungen beim Fallen und beschleunigten Sinken wurden schon erwähnt. Sie zeigen sich in geringerer Intensität auch schon beim Schaukeln, und wir wollen daher ein derartiges Experiment zuerst besprechen.

Versuch 1. Denken wir uns eine grosse gleicharmige Wage. Jeder Wagbalken hat 2^m Länge. Auf einer Wagschale befindet sich der Beobachter und ist durch Gegengewichte auf der andern aequilibrirt. Wird die Wage in Schwingungen versetzt, so führt der Beobachter fast eine reine pendelförmige Verticalschwingung aus. Darin ist also der Versuch einfacher und reiner, als derjenige, den man mit einer gewöhnlichen Schaukel ausführen kann. Nennen wir die Verticalexcursion von der Gleichgewichtslage an x, ihr Maximum a, die Schwingungsdaüer der Wage T und t die Zeit, so ist

$$x = \sin \alpha \frac{2\pi t}{T}$$

und die Beschleunigung

$$\frac{d^2 x}{dt^2} = -a \left(\frac{2\pi}{T} \right)^2 \sin \frac{2\pi t}{T}$$

Therefore, one can calculate the course of the vertical acceleration from the maximum excursion and the oscillation period. For $a = 15^{cm}$ and $T = 7$ seconds, the maximum vertical acceleration[3.17] was 12^{cm}, that is, about 0.012 of the acceleration due to gravity. However, the movements were then near the detection threshold for the subject with his eyes covered. When the excursions were somewhat larger, the subject regularly stated that he was sinking shortly before or just when reaching the highest point of the oscillation. Similarly, rising was noticed shortly before or when reaching the lowest point. Of course, the eyes were always closed. As the formula shows, the downward acceleration indeed shows a maximum at the highest point of the path of the oscillation, and the acceleration upward a maximum at the lowest point of the path of the oscillation.

One is therefore also very sensitive to changes in the magnitude of the weight acceleration,[3.18] and during vertical motion one senses acceleration rather than position or velocity. In addition to sensing acceleration in the head, it seems to me, more for these experiments than for the earlier ones, that the sensation took place throughout the body[3.19] and was particularly strong in the lowest parts.

On the basis of known mechanical principles we understand how the body on an oscillating balance senses itself and its contents as alternatingly heavier and lighter. Let us assume larger excursions, about $2a = 9^m$ and $T = 10$ seconds, as might occur on a large ship in high seas, producing a peak about 0.18 of gravity. We think it possible that such mechanical stimulation alone can lead to serious disturbances to the organsim in the form of motion sickness.[3.20]

Experiment 2. The experiment with the moving balance was initially chosen because it was easy and did not take much space. If one wished to perform similar experiments on a large Atwood[3.21] vertical accelerator, one would need a sizable room. The elevators in mine shafts are similar devices. Professor G. Schmidt[3.22] from our local Polytechnic Institute told me that by his estimate, the platform in the shaft he recently entered reached a

Man kann also aus dem Excursionsmaximum und der Schwingungsdauer die Schwankungen der Verticalbeschleunigung berechnen.
War $a = 15^{cm}$ und $T = 7$ Secunden, so war das Maximum der
Verticalbeschleunigung 12^{cm}, also rund 0.012 der Schwerebeschleunigung. Dann waren aber die Schwankungen für den
Beobachter mit bedeckten Augen an der Grenze der
Merklichkeit. Wurden die Excursionen etwas grösser, so gab
der Beobachter jedesmal an, er sinke, kurz vor dem Anlangen an
dem höchsten Punkt der Schwingung oder auf diesem selbst. Ebenso
wurde das Steigen immer kurz vor oder an dem tiefsten Punkt bemerkt, natürlich immer bei geschlossenen Augen. Wie die Formel
zeigt, ist in der That die Beschleunigung abwärts ein Maximum im
höchsten Punkt der Schwingungsbahn und die Beschleunigung aufwärts ein Maximum im tiefsten Punkt der Schwingungsbahn.
Man ist also auch für Schwankungen in der Grösse
der Schwerebeschleunigung sehr empfindlich und man
empfindet bei Verticalbewegungen nicht die Lage oder
die Geschwindigkeit, sondern die Beschleunigung. Man
hat die Empfindung der Beschleunigung auch im Kopfe, doch schien
es mir bei diesen Versuchen mehr wie bei den vorigen, dass man
die Empfindung im ganzen Körper habe und zwar in den tiefst-
gelegenen Theilen am meisten.

Wir begreifen nach den bekannten mechanischen Grundsätzen,
wie der Körper auf der schwingenden Wage, sich und seinen Inhalt
abwechselnd schwerer und leichter empfindet. Nehmen wir die
Schwankungen grösser, etwa $2a = 9^{m}$ und $T = 10$ Secunden, wie
sie etwa hei einem grossen Schiff auf hoher See vorkommen, so
betragen die Schwankungen 0.18 der Schwerebeschleunigung und wir
werden es für möglich halten, dass nun durch diese mechanischen
Eingriffe ernstliche Störungen im Organismus in Form der Seekrankheit auftreten können.

Versuch 2. Die Versuchsweise mit der schwingenden Wage
wurde zunächst der Bequemlichkeit und Raumersparniss wegen gewählt. Wollte man ähnliche Versuche auf einer grossen Atwood'-
schen Fallmaschine anstellen, so wäre hiezu ein bedeutender Raum
nöthig. Die Förderungsmaschinen in Bergwerksschachten sind nun
ähnliche Vorrichtungen. Herr Professor G. Schmidt vom hiesigen

velocity of 3 feet [/sec] in 10 seconds, corresponding to an acceleration of 0.01 of gravity. Schmidt had even sensed this acceleration clearly. Another natural scientist who was present during this discussion noted that in his experience he soon "got accustomed to" such a falling movement. This can have two interpretations. It could mean that the motion [velocity] is no longer sensed, if it becomes constant, or that even the continuous acceleration will finally no longer be sensed. We will see that both cases hold. A constant movement is never sensed, but even the sensation during a constant acceleration <u>disappears</u>[3.23] little by little.

Experiment 3. The wish to observe the phenomena described above more intensively led to another experiment. First, a falling machine was built, which could be considered a combination of Galileo's and Atwood's machines. Two wooden rails were erected in the form of a sloping plane which dropped 2^m vertically over their 22^m length. A simple cart for the subject ran on the outside of the rails. A second cart for carrying counterweights ran underneath the first cart on the inside of the rails. A rope passing over a pulley at the upper end of the rails connected the two carts. The whole device was very inexpensive and was set up within three adjoining rooms, passing through the doors. The first experiment with the apparatus, which allowed us to conduct every experiment of the Atwood machine, taught us that the enclosed subject senses every acceleration or deceleration. But even with a clearly sensed acceleration the sensation disappeared during the duration of a continuously accelerating movement. An apparent reversal of the movement sensed by the enclosed subject, when the apparatus stopped or reached constant velocity, was hardly detected and of barely noticeable duration. All these experiments suffered from the disturbance created by the <u>rumbling of the wheels</u>.[3.24] Because the experiments produced only limited results, I go into no further details. The apparatus was soon removed and a new one was substituted for it.

Polytechnicum theilte mir mit, dass nach seiner Schätzung die Schale, auf welcher er kürzlich in den Schacht eingefahren war, in 10 Secunden eine Geschwindigkeit von 3 Fuss erlangt habe, was also einer Beschleunigung von 0.01 der Schwerebeschleunigung entspricht. Schmidt hat diese Bescbleunigung noch deutlich empfunden. Ein anderer bei dieser Mittheilung anwesender Naturforscher bemerkte, er habe sich an die Fallbewegung alsbald »gewöhnt«, Dies kann einen doppelten Sinn haben. Es kann gemeint sein, dass die Bewegung nicht mehr empfunden wird, wenn sie gleichförmig geworden ist oder dass auch die bleibende Beschleuniguug schliesslich nicht mehr empfunden wird. Wir werden sehn, dass beides der Fall ist. Eine gleichförmige Bewegung wird nie empfunden, aber auch die Empfindung einer constanten Beschleunigung verschwindet nach und nach.

Versuch 3. Der Wunsch, die zuvor beschriebenen Erscheinungen in grösserer Intensität zu beobachten, führte zu einem andern Versuch. Zunächst wurde eine Fallmaschine construirt, welche als eine Combination der Galilei'schen und Atwood'schen bezeichnet werden kann. Zwei Holzschienen von 22^m Länge und 2^m Fall auf diese Länge waren als schiefe Ebene aufgestellt. Auf den Aussenseiten der Schienen lief ein einfacher Wagen für den Beobachter, auf den Innenseiten ein zweiter niederer Wagen für Gegengewichte unter dem ersten Wagen durch. Beide waren durch eine Schnur, die über eine Rolle am obern Ende der Schienen ging, mit einander verbunden. Das Ganze war mit kaum nennenswerthen Kosten sehr einfach in dem Raume von drei Zimmern durch die Thüren hindurch aufgestellt worden. Die ersten Versuche mit dem Apparate, der jedes Experiment der Atwood'schen Maschine zulässt, lehrten, dass jede Beschleunigung oder Verzögerung von dem eingeschlossenen Beobachter empfunden wurde. Aber auch bei sehr merklichen Beschleunigungen verschwand die Empfindung bei Fortdauer der gleichförmig beschleunigten Bewegung. Eine scheinbare Umkehrung der Bewegung für den eingeschlossenen Beobachter, wenn der Apparat angehalten oder die Bewegung gleichförmig wurde, trat nur in sehr geringem Maasse ein und war von kaum merklicher Dauer. Sehr störend waren bei allen Versuchen die von den Rädern herrührenden Erschütterungen. Da sich die Versuche wenig

Experiment 4. Let us imagine horizontal rails inside the frame R of our rotation device with a wheelchair for the subject placed on the rails. An assistant standing in the frame behind the chair has the task of moving it during the rotation, which itself is accomplished by another assistant. If the apparatus rotates clockwise (as seen from above), then each forward radial displacement of the subject corresponds to an acceleration to the right, each such displacement to the rear an acceleration to the left. If within the time t at an angular velocity φ, the wheelchair traverses a distance a along the rails at constant velocity, then the <u>acceleration is $a\varphi/t$</u>.[3.25] The acceleration disappears when the displacement ceases. Thus we encounter an application of the same principle that has been used as an explanation for the trade winds, changing of directional flow in rivers, etc. on a rotating earth. One has the compelling sensation that during each forward motion, after one had already forgotten the constant velocity rotation, the whole apparatus seems to fly away towards the right. That flying away motion persists, even after the displacement has already come to an end. The sensation caused by an instantaneous linear acceleration therefore includes a definite aftereffect.

It could not be decided whether the aftereffects of the movement change when the head is turned, because even with the use of a bearing to guide the head, it was not possible to rotate the head about an axis strictly parallel to the rotation axis of the apparatus. With any other rotation the subjective phenomena corresponding to the flywheel experiments were evoked so strongly that all else was disregarded.

Another thing must be mentioned concerning such types of experiments. The whole apparatus in which the displacement of the subject took place was always enclosed by paper to exclude visual influences, unless explicitly stated otherwise. We used rotation, because then one can rapidly achieve large accelerations within a limited space.

ergiebig zeigten, weshalb ich auch auf das Detail nicht eingehe,
wurde der Apparat alsbald abgeräumt und durch einen neuen ersetzt.

Versuch 4. Denken wir uns in dem Rahmen R unseres Rotationsapparates horizontale Schienen, auf den Schienen einen Rollstuhl für den Beobachter. Ein Gehilfe, der hinter dem Stuhle in dem Rahmen steht, hat die Aufgabe, denselben während der Rotation, die von einem andern Gehilfen besorgt wird, in Bewegung zu setzen. Rotirt der Apparat (von oben gesehn) im Sinne des Uhrzeigers, so entspricht jeder radialen Verschiebung des Beobachters nach vorn eine Beschleunigung nach rechts; jeder solchen Verschiebung nach hinten eine Beschleunigung nach links. Die Beschleunigung beträgt, wenn in der Zeit t bei der Winkelgeschwindigkeit φ die Schienenstrecke a gleichförmig mit dem Rollstuhl durchfahren wird, $a\varphi/t$. Die Beschleunigung verschwindet, wenn die Verschiebung aufhört. Es kommt hier, wie man sieht, dasselbe Princip zur Anwendung, welches zur Erklärung der Passatwinde, Flussabweichungen u. s. w. auf der rotirenden Erde verwendet wird. Man meint nun wirklich, dass bei jeder Vorwärtsbewegung, während man die gleichförmige Rotation längst vergessen hat, der ganze Apparat nach rechts fortfliegt. Das Fortfliegen hält noch an, wenn die Verschiebung schon beendigt ist. Die durch eine momentane Progressivbeschleunigung erregte Empfindung hat also eine merkliche Nachdauer.

Ob die Richtung des Bewegungsnachbildes sich mit einer Kopfwendung ändert, konnte nicht entschieden werden, da es selbst durch eine Axenführung nicht möglich war, den Kopf bloss um eine der Apparat-Rotationsaxe parallele Axe zu drehen. Bei jeder andern Drehung treten aber die subjectiven Analoga der Schwungmaschinenversuche so betäubend auf, dass alles übrige unbemerkt bleibt.

Was diese Versuchsweise betrifft, so muss noch Einiges bemerkt werden. Der ganze Apparat, in welchem die Verschiebung des Beobachters stattfand, war wie überall, wo nicht ausdrücklich das Gegentheil bemerkt wurde, um Gesichtsphänomene zu eliminiren, mit Papier geschlossen. Das Mittel der Rotation wurde angewandt, weil sich mit demselben bei grosser Raumersparniss plötzlich bedeutende Beschleunigungen erzielen lassen.

On a rotating sphere the movement of a rigid body from one latitude to another causes a relative angular velocity in addition to the relative linear velocity. This relative angular velocity does not occur on a rotating platform as used in our experiment. I do not know to what extent this needs to be considered with respect to the deviation of <u>railway tracks</u>.[3.26]

Experiment 5. The inner frame r is installed inside the main frame R of our rotation apparatus such that it can easily be rotated about the axis a. The enclosed subject takes his seat on the chair in r. First the frame r will be tied to R, and a fast constant velocity will be started. When the subject no longer senses the rotation, he removes the connection between R and r and commands a <u>stop</u>.[3.27] The <u>linear</u>[3.28] motion of the subject, who is positioned outside the main axis, will now suddenly cease, while the rotation of the frame r will continue according to the principle of conservation of areas. The subject immediately feels himself <u>flying away</u>[3.29] for several seconds in a direction opposite to that of the interrupted linear acceleration. Since he rotates without noticing it, the direction of the motion aftereffect also rotates together with his whole body. Unfortunately, once again, the experiment cannot be done in a convincing way with only the <u>head turning</u>.[3.30] Incidentally, one could hardly go wrong by assuming that the site of sensation for linear accelerations is in the head. I might add that such a motion aftereffect can be stopped by a quick and accurate resumption of the recently interrupted motion. I must assume that the smooth and quiet function of the apparatus was important to the success of the last experiment. However, the rotation velocity prior to the sudden stop was particularly large during the last experiment. Probably, the sensations of angular and linear accelerations are quite analogous to one another.

One must add the following comment concerning this experiment. The sensation of flying away after stopping the apparatus decays slowly. During that time, however, the frame r experiences a noticeable deceleration because of friction that is experienced as rotation in the opposite direction. It appears that finally the subjective motion beomes a spiral.

Auf einer rotirenden Kugel bringt jeder aus einem Breitengrad in den andern verschobene feste Körper nicht nur eine relative Progressivgeschwindigkeit, sondern auch eine relative Winkelgeschwindigkeit mit. Letztere fällt in einer rotirenden Ebene, wie in unserem Experiment, weg. Ich weiss nicht, wie weit dies bei Erklärung der Schienenabweichungen auf Eisenbahnen zu berücksichtigen ist.

Versuch 5. In dem Rahmen R unseres Rotationsapparates wird der Rahmen r sehr leicht um die Axe a drehbar angebraeht. Der eingeschlossene Beobachter nimmt auf dem Stuhle in r Platz. Der Rahmen r wird. an R zunächst festgebunden und eine rasche gleichförmige Rotation eingeleitet. Wenn der Beobachter letztere nicht mehr fühlt, löst er die Verbindung von R und r und commandirt Halt. Die Progressivbewegung des ausserhalb der Hauptaxe befindlichen Beobachters wird nun plötzlich gehemmt, während die Rotation des Rahmens r nach dem Flächenprincip fortdauert. Sofort meint der Beobachter entgegengesetzt der eben gehemmten Progressivbewegung einige Secunden lang fortzufliegen. Da er sich hiebei dreht, ohne etwas davon zu merken, so dreht sich also die Richtung des Bewegungsnachbildes mit seinem ganzen Körper. Leider kann der Versuch mit der blossen Kopfwendung auch hier nicht überzeugend ausgeführt werden. Man wird übrigens kaum irren, wenn man den Sitz der Empfindung von Progressivbeschleunigungen im Kopfe annimmt. Ich erwähne noch, dass ein solches Bewegungsnachbild durch rasche und geschickte Wiedereinleitung der eben unterbrochenen Bewegung ausgelöscht werden kann. Ich muss annehmen, dass der gleichmässige geräuschlose Gang des Apparates bei dem letzten Versuche besonders zum Gelingen beigetragen hat. Doch war auch die plötzlich aufgehobene Geschwindigkeit bei dem letzten Versuch am grössten. Wahrscheinlich verhalten sich die Empfindungen der Winkel- und der Progressivbeschleunigung durchaus analog.

Es muss zu diesem Versuch noch Folgendes bemerkt werden. Das nach dem Anhalten des Apparates auftretende Gefühl des Fortfliegens nimmt allmälig ab. Einstweilen hat aber der Rahmen r durch die Reibung eine merklich verzögerte Rotation angenommen, welche als Gegendrehung empfunden wird. Dadurch gewinnt es

The actual motion of the subject in the last experiment can be considered as a combination of a linear motion in a circle (without turning) and a simultaneous rotation. In general, the motion sensation disappears for all non-accelerating components of any movement.

Experiment 6. If the subject facing the rotation axis A is rotated at constant velocity, the movement may be considered as a combination of a constant sideward motion along with a constant acceleration directed towards the rotation axis. With continuation of the motion the movement sensation completely disappears. The subject has the sensation of simply being tilted backward. We conclude that for acceleration that remains constant in direction relative to the subject all motion sensation[3.31] eventually dies out.

7.

The previously described experiments have the advantage that the subject observes himself and thereby develops the greatest possible subjective confidence [in the results]. With the following experiments, however, one can show the results to several people simultaneously. However, the interpretations of the results and sensations which underlie these phenomena, and which we want to stress, remain somewhat arbitrary.

In fact, I have performed many of the previously described rotation experiments on animals as well. Even earlier, animals were passively rotated and the resulting neck distortions and nystagmic movements of the eyes and head were observed. These experiments were performed by von Graefe[3.32] (Arch Ophthalmol), Czermak[3.33] (Pflügers Archive 1873), and Breuer[3.34] (Yearbook of the Association of Physicians 1874). Breuer had already pointed out that birds made vertiginous by rotation behaved exactly as if they had semicircular canal surgery. The experiments to be described here shall next demonstrate that the same motion sensations which we know from

den Anschein, als ob die subjective Bewegung mit einer Spirale schliessen würde.

Die wirkliche Bewegung des Beobachters in dem letzten Versuch lässt sich auffassen als zusammengesetzt aus einer Progressivbewegung im Kreise (ohne Drehung) und aus einer gleichzeitigen Drehung. Es lässt sich nun die gemachte Beobachtung leicht dahin verallgemeinern, dass alle unbeschleunigten Componenten einer Bewegung für die Empfindung verschwinden.

Versuch 6. Wenn der Beobachter mit dem Gesicht der Rotationsaxe A zugewandt in gleichförmige Rotation versetzt wird, so lässt sich diese Bewegung zusammengesetzt denken aus einer gleichförmigen Seitwärtsbewegung und einer gleichförmig beschleunigten stets nach der Axe hin gerichteten Bewegung. Beim Fortdauern der Bewegung verschwindet nun das Bewegungsgefühl gänzlich. Der Beobachter hat bloss das Gefühl der Rückenlage. Daraus geht hervor, dass auch die constante gegen den Beobachter dieselbe Richtung beinhaltende Beschleunigung schliesslich keine Bewegungsempfindung mehr erregt.

7.

Haben die bisher beschriebenen Versuche den Vorzug, das man bei denselben an sich selbst beobachtet, also die grösstmögliche subjective Sicherheit erzielt, so sind die folgenden dadurch ausgezeichnet, dass man die Ergebnisse derselben mehreren Personen zugleich zeigen kann, wobei freilich die Deutung der Ergebnisse und der Schluss auf die denselben zu Grunde liegenden Empfindungen, auf die es uns hier hauptsächlich ankommt, nicht ganz frei von Willkühr ist.

Viele der beschriebenen Rotationsexperimente habe ich nämlich auch an Thieren angestellt. Schon früher sind Versuche über passive Drehung von Thieren und die dabei auftretenden Verdrehungen und nystagmatischen Bewegungen der Augen und des Kopfes angestellt worden von Graefe (Arch. f. Ophthalm. I.), Czermak (Pflügers Archiv 1873) und Breuer (Jahrb. d. Gesellschaft d. Aerzte 1874). Letzterer hat auch schon darauf aufmerksam gemacht, dass durch Drehung schwindlig gewordene Vögel sich ganz so benehmen, als wären sie an den Bogengängen operirt. Die hier zu beschrei-

introspection can be elucidated in animals.

Experiment 1. We place a pigeon on the platform of a centrifuge and place a large glass bowl with an air hole above it, taking into account the comfort of the animal. If we rotate the platform clockwise (as seen from above), the pigeon starts to turn in the opposite direction. With a faster rotation of the platform, movement of the body ceases and the pigeon turns only its head counterclockwise. It tries with its entire means to compensate for the motion imposed on it. At even higher velocity we see that the pigeon leans towards the axis and spreads its outer wing agains the glass bowl, as if to support itself. It obviously has the perception that the orientation of the enclosure is tilted. The animal remains motionless in this position as long as the strong rotation lasts. If one stops the apparatus suddenly after approximately two minutes, the pigeon starts moving uncontrollably and begins turning in the same direction as the previous rotation. If the animal were behaving according to the same rule as at the beginning of the experiment, then it would have the sensation of a forced counterrotation. If one throws the pigeon off the apparatus immediately after the stop, one creates the true picture of a pigeon whose semicircular canals had been operated upon. Although it still rotates in the same direction, after having once fallen it continues to somersault in various directions. Its legs collapse, it cannot fly, and often it falls from the table like a stone. One can also observe the well-known posture of supporting the head. A young rabbit behaved the same way as a pigeon. In the same animal, I have also observed the head twitching described by Breuer.

Experiment 2. The previous experiment is not completely decisive. The movements of the released pigeon depend not only on the position of the head during the forced rotation, but also on the head positions that the animal assumes during its subsequent movements. This occurs because the apparent rotation continues with its axis fixed to the head and moving along

benden Versuche sollen zunächst bloss zur Anschauung bringen, dass sich bei den Thieren dieselben Bewegungsempfinungen äussern, welche wir aus der Selbstbeobachtung kennen.

Versuch 1. Auf die Scheibe der Centrifugalmaschine setzen wir eine Taube und über dieselbe einen grossen Glassturz mit Luftloch, der dem Thiere alle Bequemlichkeit gestattet. Drehen wir die Scheibe (von oben gesehn) wie den Uhrzeiger, so fängt die Taube an, sich in entgegengesetztem Sinne herumzudrehen. Bei rascherer Drehung der Scheibe hört die Bewegung des Körpers auf und die Taube verdreht bloss den Kopf entgegen dem Uhrzeiger. Sie sucht also durch alle ihre Mittel die ihr aufgedrungene Bewegung zu compensiren. Bei noch grösserer Geschwindigkeit sehn wir, wie sich die Taube gegen die Axe zu neigt und den äussern Flügel, wie um sich zu stützen, gegen die Glassturzwand ausstreckt. Sie hat augenscheinlich die Vorstellung, dass die Lage des Behälters eine schiefe ist. In dieser Stellung verbleibt nun das Thier unbeweglich so lange als die heftige Rotation dauert. Hält man nach etwa zwei Minuten plötzlich den Apparat an, so beginnt die Taube krampfhafte Bewegungen und Umdrehungen in demselben Sinne, in welchem sie in Rotation versetzt worden war. Benimmt sich also das Thier nach derselben Regel, wie bei Beginn des Experimentes, so hat es die Vorstellung einer gewaltsamen Gegendrehung. Wirft man sofort nach dem Anhalten des Apparates die Taube heraus, so hat man in ihrem Benehmen das getreue Bild einer an den Bogengängen operirten Taube. Sie dreht sich zwar auch noch in demselben Sinne, überkugelt sich aber, wenn sie einmal gefallen ist, nach den verschiedensten Richtungen. Ihre Beine knicken ein, sie kann nicht fliegen und fällt häufig wie ein Stein vom Tische. Auch das bekannte Aufstützen des Kopfes kann man beobachten. Ein junges Kaninchen benahm sich ebenso wie eine Taube. Ich habe an demselben auch die von Breuer beschriebenen Zuckungen des Kopfes beobachtet.

Versuch 2. Der vorige Versuch ist nicht ganz rein. Die Bewegungen der frei gegebenen Taube hängen nicht nur von der Kopfstellung während der gewaltsamen Rotation, sondern auch von den Kopfstellungen, welche das Thier nachher bei seinen Bewegungen annimmt, ab. Denn die Scheindrehung dauert nach, ihre Axe

with it. The initial head position is easily fixed during the forced rotation. We build a kind of coffin for the rabbit in the shape of a <u>four-sided pyramid</u>.[3.35] Three of the six boards out of which the coffin is made are elongated and can be used to bolt the small apparatus to the platform of the centrifuge. One side board D can be moved by ropes and can close the coffin with the help of a hook so that it can be opened quickly. The head of the rabbit is positioned in the small space towards O, so that the abdomen of the

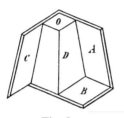

animal faces A. The space around the head is well padded with cotton wadding, so that the head position remains stationary.

Fig. 8

First we bolt A onto the platform of the centrifuge and, after a rotation in a clockwise direction, we quickly throw the animal out of the box. The animal circles in a clockwise direction spasmodically by resting on its forelimbs and hopping around with its hind legs. The head of the animal is twisted in a clockwise direction and makes the same type of movements that B r e u e r observed in birds. From all these observations one can conclude that the animal senses that it is being rotated in a counterclockwise direction.

Next we bolt B onto the platform of the centrifuge. The animal will now be rotated about the longitudinal axis of the body and head. If it is quickly thrown onto the table following the cessation of rotation, the animal persistently stumbles around its longitudinal axis for several seconds, in the same sense as it was rotated.

We now bolt C onto the platform of the centrifuge, so that the rotation is about an axis passing through the head from right to left. Our previous experience leads us to expect that the unconstrained animal will perform somersaults, thereby tumbling over forward or backward depending on the direction of rotation. In my experimental animal, which was still young and probably did not have muscle strength enough, I saw only incomplete attempts at such movements. However, I have no doubt that a strong animal

setzt im Kopfe fest und bewegt sich mit demselben. Die anfängliche Kopfstellung während der gewaltsamen Rotation lässt sich nun gut fixiren. Wir construiren uns für das Kaninchen eine Art Sarg in Form eines vierseitigen Pyramidenstutzes. Drei der sechs Brettchen, aus welchen der Sarg besteht, sind verlängert, so dass sie zum Anschrauben des kleinen Apparates auf die Scheibe der Centrifugalmaschine benützt werden können. Das eine Seitenbrettchen *D*

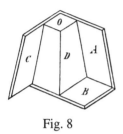

Fig. 8

ist in Bändern beweglich und wird mit Hilfe eines Hakens geschlossen, so dass man es rasch öffnen kann. Der Kopf des Kaninchens kommt nun in den engern Raum nach *O*, so dass die Bauchseite des Thieres gegen *A* hinsieht. Der Raum um den Kopf wird gut mit Watte ausgestopft, so dass die Kopfstellung nicht mehr geändert werden kann.

Schrauben wir zuerst *A* auf die Scheibe der Centrifugalmaschine und werfen nach einer Rotation im Sinne des Uhrzeigers das Thier rasch aus dem Kästchen, so dreht es sich krampfhaft im Sixme des Uhrzeigers, indem es auf den Vorderbeinen ruhend mit den Hinterbeinen. umherhüpft. Der Kopf des Thieres ist im Sinne des Uhrzeigers verdreht und macht die von B r e u e r an Vögeln beobachteten Bewegungen. Aus Allem geht hervor, dass das Thier meint, verkehrt wie der Uhrzeiger gedreht zu werden.

Wir schrauben nun *B* auf die Scheibe der Centrifugalmaschine. Das Thier wird also jetzt um die Längsaxe des Körpers und Kopfes in Rotation versetzt. Nach beendigter Drehung rasch auf den Tisch herausgeworfen, wälzt sich das Thier einige Secunden lang beharrlich um die Längsaxe und zwar in demselben Sinne, in welchem es gedreht wurde.

Schrauben wir *C* auf die Scheibe der Centrifugalmaschine, so dass die Rotation um eine von rechts nach links durch den Kopf hindurchgehende Axe ausgeführt wird, so werden wir nach den früheren Erfahrungen erwarten, dass das freigelassene Thier Purzelbäume ausführen wird, indem es nach vorne oder hinten je nach dem Rotationssinne überschlägt. An meinem Versuchsthier, welehes jedoch noch jung und wahrscheinlich nicht genug muskelkräftig

would show this phenomenon clearly.

For myself, the subjective observations I made when I was the experimental subject were of much greater value. The experiments described above, however, are very effective demonstrations for student classes.

8.

All of the experiments described thus far deal with the phenomena of passively moved humans or animals. For several reasons I placed less emphasis on experiments involving active movements and their resulting sensations. Muscular efforts during active movements by themselves lead to sensations and can induce aftereffects that can obscure and mask real motion sensations. Such experiments therefore are less clean for our purpose. Furthermore, any strong balance disturbance induces purposeful reflexive movements to regain balance, as we already have seen nicely with the rotated animals. Such movements also occur with illusory motion. Even if they should not cause the subject to fall over in this case, they induce inappropriate reactions resulting in stronger or weaker balance disturbances, which complicate the observed sensations. In the purest form of these experiments one is moved passively with the body supported. Every tendency towards movement is then much weaker, and when it occurs it has no visible effect and therefore is usually not noticed. For these reasons, and because I do not wish to enter into discussions about pure physiological matters[3.36], I confined myself almost entirely to repeating the experiments previously done by Purkyně. The most important aspects are summarized below.

If one stands erect and makes several turns about the body's longitudinal axis, visible objects that at first appear stationary, start to move in the opposite direction. If one suddenly stops after a prolonged rotation, the visible objects continue with their motion. Furthermore every object which

war, habe ich nur umvollkommene Versuche dieser Bewegung gesehn. Ich zweifle jedoch nicht, dass ein kräftiges Thier diese Erscheinung rein zeigen wird.

Für mich haben nun zwar die an mir selbst angestellten Beobachtungen einen viel grösseren subjectiven Werth. Die eben
beschriebenen Versuche eignen sich aber vorzüglich zu Collegien-
demonstrationen.

8.

Alle bisher beschriebenen Versuche behandeln die Erscheinungen an passiv bewegten Menschen und Thieren. Auf Versuche
über active Bewegungen und deren Empfindung habe ich aus mehreren Gründen ein geringeres Gewicht gelegt. Die Muskelanstrengungen bei der activen Bewegung führen an sich schon Empfindungen
mit sich und können Nachempfindungen veranlassen, welche die
eigentlichen Bewegungsempfindungen zu überdecken und zu stören
vermögen. Solche Versuche sind daher für unsern Zweck weniger rein.
Ferner löst jede stärkere Gleichgewichtstörung reflectorisch zweck-
mässige Bewegungen zur Erhaltung des Gleichgewichtes aus, wie
wir dies sehr schön an gedrehten Thieren sehen können. Solche
Bewegungen treten nun auch bei den Scheinbewegungen auf und
werden dann, wenn sie auch den Beobachter nicht gerade nieder-
werfen, als in diesem Fall unzweckmässige Bewegungen doch mehr
oder weniger starke Gleichgewichtsstörungen hervorbringen, welche
die zu beobachtenden Empfindungen compliciren. Am reinsten fallen die Versuche aus, wenn man bei durchaus unterstütztem Körper
passiv bewegt wird, wobei jene Bewegungstendenzen schon sehr viel
schwächer auftreten, und wo sie vorhanden sind, keinen sichtbaren
Erfolg haben, so dass sie in der Regel gar nicht bemerkt werden.
Aus diesen Gründen und weil ich mich auf rein Physiologisches
nicht einlassen wollte, habe ich mich fast darauf beschränkt, gele-
gentlich die schon von Purkyně angestellten Versuche zu wieder-
holen. Hierüber mag hier das Nothwendigste in aller Kürze folgen.

Wenn man sich in verticaler Stellung um die Längsaxe des
Körpers mehreremal umdreht, so beginnen die sichtbaren Gegenstände, welche anfangs ruhend erscheinen, bald eine Bewegung in
entgegengesetzter Richtung. Bleibt man nach längerer Drehung

one touches, and even our body, with all its components, seems to move in a direction opposite to the previous active rotation. These are the phenomena of visual and tactile vertigo. The axes of the illusory rotations of both the visual and the tactile vertigo are determined by that axis about which the head had been rotated, as already known by Darwin and Purkyně. After cessation of the active rotation, the axis of the illusory movement retains its orientation in the head and follows every head movement.

Purkyně apparently assumed that the prolonged activity of the postural muscles, in an attempt to rotate the body, continues for some time in an "unconscious manner" after the rotation ceases. Similarly, the prolonged activity of the eye muscles, in an attempt to maintain fixation, continues unconsciously after the rotation ceases. Vertigo results from ascribing the object motion to the sensory effect of these unconscious movements.

Breuer has a different view of the matter. He assumes that the sensation of motion causes unconscious reflexive movements, which occur at the beginning as well as at the end of the rotation. Purkyně also floats the notion that [head] movements act directly on the brain, and thereby produce involuntary movements. In fact, eye movements occurring in the course of experiments that we will discuss somewhat later can probably only be understood according to Breuer's interpretation.

The interpretation first mentioned by Purkyně cannot be completely ruled out, if one does not completely deny the power of habits. The assumption of reflex movements, however, is not so different from this view. Reflex movements can always be interpreted as caused by habit, even if this habit is attributed to a much older and greater historical law (transcending the individual). For reasons, which I will have an opportunity to discuss later, I must adhere to Breuer's interpretation.

plötzlich stehen, so setzen die sichtbaren Gegenstände ihre Bewegung fort, aber auch jeder tastbare Gegenstand, den man anfasst, so wie überhaupt unser Körper mit seinem ganzen Inhalt scheint sich in einem der ausgeführten activen Drehung entgegengesetztem Sinne zu drehen. Dies sind die Phänomene des Gesichts- und des Tastschwindels. Die Axen der scheinbaren Drehung sind sowohl für den Gesichts- als für den Tastschwindel bestimmt durch jene Axe, um welche der Kopf gedreht wurde, wie Darwin und Purkyně schon wussten. Nach Beendigung der activen Rotation behält die Axe der Scheinbewegung ihre Lage im Kopfe bei und macht jede Bewegung des Kopfes mit.

Purkyně scheint nun anzunehmen, dass die längere Zeit ausgeführte Bewegung der Muskel und der Augen, welche erstere durch das Bestreben sich zu drehen, letztere durch das Bestreben zu fixiren, bedingt ist, längere Zeit auch nach dem Stehenbleiben »bewusstlos« fortgesetzt wird. Indem der Effect dieser bewusstlosen Bewegungen den Objecten zugeschrieben wird, entstehn die Schwindelbewegungen.

Breuer hat eine andere Auffassung der Sache. Er nimmt an, dass die bewusstlosen Bewegungen reflectorisch durch die Bewegungsempfindungen erregt werden, welche sowohl beim Beginn, wie beim Ende der Drehung auftreten. Auch bei Purkyně taucht flüchtig die Annahme auf, dass die Bewegung direct auf's Hirn wirkt und von hier aus die unbewussten Bewegungen erregt werden. In der That lassen sich die bei den Versuchen auftretenden Augenbewegungen, von welchen wir später noch sprechen werden, wohl nur nach der Breuer'schen Auffassung verstehn. Die Purkyně'sche ersterwähnte Auffassung wird sich nicht ganz bestreiten lassen, wenn man nicht die Macht der Gewohnheit überhaupt bestreiten will. Doch ist die Annahme von Reflexbewegungen nicht gar so verschieden von dieser Auffassung. Die Reflexbewegungen werden sich immer ansehn lassen als hervorgerufen durch Gewohnheit, wenn auch durch eine Gewohnheit von viel älterem und grösserem (über das Individuum hinausreichendem) historischen Recht. Ich muss mich aus Gründen, welche vorzubringen die Gelegenheit nicht fehlen wird, der Breuer'schen Auffassung anschliessen.

Flourens' Experiment.

1.

Some of Flourens' experiments have been repeated by Dr. Kessel in my laboratory. I therefore know these phenomena well enough from personal observation to able to state that they are very similar to those observed in rotated animals. However, because I have never performed vivisection myself, I must limit myself in this chapter to reporting on aspects of others' work most important for us.

In the course of investigating the conditions underlying the process of hearing, <u>Flourens</u>[4.1] destroyed all parts of the hearing organ in sequence. He found that the pigeon's ability to hear is relatively unaffected by removal of the eardrum and the ossicles in the middle ear; however, it was considerably disturbed by removing the stapes. Opening the <u>middle ear</u>[4.2] and even destroying its nerves does not completely eliminate hearing. Absolute deafness only occurs if the cochlear nerve is destroyed. Although the middle ear and cochlea seem to be insensitive to pain, the semicircular canals prove to be very sensitive. Hearing seems to be unaffected by surgical interference with the semicircular canals. However, Flourens was very surprised to find striking movements of the animals after <u>cutting the semicircular canals</u>.[4.3]

Each time the contents of the horizontal semicircular canal were cut or pierced with a needle, the animal seemed to experience considerable <u>pain</u>[4.4], and it made strong movements of the head in a horizontal direction from right to left and from left to right. This pendular movement of the head generally subsides; however, it reoccurs each time the animal tries to move. If the animal started to walk, its movements became more and more uncoordinated, until it finally succumbed to vertigo. Occasionally spontaneous rotational movements of the animal took place about the vertical axis.

.... "once the animal started to move, its head resumed shaking, and that shaking of the head got worse with movements of the body; all locomotion, all regular movements, finally became impossible, a little like losing

Der Flourens'sche Versuch.

1.

Einige der Flourens'schen Versuche sind von Herrn Dr. Kessel in meinem Laboratorium wiederholt worden. Ich kenne also die Erscheinungen so weit aus eigener Anschauung, um die grosse Aehnlichkeit derselben mit den an gedrehten Thieren beobachteten zu constatiren. Doch muss ich mich, da ich nie selbst Vivisectionen ausgeführt habe, in diesem Capitel darauf beschränken, das für uns Wichtigste aus den Arbeiten anderer zu referiren.

Flourens hat bei Gelegenheit einer Untersuchung über die Bedingungen der Hörfähigkeit nach einander alle Theile des Gehörorganes zerstört. Er hat gefunden, dass die Hörfahigkeit einer Taube wenig beeinträchtigt wird durch Abtragung des Trommelfells und der Gehörknöchelchen, sehr merklich hingegen durch Entfernung des Steigbüigels. Blosslegung des Vorhofs und selbst Zerstörung der Nerven desselben vernichtet das Gehör nicht vollständig. Absolute Taubheit tritt nur bei Vernichtung des Schneckennerven ein. Vorhof und Schnecke scheinen gegen Schmerz unempfindlich, wogegen sich die Bogengänge sehr empfindlich erweisen. Das Gehör scheint durch Eingriffe in die Bogengänge nicht afficirt zu werden. Dagegen fand sich Flourens sehr überrascht durch die auffallenden Bewegungen der Thiere bei Durchschneidung der Bogengänge.

Jedesmal wenn der Inhalt eines horizontalen Bogenganges durchschnitten oder mit einer Nadel gestochen wird, scheint das Thier lebhaften Schmerz zu empfinden und es treten heftige Bewegungen des Kopfes in horizontaler Richtung von rechts nach links und von links nach rechts auf. Diese Pendelbewegung des Kopfes beruhigt sich oft, tritt aber bei jedem Bewegungsversuch des Thieres von neuem auf. Fing das Thier an zu gehn, so wurden seine Bewegungen immer ungeordneter, bis es endlich ganz dem Schwindel unterlag. Zuweilen treten auch spontane Drehbewegungen des Thieres um die verticale Axe auf.

.... »dès que l'animal commençait à marcher, la tète recommençait à s'agiter; et cette agitation de la tète s'accroissant avec les mouvements du corps, toute démarche, tout mouvement régulier,

one's equilibrium and movement stability when turning about several times or shaking the head violently"*). One sees by how little Flourens misses our interpretation.

Because of an unmistakable similarity between these phenomena and those that Flourens obtained after certain cerebellar lesions, he looked very carefully to see whether such lesions had occurred. Because of its importance, I will cite the relevant passage**).

"The striking similarity of that last part of the phenomena to the phenomena which follow cerebellar lesions could lead to the belief that one had created a lesion of that organ, if not directly, then at least indirectly. I therefore examined the cerebellum with the greatest care; it appeared perfectly intact.

To establish even more precisely the independence of the phenomena that I have described from all lesions, be they direct or indirect, of the cerebellum or any other part of the brain, I resorted to the following precautions. With great care I exposed the semicircular canals in a pigeon; then I cut the horizontal canals on both sides.

I chose this canal as the subject of my experiments because it is the one that is the furthest away from the brain, particularly from the cerebellum; and in cutting it I placed full attention to avoiding the least shaking which could have extended to the walls of the skull.

Despite these precautions, the cutting of the two canals was followed by precipitate shaking of the head.

Once this shaking had stopped, one could always reinitiate it either by poking the internal walls of the canals with a needle or by stimulating the animal to move.

The shaking was always the strongest in the beginning; then it diminished, and finally little by little it disappeared completely. But this was only true as long as the animal remained at rest; when it walked, to the contrary, the shaking was always more prominent when the animal

---

*) Researches. 2nd edition. p. 446.
**) p. 447.

finissaient par devenir impossibles, à peu près comme on perd l'equilibre et la stabilité de ses mouvements quand on tourne quelque temps sur soi-même, ou qu'on secoue violement la tète«*). Man sieht, wie hart Flourens an unserer Auffassung vorbeigeht.

Da eine Aehnlichkeit dieser Erscheinungen mit denjenigen, welche Flourens nach gewissen Kleinhirnverletzungen erhalten hatte, unverkennbar war, so untersuchte er sehr genau, ob solche Verletzungen eingetreten waren. Der Wichtigkeit dieses Punktes wegen will ich die darauf bezüglichen Worte anführen**).

»La rassemblance frappante de cette dernière partie du phenomene avec les phénomènes qui suivent les lésions du cervelet pouvait faire croire à quelque lésion, sinon directe, du moins indirecte de cet organe. J'examinai donc le cervelet avec le plus grand soin; il parut dans un état d'integrité parfaite.

Pour établir, avec plus de précision encore, l'indépendance du phénomène que je décris de toute lésion, du moins directe, soit du cervelet, soit de toute autre partie de l'encéphale, j'eus recours aux précautions suivantes. Je mis bien exactement à nu les canaux semicirculaires sur un pigeon; puis je coupai le canal horizontal des deux côtes.

Je choisis ce canal pour sujet de mon expérience, parce qu'il est le plus éloigné de l'encéphale, surtout du cervelet; et en le coupant je mis toute mon attention à éviter la moindre secousse qui eû pu se communiquer aux parois du cràne.

Malgré ces précautions, la section des deux canaux fut suivie du branlement impétueux de la tète.

Quand ce branlement s'arrètait, ou pouvait toujours le reproduire, soit en piquant avec la pointe d'une aiguille les parois internes des canaux, soit en excitant l'animal à se mouvoir.

Le branlement était toujours plus vif au moment où il commençait; puis il allait en se ralentissent, et finissait peu à peu par cesser tout-à-fait. Mais les choses ne se passaient ainsi qu'autant que l'animal restait en repos; quand il marchait, au contraire, le

---

*) Recherches. 2e édition. p. 446.
**) p. 447.

attempted to walk faster.

I have assured the absence of any lesion or even any direct injury of the cerebellum; but there still remains one other possible cause of a lesion to be examined.

In breaking the semicircular canals with scissors, as I have done thus far, one inevitably cuts the small artery which runs on the outer side; and this rupture quickly leads to an effusion of blood which rapidly reaches the cerebellum, the medulla oblongata, and the whole bony structure of the posterior walls of the skull. It was therefore important to avoid the complication that could result from such bleeding.

Therefore, after the semicircular canals were exposed in a pigeon, I opened one of the canals, the horizontal, on the opposite side of the artery, consequently without rupturing the artery.

All bleeding being avoided this way, I poked the inner parts of the canal; pain and agitation of the head soon followed, in the same manner as in the previous experiments; although now the agitation was much less than in the cases where the canal was completely severed."

The described consequences of the operation on a pigeon operated on November 15, 1824, were evident until May 17, 1825. Every time the animal was stimulated to move, pendular movements of the head and rotations were seen. The animal was sacrificed and examination of the brain proved it to be completely intact. One can see that Flourens has carefully experimented and observed.

Cutting the horizontal canals (both sides) causes violent horizontal side-to-side movements of the head. Strong eye movements always went hand in hand with these head movements.

If both lower vertical [posterior] canals were cut, the head movements were in an up-down direction, with the head often tilted to one side or the other. The ability to fly is lost, and the eye movements are as violent as in the earlier case. The animal frequently somersaults and falls on its back.

branlement était toujours d'autant plus vif que l'animal cherchait à marcher plus vite.

J'avais constaté l'absence de toute lésion ou plutôt de toute blessure directe du cervelet; mais il restait une cause particulière de lésion à examiner encore.

En rompant avec des ciseaux, comme je l'avais fait jusqu'ici, les canaux semi-circulaires, on rompt inévitablement la petite artère qui rampe sur leur côté externe; et cette rupture amène bientôt un épanchement de sang qui gagne rapidement le cervelet, la moelle allongée et toute la cellulosité osseusse des parois postérieures du crane. Il importait donc d'éviter la complication qui pouvait résulter le cet épanchement.

A cet effet, les canaux semi-circulaires etant mis à nu sur un pigeon, j'ouvris l'un de ces canaux, l'horizontal, par le côté opposé à celui qu'occupe l'artère, et sans ouvrir l'artère par consequent.

Tout epanchement étant évité ainsi, je piquai les parties internes de ce canal; la douleur et l'agitation de la tète suivirent aussitôt, et de la même manière que dans les expériences precédents; á cela près, néanmoins, que l'agitation fut bien moindre dans ce cas que dans le cas de la rupture complete du canal.»

Die beschriebenen Folgen der Operation erhielten sich nun an einer am 15. November 1824 operirten Taube bis zum 17. Mai 1825. Jedesmal wenn das Thier zu Bewegungen gereizt wurde, zeigten sich die Pendelbewegungen des Kopfes und die Drehungen. Das Thier wurde getödtet und die Untersuchung des Gehirns erwies dasselbe als vollkommen unversehrt. Man sieht hieraus, dass Flourens sorgfältig experimentirt und beobachtet hat.

Durchschneidung der Horizontalcanäle (beiderseits) bringt heftige horizontale Hin- und Herbewegungen des Kopfes hervor. Mit diesen Kopfbewegungen gingen auch stets starke Augenbewegungen Hand in Hand.

Wurden beiderseits die untern verticalen Canäle durchschnitten, so traten die Kopfbewegungen auf- und abwärts ein, wobei der Kopf oft nach der einen oder andern Seite sich neigte. Das Flugvermögen ist verloren, die Augenbewegung wie in dem frühern Falle heftig. Das Thier überkugelt sich häufig, indem es auf den Rücken fällt.

After cutting both upper [anterior] canals, violent up-and-down head movements appear and the pigeon frequently tumbles forward.

Merely opening the bony canals has no effect. The movements occur only when the membranous canals are pierced with a needle, however, they are weaker than in the case of cutting the canals.

"In over twenty experiments on these canals, I was always certain of the complete integrity of the cerebellum"*).

Flourens also did experiments with rabbits. Cutting of the left horizontal canal induced movements of the head from right to left and vice versa, which reoccurred for each movement stimulation. Violent eye movements always accompanied it. At rest the head is turned to the left. The animal frequently rotates towards the left about a vertical axis. If the operation is also performed on the other side, the rotation occurs sometimes to the right and at other times to the left. At rest the head position then appears normal. The operation on the right side alone induces a rotation to the right.

Cutting of both posterior vertical canals induces backward tumbling. Cutting of the anterior vertical canals induces forward tumbling.

The fact that the canals can be cut without a noticeable deficit in hearing allows one to assume that the nerves concerned are not associated with hearing. Whether stimulation of these nerves produces pain cannot easily be decided on the basis of Flourens' remarks, because he does not state how the animals express their pain. The violent motions and struggling can have totally other origins. If pain is present, it could come from nerves other than those that stimulate movement. The most natural assumption is that these nerves, stimulated according to their specific energy[4.5], trigger motion sensation. We also see that, although Flourens does not discuss this matter in further detail, and is rather inclined to a painfully increased hearing sensitivity, he finds no better picture than rotational vertigo to describe the phenomena.

---

*) p. 466.

Nach Durchschneidung der beiden obern verticalen Canäle erscheinen heftige Kopfbewegungen auf und ab und die Taube überkugelt häufig nach vorn.

Blosees Aufbrechen der knöchernen Canäle hat keine Wirkung. Erst wenn man die häutigen Canäle mit einer Nadel sticht, treten die Bewegungen ein, wiewohl schwächer als bei Durchschneidung der Bogengänge.

»Dans plus de vingt expériences sur ces canaux, je me suis constamment convaincu de l'integrité complète et absolue du cervelet«*).

Auch an Kaninchen hat Flourens experimentirt. Durchschneidung des linken horizontalen Bogganges brachte Bewegungen des Kopfes von rechts nach links und umgekehrt hervor, die sich bei jeder Aufreizung zur Bewegung wiederholten. Hiebei immer heftige Augenbewegungen. In der Ruhe ist der Kopf nach links verdreht. Das Thier dreht sich häufig um die Verticalaxe nach links. Wird die Operation auch auf der andern Seite ausgeführt, so tritt die Drehung bald nach rechts, bald nach links ein. In der Ruhe ist dann die Kopfstellung normal. Die Operation auf der rechten Seite allein bringt Rechtsdrehung hervor.

Durchschneiden der beiden hintern Verticalcanäle bringt Ueberschlagen nach hinten hervor. Durchschneiden der vordern Verticalcanäle gibt Ueberschlagen nach vorn.

Der Umstand, dass die Bogengänge durchschnitten werden können, ohne eine merkliche Gehörsstörung hervorzubringen, legt die Vermuthung nahe, dass die betreffenden Nerven mit dem Hören nichts zu thun haben. Ob die Reizung dieser Nerven Schmerz verursacht, lässt sich nach den Angaben von Flourens schwer entscheiden, da er nicht angibt, wodurch die Thiere den Schmerz äussern. Die heftigen Bewegungen und das Sträuben können ganz andere Ursachen haben. Sind Schmerzen vorhanden, so können diese auch von andern Nerven herrühren als denjenigen, welche die Bewegungserscheinungen anregen. Die natürlichste Annahme ist die, dass diese Nerven vermöge ihrer specifischen Energie gereizt Bewegungsempfindungen erregen. Wir sehen auch, dass Flourens,

---

*) p. 466.

2.

All of the important points of Flourens' experiments were confirmed by Harless[4.6], Czermak[4.7], Brown-Séquard[4.8], and Vulpian.[4.9] After cutting the auditory nerve in mammals and frogs Brown-Sequard again observed twisting and yaw movements, a finding that Schiff[4.10] disputed but which was noted again by Goltz.

Vulpian explains the phenomenon by assuming a strong excitation of the auditory nerve, which frightens the animal. Brown-Sequard and later Löwenberg also considered an excitation of the auditory nerve that would then induce the strange movements reflexively.

Goltz proposed a substantially new interpretation of Flourens' phenomena. Relying upon his experiments with frogs in which he cut the auditory nerve, and upon several clever modifications of Flourens' experiments, he came to the following conclusion: "Whether the semicircular canals are organs of hearing remains to be seen. In addition, however, they constitute an apparatus that serves to maintain balance. They are, so to speak, sense organs for the equilibrium of the head and thereby for the whole body." As one sees, this is the very hypothesis, which I also support with some modifications, and to which I was led through totally different experiments and considerations.

Goltz now develops a somewhat more detailed hypothesis of the function of this organ of equilibrium in the following words: "I have thought of the following as a possibility. We will assume that the appropriately arranged nerve endings which exist in the ampullae may be excited by pressure or stretching similar to the nerves of the skin surface that serve to sense pressure. According to well-known physical laws, the fluid found in the canals (endolymph) would expand the furthest the lowermost parts of the wall. Depending on the position of the head, the distribution of the fluid pressure[4.11] will vary, and a distinct pattern of nerve excitation would correspond to each head position.

obwohl er hierauf nicht eingeht und sogar eher an schmerzhaft erhöhte Empfindlichkeit des Gehörs denkt, kein besseres Bild zur Beschreibung der Erscheinungen findet, als den Drehschwindel.

2.

Die Versuche von Flourens sind in allen wichtigen Punkten von Harless, Czermak, Brown-Sequard und Vulpian bestätigt worden. Brown-Sequard hat bei Säugethieren und Fröschen mit durchschnittenem Hörnerv Roll- und Drehbewegungen beobachtet, die Schiff bestritten, Goltz aber wieder bemerkt hat.

Vulpian nimmt zur Erklärung der Erscheinungen eine heftige Erregung der Gehörnerven an, die erschreckend auf das Thier wirkt. Brown-Sequard und später Löwenberg haben ebenfalls an Reizung des Gehörnerven gedacht, welche dann reflectorisch die sonderbaren Bewegungen auslöst.

Goltz hat eine wesentlich neue Auffassung der Flourens'schen Phänomene gefunden, indem er, gestützt auf seine Versuche an Fröschen mit durchschnittenen Gehörnerven und auf mehrere sinnreiche Abänderungen der Flourens'chen Versuche den Satz ausspricht: »Ob die Bogengänge Gehörorgane sind, bleibt dahingestellt. Ausserdem aber bilden sie eine Vorrichtung, welche der Erhaltung des Gleichgewichts dient. Sie sind so zu sagen Sinnesorgane für das Gleichgewicht des Kopfes und mittelbar des ganzen Körpers« Wie man sieht, ist dies diejenige Vorstellung, welche auch ich mit einigen Modifictionen glaube festhalten zu müssen und auf die ich durch ganz andere Experimente und Ueberlegungen geführt worden bin.

Goltz entwickelt nun eine etwas detaillirtere Vorstellung von der Wirkungsweise dieses Gleichgewichtsorgans in folgenden Worten: »Ich habe mir folgende Einrichtung als möglich gedacht. Wir wollen annehmen, dass die in den Ampullen vorhandenen Nervenendigungen in ähnlicher Weise geeignet sind, durch Druck oder Dehnung erregt zu werden, wie etwa die dem Drucksinne dienenden Nerven der äussern Haut. Die in den Bogengängen befindliche Flüssigkeit (Endolymphe) wird nach bekannten physikalischen Gesetzen diejenigen Abschnitte der Wandung am stärksten anspannen, welche am meisten nach abwärts gelegen sind. Je nach der Stel-

3.

We should now mention some work that was critical of Flourens' experiments and the theories derived from them. Schklarewsky points out a special outgrowth of the cerebellum located between the semicircular canals, which could easily be damaged while performing Flourens' experiments and could cause the observed phenomena. Böttcher declares this outgrowth to be the vestibular aqueduct, based upon his own and Hasse's*)[4.12] anatomical investigations. The interpretations that Cyon[4.13] arrives at in his work combine those of Goltz with some older concepts.

Cyon concludes from his experiments that the semicircular canals are organs of space which deliver "unconscious sensations" to us, from which we draw "unconscious deductions" about head position. Cyon believes that the disturbances following cutting of the canals result from anomalous hearing sensations. The latter seems quite unlikely when one considers that similar phenomena occurring with rotational vertigo are not accompanied by acoustic sensations at all.

Based on numerous carefully performed experiments in pigeons, Curschmann[4.14] finds all of the previously proposed theories about Flourens' phenomena to be indefensible; namely the concept of the canal as a sense organ. He does not consider the phenomena to be the results of a stimulus but rather the results of a functional deficiency without, however, being able to supply a comprehensive theory himself.

The single publication that impressed me most, both by the number of experiments and the decisiveness of the language, is the one by Böttcher[4.15], because it rejects almost everything that had been considered correct about Flourens' experiments. I will quote those sentences which Böttcher himself considered important. Disturbances of movement are discussed first:

---

*) Hasse, anatomical studies. no 4. p. 765.

lung des Kopfes wird die Vertheilung des Druckes der Flüssigkeit wechseln, und einer jeden Kopfhaltung wird demgemäss immer eine bestimmte Form der Nervenerregung entsprechen.«

3.

Wir haben nun noch einiger Arbeiten zu erwähnen, welche sich die Kritik des F l o u r e n s 'schen Versuches und der daran geknüpften Theorieen zur Aufgabe gemacht haben. S c h k l a r e w s k y weist auf einen zwischen den Bogengängen befindlichen Kleinhirnfortsatz hin, der bei dem F l o u r e n s 'sehen Versuch sehr leicht mit verletzt werden und die beobachteten Erscheinungen bedingen kann. Letzteren Kleinhirnfortsatz erklärt B ö t t c h e r gestützt auf seine und H a s s e's*) anatomische Untersuchungen für den aquaeductus vestibuli. Die Anschauungen, zu welchen C y o n in seiner Arbeit gelangt, setzen sich aus jenen von G o l t z und aus älteren Vorstellungen zusammen.

C y o n schliesst aus seinen Versuchen, dass die Bogengänge Raumorgane seien, welche uns "unbewusste Empfindungen" liefern, aus welchen wir »unbewusste Schlüsse« über die Kopfstellung ziehn. C y o n hält die Störungen bei Durchschneidung der Bogengänge für Folgen anomaler Gehörsempfindungen. Letzteres wird man sehr unwahrscheinlich finden, wenn man bedenkt, dass die beim Drehschwindel auftretenden ganz ähnlichen Phänomene keineswegs von Gehörsempfindungen begleitet sind.

C u r s c h m a n n findet auf Grund zahlreicher sorgfältig angestellter Versuche an Tauben, dass alle bis dahin aufgestellten Theorieen der F l o u r e n s 'schen Erscheinungen, namentlich die Auffassung der Canäle als Sinnesorgane, unhaltbar seien. Er hält die Phänomene nicht für Folgen eines Reizes, sondern für Folgen eines Functionsausfalles, ohne jedoch selbst eine umfassende Theorie geben zu können.

Diejenige Arbeit, welche sowohl durch die Zahl der Versuche als auch durch die Entschiedenheit der Sprache am meisten imponiren möchte, indem sie so ziemlich Alles in Abrede stellt, was bisher über den F l o u r e n s 'schen Versuch als richtig angenommen wurde, ist die von B ö t t c h e r. Ich will diejenigen Sätze, die B ö t t -

---

*) Hasse, anatomische Studien. 4. Heft. S. 765.

"1. Twisting of the head to one side and lowering it to the floor, so that the crown touches the floor, and the beak seems to be directed more or less behind."– "This disturbance can be induced by a unilateral operation, if the lesion is a deep one.".... – "On the other hand, head twisting is never seen if one operates cautiously." It is further noted, "that head twisting often developed only over time." In the majority of these cases Böttcher "found changes inside the skull, which had to be directly related to previous signs of illness." It is also mentioned "that the head twist often develops quite suddenly."

"2. Racetrack movements and fore-and-aft movements about a transverse axis" – "I have proven that after cutting the canals on both sides the initial occurrence of movement disturbance can totally disappear." – "If the loss of the semicircular canals and the resulting vertigo had been the cause of the aforementioned movement disturbances, they would have had to continue." – "As a further result of my investigation might I stress the fact that the movement disorders of the extremities always manifest themselves on the side on which the sectioning of the semicircular canal apparatus had taken place." – "Furthermore I must state that the movement disorders under discussion may often be present frequently for motion on the ground and during flight, but are also frequently observed only during motion on the ground, or only during flight." – "If in one case the surgical intervention only affects the movement on the ground, and for another case only the ability to fly, then this disturbance cannot result from vertigo, because then it should affect movements during motion on the ground in the same way as during flight." – "Finally it should be noted that the manner of movement disturbance does not solely depend on which canals were cut, but also where these cuts had been made."

"3. The pendular movements of the head can be a temporary phenomenon." – "In other cases the pendular movements remain while the

cher selbst für wichtig erkennt, herausheben. Unter den Bewegungsstörungen wird zuerst besprochen:

»1. Die einseitige Verdrehung des Kopfes und die damit ver-
knüpfte Senkung desselben auf den Boden, so dass der Scheitel den
Boden berührt und. der Schnabel mehr oder weniger nach hinten
gerichtet erscheint.«–»Diese Störung kann durch einseitige Operation erzeugt werden, wenn der Eingriff ein tiefer ist«.... –»Dagegen zeigt sich die Kopfverdrehung niemals, wenn mit Vorsicht
operirt wird.« Es wird ferner bemerkt, »dass die Kopfverdrehung
sich häufig erst in der Folge entwickelte.« In der Mehrzahl dieser
Fälle hat Böttcher »Veränderungen in der Schädelhöhle feststellen
können, die mit den vorhergegangenen Krankheitserscheinungen in
directe Beziehung gebracht werden müssen«. Auch wird erwähnt,
»dass die Kopfverdrehung sich öfters ganz plötzlich entwickelt«.

»2. Die Reitbahnbewegungen und die Bewegungen nach vorn
und hinten um die Querachse« – »Ich habe nachgewiesen, dass
nach Durchschneidung des Bogenapparates auf beiden Seiten die
Anfangs auftretenden Bewegungsstörungen gänzlich schwinden können.« – »Wäre der Verlust der Bogengänge und ein dadurch ver-
ursachter Schwindel die Ursache obengenannter Bewegungstörungen,
so hätten diese andauern müssen.« – »Als ferneres Ergebniss meiner
Untersuchung darf ich die Thatsache hervor heben, dass die Be-
wegungsstörung immer an den Extremitäten der Seite sich geltend
macht, auf welcher die Durchschneidung des Bogenapparates vorgenommen wurde.« – »Weiterhin habe ich geltend zu machen, dass
die in Rede stehenden Beweguugsstörungen zwar bald beim Gange
und Fluge zugleich vorhanden sind, bald aber auch nur beim Gange
oder nur beim Fluge beobachtet werden.« – »Wenn nun einerseits
durch den operativen Eingriff' nur das Gehvermögen und anderseits
nur das Flugvermögen leidet, so kann diese Störung nicht von einem
Schwindel abhängen, weil ein solcher die Bewegungen beim Gange
und Fluge in ganz gleicher Weise beeinträchtigen müsste.« – »Endlich ist noch darauf aufmerksam zu machen, dass die Art der Bewegungsstörungen nicht nur davon abhängt, welche Bogengänge,
sondern auch davon, wo diese durchschnitten werden.«

»3. Die Pendelbewegungen des Kopfes können eine vorüber-
gehende Erscheinung sein.« – »Die Pendelbewegungen sind in an-

head stays continuously erect." – "If a head twist occurs, the pendular movements usually cease. "

Concerning the theory of Goltz, Böttcher notes:
"If his hypothesis were right, then after cutting the semicircular canal apparatus on both sides in every case a permanent disturbance of balance would remain." – "Despite unchanged head position, however, movement disturbances of the trunk do nevertheless occur." – "The balance disturbances of the trunk cannot be derived from balance disturbances of the head." – "The balance disturbances of the trunk can disappear completely." – "The continuity of the semicircular canals, i.e., of corresponding canals on both sides, can be disrupted without necessarily resulting in any balance disturbance of the trunk." – "If, as Goltz assumes, vertigo were the cause of racetrack movements and of rotations about the transverse axis, this disturbance would express itself in the same manner for the use of wings as for the use of legs." Concerning the head twist Böttcher says: "It is therefore neither a direct consequence of cutting the semicircular canals nor the cause of the movement disturbances of the trunk."

According to Böttcher, the racetrack movements and the pendular movements depend upon the cutting of the semicircular canal itself. Specifically for racetrack movements the outcome of the surgery is directly related to the pull exerted on the membranous canals. Böttcher believes that this pull generates changes within the brain that cause the movement phenomena. Böttcher also attributes the pendular movements due to damage of the central organ, because he feels he is able to rule out other explanations.

Thus we see that the main factors for Böttcher are brain lesions, which Flourens disputes. Böttcher pointed out that brain lesions are the probable cause of head twisting. The conclusion stated in his treatis, that such lesions are also the cause for the turning and pendular movements of importance to us, is bare speculation. His deductions are generally somewhat

dern Fällen bleibend bei unausgesetzt gerader Kopfhaltung.« -
»Wenn eine Kopfverdrehung sich einstellt, hören die Pendelbewegungen meist auf.«

Ueber die G o l t z ' sche Theorie bemerkt B ö t t c h e r :

»Wenn die Hypothese desselben richtig wäre, so müssten
nach Durchschneidung des Bogenapparates auf beiden Seiten in jedem
Fall bleibende Gleichgewichtsstörungen vorhanden sein.« – »Trotz
unveränderter Kopfstellung kommen nun aber doch Bewegungsstörungen des Rumpfs vor.« – »Die Gleichgewichtsstörungen des
Rumpfs können aber nicht aus Gleichgewichtsstörungen des Kopfs
abgeleitet werden.« – »Die Gleichgewichtsstörungen des Rumpfs
können sich vollständig verlieren.« – »Es kann also die Continuität
der Bogengänge, und zwar gleichnamiger Bogengänge, auf beiden
Seiten unterbrochen sein und braucht darum dennoch keine Gleichgewichtsstörung des Rumpfs zu bestehen.« – »Wenn ein Schwindel,
wie G o l t z annimmt, die Ursache der Reitbahnbewegungen und der
Drehungen um die Querachse wäre, so müsste sich diese Störung
beim Gebrauch der Flügel und der Beine stets in gleicher Weise
äussern.« – Von der Kopfverdrehung sagt B ö t t c h e r : »Sie ist demnach weder eine unmittelbare Folge der Bogengangdurchschneidung
noch auch kann sie die Ursache der Bewegungsstörungen des
Rumpfs sein.«

Die Reitbahnbewegungen und die Pendelbewegungen hängen
nach B ö t t c h e r mit der Durchschneidung der Bogengänge selbst
zusammen, und zwar steht, was die Reitbahnbewegungen betrifft, der
Erfolg der Operation in geradem Verhältnisse zu der Zerrung, welche
auf die häutigen Bogengänge ausgeübt wird. Durch diese Zerrung
werden, wie B ö t t c h e r glaubt, Veränderungen im Hirn hervorgebracht, welche die Ursachen der Bewegungserscheinungen sind. Auch
die Pendelbewegungen schiebt B ö t t c h e r auf eine Mitleidenschaft
des Centralorgans, indem er meint andere Erklärungen ausschliessen
zu können.

Wie wir also sehn, spielen bei B ö t t c h e r Hirnverletzungen,
welche F l o u r e n s in Abrede stellt, die Hauptrolle. Störungen im
Hirn hat B ö t t c h e r als wahrscheinliche Ursache der Kopfverdrehung
aufgewiesen. Dass solche Störungen auch Ursache der für uns wichtigen Drehungen und Pendelbewegungen seien, ist, wie man aus seiner

arbitrary. It is difficult to see why permanent balance disturbances would have to occur after canal surgery, as predicted by Goltz' theory. Couldn't the nerve endings retain their sensitivity in one case and lose it in another, so that in the latter case the disturbances caused by irritation would also disappear? Do the movement disorders really occur in such a regular and machine-like fashion that one should be surprised to see them more developed during locomotion in one case and during flight in another? As for the persistent denial of the existence of vertigo after semicircular canal operation, I merely invite the reader to place a pigeon or a rabbit on a centrifuge and rotate it. The experiment is easily performed and conclusive.

If the function of the ampullar nerves is to react to mechanical stimuli, then quite obviously the success of the operation will be directly related to any pull on the canals. Obviously the ampullar nerves would not be stimulated if one could cut the semicircular canals without any mechanical effect on them. Since it is difficult to separate mechanical irritation of the ampullae from disturbances of the brain, one must acknowledge that the experiment is not clean. However, it is arbitrary to conclude that of the two manipulations, pull on the nerve and propagation of this pull to the brain, only the latter is effective. Admitting that pendular and rotational movements are directly related to cutting of the semicircular canals, even after consideration of Böttcher's work, the most natural explanation still seems to me to be the one I have proposed, namely that the ampullar nerves respond to every stimulus with a rotation perception corresponding to their specific energy.

As for Böttcher's remark about my discourse, that I would assume the existence of a balance organ, and that it therefore would be pointless to follow the conclusions drawn from, I simply have to reply the following in opposition. In this case he seems to have ignored the fact that vivisection is neither the only nor the best method to learn about physiological evidence or to test theories of it. Otherwise he would have noticed that my work consists mostly of experiments that would retain a certain value, even if

Abhandlung ersehn kann, eine blosse Vermuthung. Auch sonst sind seine Deductionen nicht frei von Willkührlichkeit. Es ist schwer einzusehn, warum bleibende Gleichgewichtsstörungen nach der Bogengangoperation eintreten müssten wenn die Goltz'sche Theorie richtig wäre. Können denn nicht die Nervenendapparate in einem Falle ihre Empfindlichkeit behalten, im andern dieselbe einbüssen, so dass im letzteren Falle auch die von Reizungen herrührenden Störungen verschwinden? Treten denn die Bewegungsstörungen überhaupt so regelmässig und maschinenmässig auf, dass es überraschen muss, sie einmal mehr im Gange, ein anderes mal mehr im Fluge entwickelt zu sehn? Was das fortwährende Leugnen des Schwindels nach der Bogengangoperation betrifft, so möchte ich den Leser bitten eine Taube oder ein Kaninchen auf der Centrifugalmaschine in Rotation zu versetzen. Das Experiment ist leicht ausgeführt und spricht deutlich.

Sind die Ampullennerven bestimmt, auf mechanische Reize zu reagiren, so ist es ganz selbstverständlich, dass der Erfolg der Operation im geraden Verhältnisse zur Zerrung der Bogengänge stehn wird. Natürlich würden auch die Ampullennerven nicht gereizt, wenn man die Bogengänge ganz ohne mechanische Wirkung auf erstere durchschneiden könnte. Wenn von mechanischen Ampullenreizungen Störungen im Hirn schwer zu trennen sind, so ist anzuerkennen, dass das Experiment unrein ist. Willkührlich aber ist es, von den beiden Eingriffen, Zug am Nerv und Fortpflanzung dieses Zuges ins Hirn, bloss den letzteren als wirksam zu betrachten. Wird einmal zugegeben, dass Pendel- und Drehbewegungen mit der Durchschneidung der Bogengänge selbst zusammenhängen, so scheint mir auch nach der Böttcher'schen Arbeit die natürlichste Ansicht noch immer die von mir aufgestellte, dass die Ampullennerven vermöge ihrer specifischen Energie jeden Reiz mit einer Drehempfindung beantworten.

Was die Bemerkung Böttchers über meine Abhandlung betrifft, dass ich mit dem Gleichgewichtsorgan wie mit einer gegebenen Grösse operire und es demnach keinen Zweck hätte den daran geknüpften Betrachtungen zu folgen, so habe ich dagegen nur Folgendes zu bemerken. Er scheint hiebei nicht beachtet zu haben, dass Vivisectionen weder das einzige noch das beste Mittel sind, physiologische Thatsachen kennen zu lernen oder deren Theorien zu

Flourens' experiment had never existed. It leads to a general theory, whose main points would not be effected, even if one withdrew the assignment of the semicircular canals as the organ of equilibrium. Nevertheless, these experiments support the existence of a balance organ and are not as irrelevant to the interpretation of Flourens' experiments as Mr. Böttcher has supposed.

A paper by Berthold[4.16] describing very careful experiments appeared at the same time as Böttcher's publication. Berthold observed Flourens' phenomena by cutting the membranous semicircular canals with a thread sling after having opened the bony canals. Even before cutting [of the membranous semicircular canals], one sometimes observes bending of the legs on the operated side when the animal walks. Berthold reaches the following conclusion: "The semicircular canals have the function of movement coordination by means of reflexes. They perform this work in connection with two sensory organs, the senses of vision and touch. It has not yet been demonstrated whether the semicircular canals play a role in hearing."

Since I cannot repeat the relevant experiments I must rely mainly on the literature concerning this subject. There is general agreement among researchers concerning the main argument, namely the connection between pendular and rotational movements and the cutting of the semicircular canals. Other aspects of the experiment, whose performance requires careful attention, vary considerably and thereby influence any interpretation of the phenomena. Only that main argument was important to me, and I tried to interpret it in the most natural way. This is the same interpretation reached by Breuer and Brown. I therefore still feel justified in maintaining my hypothesis, even after consideration of Böttcher's work. The status of this episode is that in this field, in which researchers cannot agree on any theories even after 50 years of experiments, Flourens' experiment cannot be taken for proof, but rather as a hint.

prüfen. Sonst würde er bemerkt haben, dass meine Arbeit grösstentheils Versuche enthält, welche für sich schon einen gewissen Werth behalten, auch wenn der Flourens'sche Versuch gar nicht existirt, und die zu einer allgemeinen Theorie führen, an welcher die Hauptsache durch das Aufgeben des Gleichgewichtsorgans in den Halbcirkelcanälen gar nicht afficirt wird. Doch sprechen diese Versuche für das Gleichgewichtsorgan und werden daher auch für die Beurtheilung des Flourens'schen Experimentes nicht so gleichgültig sein, als Herr Böttcher meint.

Mit der Böttcher'schen Arbeit gleichzeitig erschien eine Abhandlung von Berthold, welche sehr sorgfältig angestellte Experimente enthält. Berthold hat die Flourens'schen Erscheinungen beobachtet, indem er nach aufgebrochenen knöchernen Bogengängen die häutigen mit Hilfe einer Fadenschlinge durchschnitten hat. Beim Gehen der Thiere beobachtet man schon vor der Durchschneidung ab und zu ein Einknicken der Beine auf der operirten Seite. Der Schluss zu dem Berthold kommt, ist folgender: »Die Bogengänge haben sonach die Function, die Coordination der Bewegungen auf dem Wege des Reflexes zu vermitteln. Sie leisten diese Arbeit in Verbindung mit zwei Sinnesorganen, mit dem Gesichts- und dem Gefühlssinn. Ob die Bogengänge zum Hören dienen, ist bis jetzt nicht festgestellt.«

Aus der Literatur des Gegenstandes, an welche ich mich hauptsächlich halten muss, da ich die hieher gehörigen Versuche nicht nachmachen kann, geht hervor, dass in Betreff der Hauptsache, des Zusammenhangs der Pendel- und Drehbewegungen mit der Durchschneidung der Bogengänge unter den Forschern grosse Uebereinstimmung herrscht. Andere Umstände des Versuches, dessen Ausführung überhaupt sehr viel Aufmerksamkeit erfordert, variiren ausserordentlich und damit auch die Deutung der Erscheinungen. Auf jene Hauptsache allein, deren natürlichste Deutung ich versucht habe und auf welche auch Breuer und Brown verfallen sind, auf diese allein kommt es mir an. Ich glaube mich demnach auch nach der Böttcher'sehen Arbeit noch berechtigt, meine Hypothese festzuhalten. Als Beweis ist bei dem Zustand des Capitels, der sich in der Nichtübereinstimmung der Forscher in allem Theoretischen selbst nach 50jährigem Experimentiren ausspricht, der Flourens'sche Versuch nicht zu gebrauchen, wohl aber als Fingerzeig.

Phenomena Reminiscent of Flourens' Experiment.

1.

Pathological cases were reported which vividly remind us of the phenomena of Flourens'[5.1] experiments. Menière[5.2] studied several peculiar forms of impaired hearing, tinnitus, vomiting, vertigo and rotation movements, and established exuded matter in the canals as their cause*). Such phenomena have also been observed in animals.

Recently Sigm Exner[5.3] also reported a disease he had observed in rabbits that is analogous to Menière's disease.

"The signs of this disease and the autopsy results are used to prove that those signs of disease that can be observed in animals with lesioned canals may also exist without any influence on the cerebellum. This occurs with respect to the recent conjecture that vertigo, which Flourens demonstrated in operated animals, resulted from lesions of the cerebellum**)."

In the paper referred to earlier, Berthold also referred to two cases of pendular head movements in humans. Because those observations are without autopsy reports, no conclusions can be drawn from them.

2.

Phenomena similar to those seen in Flourens' experiment may sometimes actually be solely by disease of the canals alone. This is important for us. If the canals are end organs we should not be surprised that the same or similar phenomena can also be produced by effects on the central organ itself. In this regard, Brücke's***)[5.4], discussion of the case of one of Türk's[5.5] patients should probably be described. She felt as though the bed were lifted and turned, so that in fright she rotated toward the

---

*) v. Tröltsch, Otology. 5th ed. 1873. p. 489.

**) Newsletter of the Vienna Academy. 1874. no. 20.

***) Physiology p. 59.

Erscheinungen, welche an den Flourens'schen Versuch erinnern.

1.

Es sind pathologische Fälle bekannt geworden, welche lebhaft an die Erscheinungen des Flourens'schen Versuchs erinnern. Menière hat die eigenthümlichen Formen von Schwerhörigkeit mit Ohrensausen, Erbrechen, Schwindel und Drehbewegungen studirt und als Ursache derselben Exsudat in den Bogengängen nachgewiesen*). Solche Phänomene sind auch bei Thieren beobachtet worden.

In neuester Zeit hat auch Sigm. Exner über eine von ihm beobachtete, bei Kaninchen auftretende Krankheit berichtet, welche ein Analogon der Menière'schen Krankheit bildet.

»Die Erscheinungen dieser Krankheit und die Sectionsbefunde werden benützt, um nachzuweisen, dass Krankheitserscheinungen, wie diejenigen sind, welche man an Thieren mit verletzten Bogengängen beobachtet, auch vorhanden sein können, ohne Affection des Kleinhirns. Es geschieht dies mit Bezug auf die in jüngster Zeit aufgestellten Vermuthungen, dass die Schwindelerscheinungen, welche nach Flourens operirte Thiere zeigen, von Kleinhirnverletzungen herrühren**).«

Zwei Fälle von Pendelbewegung des Kopfes bei Menschen hat auch Berthold in seiner früher erwähnten Arbeit publicirt. Da diesen Beobachtungen aber keine Sectionsbefunde beiliegen, so lassen sich aus denselben auch keine Schlüsse ziehn.

2.

Aehnliche Erscheinungen wie beim Flourens'schen Versuch scheinen also wirklich mitunter durch blosse Bogengangerkrankungen bedingt zu sein. Dies ist für uns wichtig. Sind aber die Bogengänge End organe, so dürfen wir uns nicht wundern, dass dieselben oder ähnliche Phänomene auch noch durch Affection der Centralorgane selbst erregt werden können. Hieher gehört wahrscheinlich der von Brücke***) erwähnte Fall einer Kranken bei Türk, welche das

---

*) v. Tröltsch, Ohrenheilkunde. 5. Aufl. 1873. S. 489.
**) Anzeiger d. Wiener Akademie. 1874. No. 20.
***) Physiologie S. 59.

opposite side. Brücke also mentions that patients with cerebellar degeneration feel that they must constantly move forward or backward in order avoid falling.

Despite freely expressed views to the contrary, many movement sensations of different kinds can originate directly in the central organ. These include the circling movements of animals after a lesion of the superior colliculus (Wundt[5.6], Physiological Psychology, p. 169), as well as falling forward and backward after cutting the anterior or posterior region of the vermis (Wundt, cited above, p. 207), and finally the observations by Magendie[5.7], Schiff[5.8] and Mitschell[5.9] concerning continuous forward movement after lesion of the lateral geniculate body (Henle[5.10], Neurology, 1871, p. 304)

3.

First Ritter*)[5.11] and later Purkyně[5.12] directed electrical current from ear to ear.[5.13] The latter thought he was moving from the zinc toward the copper electrode. I repeated the experiment with a Smee's battery[5.14] of six elements (in series) and fully confirmed it. I believed myself moving from the zinc to the platinum electrode.

Hitzig**)[5.15] repeated the Purkyně experiments in a very careful manner. As Hitzig observed, if a current traverses the head (preferably from one mastoid process to the other), one senses a tilting and sinking of the head or the body towards the side of the cathode when the circuit is completed.[5.16] With stronger currents this sensation may be demonstrated objectively, as the subject is drawn towards the anode upon completion of the circuit. Introspectively the subject perceives the movement as voluntary, provoked by the sensation of tilting and the attempt to maintain balance. If the eyes are open, illusory movements[5.17] of visual objects occur which, according to Hitzig, are the result of involuntary eye movements caused by

---

*) Hufelands Journal for Practical Medicine vol. 17. no 3., p. 34, 72.
**) On the disturbances of muscle innervation and the perception of space with galvanisation of the head, Dubois Archive 1871.

Gefühl hatte, als ob das Bett aufgehoben und umgedreht würde und die sich mit Angst nach der entgegengesetzten Seite wältzte. Dass Kranke bei Kleinhirndegenerationen meinen, sich constant vor- oder rückwarts bewegen zu müssen, um nicht zu fallen, führt Brücke ebenfalls an. Die Manègebewegungen der Thiere nach Verletzung der Sehhügel (Wundt, physiologische Psychologie S. 169), sowie das Vor- und Rückwärtsfallen derselben nach Durchschneidung der vorderen oder hinteren Gegend des Wurmes (Wundt a. a. O. S. 207), endlich die von Magendie, Schiff und Mitschell beobachtete rastlose Vorwärtsbewegung nach Verletzung der Streifenhügel (Henle Nervenlehre 1871. S. 304), gegen deren Auffassung freilich Bedenken laut geworden sind, alle diese Erscheinungen sprechen dafür, dass Bewegungsgefühle verschiedener Art auch direct in den Centralorganen erregt werden können.

3.

Schon Ritter*) und später Purkyně haben den electrischen Strom von Ohr zu Ohr durchgeleitet. Letzterer meinte dabei sich vom Zinkpol gegen den Kupferpol zu bewegen. Ich habe den Versuch mit einer Smee'schen Batterie von sechs Elementen (hintereinander) wiederholt und vollständig bestätigt gefunden. Ich glaubte mich vom Zink zum Platin zu bewegen.

Die Purkyně'schen Versuche sind von Hitzig in sehr sorgfältiger Weise wiederholt worden**). Wie Hitzig beobachtet hat, empfindet man, wenn ein Strom (am besten von einer Fossa mastoidea zur andern) quer durch den Kopf geht, bei Kettenschluss ein Umsinken des Kopfes oder Körpers nach der Kathodenseite hin. Bei stärkern Strömen äussert sich diese Empfindung auch objectiv, indem die Versuchsperson bei Kettenschluss nach der Anodenseite hinschwankt. Der Selbstbeobachtung fähige Personen bezeichnen diese Bewegung als eine willkührliche, die hervorgerufen wird durch die Empfindung des Umsinkens und durch das Bestreben im Gleich-

---

*) Hufelands Journal für praktische Heilkunde. Bd. 17. Heft 3. S. 34, 72.
**) Ueber die beim Galvanisiren des Kopfes entstehenden Störungen der Muskelinnervation und der Vorstellungen vom Verhalten im Raume, Dubois Archiv 1871.

stronger currents.

Hitzig explains this illusion by asymmetric effects of the current on the cerebellum. Breuer and I first attributed it to stimulation of the labyrinth. I believe that Breuer still maintains this view.

The assumption of a labyrinthine effect causes difficulty for me, however, when I consider that only small variations in perception occur with significant displacement of the electrodes. I tried, without noticeable success, to place both electrodes close to one labyrinth so that the current flowing through the brain would be significantly weakened. I am not in a position to say that the cerebellum plays no role in these phenomena, or perhaps even causes them by itself. The experiment is far too complicated to serve our purposes.

I now permit myself to describe another experiment closely related to that of Hitzig. About two years ago, the institute's laboratory assistant told me that he had seen fish anesthetized by the electric current created by an Escamoteur.[5.18] I immediately repeated the experiment on several fish with Dr. Kessel, whereby I included the whole water basin in the circuit of a Ruhmkorff apparatus.[5.19] Indeed, after a few shocks the fish turned onto their backs, most became tetanic and recovered some time later. At first it seemed as if the fish would turn in a specific sense according to the Ruhmkorff-polarity. However, this result was not consistent, and since the matter seemed to involve many questions of nerve physiology, the experiments were again discontinued.

Some time ago, however, I found very consistent results with small specimens of Cobitis barbatula L.[5.20] subjected to constant current. The approximately 7^{cm} long fish are placed in a small glass whose water contains a little table salt. Immediately after passing current through the head by means of two platinum plates the fish become very calm and lay on their back such that the back is always turned down towards the zinc electrode of

gewicht zu bleiben. Sind die Augen offen, so treten Scheinbewegungen der Gesichtsobjecte auf, welche nach Hitzig Folgen der bei stärkeren Strömen sich einstellenden unwillkührlichen Augenbewegungen sind.

Hitzig erklärt diese Erscheinungen aus unsymmetrischen Affectionen des Kleinhirns durch den Strom. Breuer und ich haben zunächst an eine Reizung des Labyrinths gedacht. Wie ich glaube, hält Breuer diese Ansicht noch fest.

Die Annahme einer Labyrinthaffection bietet mir aber Schwierigkeiten, wenn ich bedenke, wie geringe Variationen in der Empfindung bei bedeutender Versetzung der Electroden auftreten. Ich habe ohne merklichen Erfolg versucht, beide Electroden dem einen Labyrinth so nahe zu bringen., dass die durchs Hirn gehenden Stromschleifen sehr abgeschwächt werden. Ich bin nicht in der Lage zu sagen, dass hiebei das Kleinhirn nicht mitspielt oder vielleicht sogar allein die Erscheinungen bedingt. Der Versuch ist überhaupt zu complicirt, um für unsern Zweck verwerthbar zu sein.

Ich erlaube mir hier noch ein Experiment zu beschreiben, welches zu dem Hitzig'schen in naher Beziehung steht. Vor etwa zwei Jahren erzählte der Laborant des Institutes, er habe bei einem Escamoteur Betäubung von Fischen durch den electrischen Strom gesehn. Ich wiederholte das Experiment sofort mit Herrn Dr. Kessel an mehreren Fischen, indem ich den ganzen Wasserbehälter in den Ruhmkorff'schen Apparat einschaltete. Wirklich legten sich die Fische nach wenigen Schlägen auf den Rücken, bekamen meist Tetanus und erholten sich nach einiger Zeit wieder. Anfangs schien es, als ob die Fische in einem durch die Ruhmkorff-Pole bestimmten Sinne sich umlegen würden. Da sich aber dies nicht constant zeigte und die Sache zudem in viele Fragen der Nervenphysiologie einzugreifen schien, wurden die Experimente wieder aufgegeben.

Vor einiger Zeit erhielt ich nun an kleinen Exemplaren von Cobitis barbatula L. bei Anwendung des constanten Stromes sehr gleichförmige Resultate. Die etwa 7^{cm} langen Fische befinden sich in einem kleinen Glasgefäss, dessen Wasser eine Spur Kochsalz enthält. Sofort beim Durchleiten des Stromes durch den Kopf mit Hilfe zweier Platinplatten werden die Fische sehr ruhig und

Smee's battery (consisting of six elements in series). Right after interruption of the current the fish are once again healthy and lively. Some time after frequent repetition of the experiment, however, they spontaneously show signs of vertigo and expire. As I said earlier I cannot draw conclusions from this complicated experiment but I believe that it is not completely useless to mention it.

Comparison of Movement Sensations with Other Sensations.

1.

Among the natural phenomena, a certain group, which we designate as our body, assumes special subjective importance for us. The phenomena can be divided into elements that we call sensations, as they are connected to specific processes of the body and may be considered caused by them. Sensations may possibly exhibit a general natural dependence on the sensors, often at a subjective level. As commonly put, stimulus leads to sensation.

Knowledge of the context of natural phenomena, or of the rules governing how they reoccur, allows us to avoid fully observing the phenomena each time. The issue is to derive the whole phenomenon from only a portion of it. In describing the context of the phenomena we try to be as complete as possible once and for all. That is the essence of natural science. To that end one has to separate the phenomena into elements that are easily understandable and producible, and which can be combined to restore the [original] phenomenon. A phenomenon can thus be described by the simple enumeration of these elements, since counting is merely the simplest form of description. The rule then consists of specification of the quantitative relationships among different elements of the phenomenon. For example, in providing sufficient work to decompose water, we say that for [a mass of] p kilograms that sinks one meter because of its weight, q cubic centimeters of oxyhydrogen gas will appear. The mechanical and the

legen sich auf den Rücken, indem der Rücken constant gegen den Zinkpol der angewandten (aus sechs Elementen hintereinander bestehenden) Smee'schen Batterie umsinkt. Gleich nach der Unterbrechung des Stromes befinden sich die Fische wieder wohl und munter. Nach mehrmaliger Wiederholung des Experimentes zeigen sie jedoch später spontan Schwindelerscheinungen und gehen zu Grunde. Folgerungen kann ich, wie gesagt, aus diesem complicirten Experiment nicht ziehen, glaube aber, dass die Erwähnung desselben nicht ganz unnütz sein wird.

Vergleichung der Bewegungsempfindungen mit andern Sinnesempfindungen.

1.

Unter den Naturerscheinungen hat eine gewisse Gruppe, die wir als unsern Körper bezeichnen, für uns eine besondere subjective Wichtigkeit. Die Erscheinungen lassen sich in Elemente zerlegen, die wir, insofern sie als mit bestimmten Vorgängen des Körpers verbunden und durch dieselben bedingt angesehn werden können, Empfindungen nennen. Vermöge des allgemeinen Naturzusammenhanges greifen die Vorgänge oft in jenes subjective Gebiet über. Der Reiz erregt Empfindung, wie man zu sagen pflegt.

Die Kenntniss des Zusammenhanges der Naturerscheinungen, oder der Regeln, nach welchen sich wiederholende Erscheinungen erfolgen, setzt uns in den Stand, die jedesmalige vollständige Beobachtung der Erscheinungen zu ersparen. Es handelt sich also darum, zu einem gegebenen Erscheinungstheil den Rest abzuleiten. Bei der Beschreibung des Zusammenhanges der Erscheinungen suchen wir so viel wie möglich auf einmal und ein für allemal anzugeben. Darin besteht eben die Naturwissenschaft. Zu diesem Zwecke hat man die Erscheinungen in möglichst leicht verstehbare und herstellbare Elemente zu zerlegen, aus deren Addition dann die Erscheinung besteht. Ein Phänomen lässt sich dann durch die blosse Abzählung dieser Elemente beschreiben, denn das Zählen ist eben das einfachste Beschreiben. Die Regel besteht dann in der Angabe des Zahlenzusammenhanges der Erscheinungselemente der einen Art

chemical aspects of the phenomenon are divided into equal, countable For example, in providing sufficient work to decompose water, we say that for [a mass of] p kilograms that sinks one meter because of its weight, q cubic centimeters of oxyhydrogen gas will appear. The mechanical and the chemical aspects of the phenomenon are divided into equal, countable elements and the quantitative connection between the two is given.

Recent physics provided the insight that all different types of phenomena are connected to one another by common mechanical principles. The mechanical work[6.1] corresponding to a certain phenomenon therefore reappears in all similar phenomena.

2.

If one could maintain a sensory organ in an undisturbed state, then a constant stimulus would correspond to a constant sensation. Let us call the stimulus delivering an amount of work in unit time A, and the corresponding sensory process work for a unit time a. A bundle of light shining onto the pupil, for example, would deliver the mechanical energy A in unit time. After all refraction and reflection, only the fraction α would remain in the retina. It would then release the work a associated with the sensation, which in turn could be much greater than α. This is because the work leading to sensation is not done by the stimulus itself. Rather, the stimulus only releases the work[6.2] of other forces, which reside in the sensory organs. The sensory organs therefore operate according to a principle that is very different from that of most physical devices. They function like a relay.[6.3] This explains the high sensitivity of the sensory organs.

Provided that the organ remains unchanged, a constant stimulus level would correspond to a constant level of sensation. Such simple relationships are not what we find, however. On the one hand we see as a rule that the released sensory process continues after the stimulus dies out, because the stimulus does not supply the sensory work, but merely enables the work to be released. This extended action produces the so-called positive afterimage. On the other hand, the energy supply of the organ is exhausted during the stimulation, and the sensitivity of the organ is reduced. Therefore,

mit jenen der andern Art. Wenn wir z. B. die Arbeit angeben, welche zur Wasserzersetzung nothwendig ist, so sagen wir, für p Kilogramme, welche ein Meter tief durch die Wirkung der Schwere sinken, erscheinen q Cubikcentimeter Knallgas. Der mechanische und der chemische Theil der Erscheinung ist in gleiche abzählbare Elemente zerlegt und der Zahlenzusammenhang zwischen beiden ist angegeben.

Die neuere Physik hat die Einsicht gewonnen, dass alle Erscheinungsarten mit einander und demnach auch mit mechanischen zusammenhängen. Die mechanische Arbeit, welche einer Erscheinung entspricht, gibt also ein in allen Gebieten verwendbares Erscheinungsmerkmal ab.

2.

Könnte man ein Sinnesorgan in unveränderlichem Zustande erhalten, so würde einem constanten Reize eine constante Empfindung entsprechen. Liefert der Reiz in der Zeiteinheit die Arbeit A, so entspricht für den Empfindungsvorgang der Zeiteinheit die Arbeit a. Ein auf die Pupille fallendes Lichtbündel würde z. B. in der Zeiteinheit die mechanische Arbeit A liefern, davon würde nach allen Brechungen und Reflexionen in der Netzhaut der Bruchtheil α verbleiben und daselbst eine Empfindungsarbeit a auslösen, welche auch vielmal grösser sein kann als α. Die Empfindungsarbeit wird. nämlich nicht vom Reize geleistet, sondern der Reiz ist bloss der Anlass zr Auslösung der Arbeit anderer Kräfte, welche in den Sinnesorganen vorhanden sind. Die Sinnesorgane wirken also nach einem Princip, welches von jenem der meisten physikalischen Apparate sehr verschieden ist, sie wirken nach Art eines Relais. Hierauf beruht eben die hohe Empfindlichkeit der Sinnesorgane.

Einem constanten Reizstrom würde also bei unverändert erhaltenem Organ ein constanter Empfindungsstrom entsprechen. So einfache Verhältnisse finden wir jedoch nicht vor. Einmal sehen wir in der Regel den ausgelösten Empfindungsvorgang fortdauern, wenn der Reiz erlischt, eben weil der Reiz nicht die Empfindungsarbeit, sondern bloss die Auslösungsarbeit verrichtet. Diese Fortdauer gibt das sogenannte positive Nachbild. Ferner wird durch den Reiz der Arbeitsvorrath des Organs erschöpft und das

as a rule, a constant stimulus produces a declining sensation. With the exhaustion of the organ other processes come into play, which serve to return the organ to its original state. These latter processes are the source of other kinds of after-images.

Plateau*)[6.4], who earned special credit for work in this field, believed that a general rule could be postulated covering all these processes, which he formulated as follows:

"When an organ is subjected to prolonged excitation, it generates a resistance that grows with the duration of that excitation. If the source of the excitation is then suddenly removed, it tends to regain its normal state by a process analogous to that of a spring which, once stretched away from its equilibrium length and released, returns to its original length by a series of decreasing oscillations. During that return course it passes the equilibrium state alternately from the two opposite directions. That is to say that, at the moment the organ is no longer under the influence of the excitation, it initially moves rapidly toward its normal state. Because of the acquired speed, however, it overshoots that normal state to appear momentarily in the opposite state, after which it returns once again towards the normal state, and again tends to overshoot, but with less intensity. It passes the opposite state again, a second time, but even more weakly and so forth until, finally, the normal state is fully attained.

These successive states of the organ result in a sequence of alternatively opposing phases of sensation, which decrease in intensity. Phases which are in the same sense as the original sensation may be called positive phases, whereas the others of opposite sense may be called negative phases."

---

*) Essay concerning a general theory of visual appearances following the viewing of colored objects, etc. Mem. Acad. Brussels. vol. 8.

Organ dadurch in einen Zustand geringerer Empfindlichkeit versetzt. Desshalb entspricht einem constanten Reize in der Regel eine an Stärke abnehmende Empfindung. Mit der Erschöpfung des Organs treten aber Folgezustände ein, durch welche das Organ seinen ursprünglichen Zustand wieder zu gewinnen sucht. Diese Folgezustände sind die Ursachen anderer Arten von Nachbildern.

Plateau*), der sich um dieses Capitel besonders verdient gemacht hat, war der Meinung, dass für alle diese Vorgange ein allgemeines Gesetz aufgestellt werden könne, welches er folgendermaassen formulirt hat:

»Lorsque un organe est soumis a une excitation prolongée, il oppose une résistance qui croît avec la durée de cette excitation. Alors s'il vient a ètre subitement soustrait â la cause excitante, il tend à regagner son état normal par une marche analogue à celle d'un ressort qui, écarté de sa forme d'equilibre et abandonné ensuite à lui mème, revient à cette forme par des oscillations decroissantes en vertu desquelles il la dépasse alternativement en deux sens opposés. C'est-à-dire que l'organe, au moment où il cesse d'etre sous l'influence de la cause excitante, marche d'abord rapidement vers son état normal; mais qu'emporté par une sorte de vitesse acquise, il dépasse cet état normale pour se constituer momentanément dans un état opposé; puisqu'il revient de nouveau vers l'état normal, et tend encore à dépasser celui-ci en reprenant, mais avec moins d'intensité, l'état correspondant à l'excitation, pour repasser une seconde fois, mais plus faiblement aussi, à l'état contraire, et ainsi de suite, jusqu'à ce que l'état normal soit définitivement atteint.

De ces états successifs de l'organe resulte une suite de phases de la sensation alternativement opposées, et décroissantes en intensité, phases dont les unes sont de la mème nature que la sensation primitive et peuvent être appelées les phases positives, tandis que les autres sont d'une nature contraire et peuvent être appelées les phases négatives.«

---

*) Essai d'une théorie génerale comprenant l'ensemble des apparences visuelles qui succèdent à la contemplation des objects colorés ect. Mém. de l'Acad. de Bruxelles. T. 8.

3.

Plateau supported his principle with many examples, some of which are very convincing. However, to me his formulation does not seem to be generally valid without modification. There is no doubt that in many cases oscillatory aftereffects are produced by the stimulated organ. However, the assumption that they are always oscillatory goes too far and is not supported by facts. It also seems not entirely justified to me to name the phases positive and negative.

If two things A and B are indicated as positive and negative relative to each other, then it is understood that A can be partly or totally cancelled by adding B. This relation is indeed found between some sensations and their after-images, but not between all of them. Such relations exist, e.g., between the sensation of motion and its aftersensation, which is completely similar except that its motion is in the opposite sense. It does not exist, e.g., between the sensations black and white, one of which can be the after-image of the other*). The two sensations are indeed dissimilar and do not neutralize each other, but like two different colors yield a mixed color, in this case gray. Especially in this case, therefore, the words positive and negative are inappropriate. It is sufficient to state that the aftereffects of a stimulation can vary greatly, and can trigger very different sensations, which in some cases can be oscillatory and can oppose each other as + and −.

The most general phenomenon in the field of sensation processes seems to be the outlasting duration of the sensation. The occasional prolongation of sensation after cessation of the stimulus is understandable if one considers that the sensory process is only an indirect effect of the stimulus process and that it runs by itself. This is the reason why intermittent light, acoustic, and haptic stimuli fuse to produce continuous sensations.

---

*) Mach, About intermittent light stimulation, Archiv von Reichert u. Dubois 1865.

3.

Plateau hat sein Princip durch viele zum Theil sehr schlagende Beispiele belegt. Doch scheint mir dasselbe in dieser Form ausgesprochen ohne Modification nicht allgemein gültig zu sein. Es ist nicht zu bezweifeln, dass die Folgezustände des erregten Organs in vielen Fällen oscillatorisch verlaufen. Dagegen geht die Annahme, dass der Hergang immer oscillatorisch ist, zu weit und wird durch die Thatsachen nicht bestätigt. Auch scheint mir die Benennung positive und negative Phasen nicht durchaus bezeichnend.

Wenn zwei Dinge A und B als positiv und negativ im Verhältniss zu einander bezeichnet werden, so versteht man darunter, dass A durch die Hinzufügung von B ganz oder theilweise vernichtet werden kann. Dieses Verhältniss besteht nun zwischen manchen Empfindungen und ihren Nachbildern, aber nicht zwischen allen. Dieses Verhältniss besteht z. B. zwischen der Empfindung einer Bewegung und dem Nachbilde derselben, welches eine vollständig ähnliche nur dem Sinne nach entgegengesetzte Bewegung ist. Es besteht aber z. B, nicht zwischen den Empfindungen Schwarz und Weiss, von welchen ebenfalls die eine das Nachbild der andern sein kann*). Beide Empfindungen sind einander durchaus unähnlich und beide zusammen heben sich nicht auf, sondern geben wie zwei verschiedene Farben eine Mischfarbe, sie geben Grau. Gerade in diesem Falle ist also die Bezeichnung positiv und negativ nicht passend. Wir wollen. uns daher damit begnügen, zu sagen, dass die Folgezustände einer Erregung sehr mannigfalti sein und sehr verschiedene Empfindungen auslösen können, welche zuweilen auch oscillatorisch wechseln und in dem Gegensatze von + und − zu einander stehn können.

Die allgemeinste Erscheinung im Gebiete der Empfindungsvorgänge scheint die Nachdauer der Empfindung zu sein. Das zeitweilige Fortbestehn der Empfindung nach dem Erlöschen des Reizes ist begreiflich, wenn man bedenkt, dass der Empfindungsvorgang nur eine mittelbare Folge des Reizvorganges ist und dass er für sich abläuft. Hierauf beruht es, dass intermittirende Licht-, Schall

---

*) Mach, über intermittirende Lichtreize, Archiv von Reichert u. Dubois 1865.

The exhaustion of the organ during prolonged continuous stimulation is another very general attribute, although the explanations given for this exhaustion are often not sufficient. We see this phenomenon clearly for the senses of touch and vision. It is difficult to demonstrate it for the acoustic sense. One seems to hear a constant tone for several hours at <u>constant intensity</u>.[6.5] However, there are observations that indicate such exhaustion for the hearing organ as well.

If one holds down a key of a harmonium and pays attention to the constant tone for about half an hour, no gradual weakening of the sound is noted, however, one overtone after the other comes out very clearly. This can only be explained by exhaustion to the primary tones, which had previously received all one's attention*).

If one strikes a table in a room with a hammer, the sound is very brief, and one seems to perceive absolute stillness right after the blow. We can modify the experiment, however. An assistant strikes the table with a hammer while we block both outer ears by pressing with our fingers. If we open the ear 0.5 - 1.0 seconds after the blow then we hear the blow newly created. After the blow we can quickly open and close the ear for several times. With every opening a new blow is heard that, of course, becomes weaker the longer the opening occurs after the blow. This is explained by the slowly attenuating acoustic waves in the room, which will be detected only by a non-fatigued organ or if the organ had a short time to recover.

Teleologically it is easily understood why all acoustic processes serving time perception must proceed very rapidly, even though their physiological explanation might be rather difficult. Consequently, true after-images of sound cannot be observed directly at all. Their existence probably occurs

---

*) M a c h, Introduction to Helmholtz' theory of music. Graz 1866. p. 29.

und Tastreize sich zu einem continuirlichen Empfindungsstrom zusammensetzen.

Die Erschöpfung des Organs durch einen länger anhaltenden Reiz ist auch eine sehr allgemeine Erscheinung, wenngleich die Erklärungen, die man auf diese Erschöpfung gegründet hat, nicht immer ausreichend sein mögen. Wir sehn die Erscheinung deutlich am Tast- und Gesichtssinn. Am Gehörsinn ist sie schwer nachweisbar. Man meint einen constanten Ton stundenlang in constanter Intensität zu hören. Doch gibt es Beobachtungen, welche auf eine solche Erschöpfung auch beim Gehörorgan deuten.

Klemmt man eine Taste des Harmoniums fest und beobachtet den constanten Ton durch etwa eine halbe Stunde, so kann man zwar keine allmälige Abschwächung des Klanges wahrnehmen, aber ein Oberton nach dem andern tritt jetzt in voller Deutlichkeit hervor, was sich doch nur aus einer Erschöpfung für jene Partialtöne gut erklären lässt, welchen die Aufmerksamkeit früher zugewendet war *).

Führt man im Zimmer mit dem Hammer einen Schlag auf den Tisch, so ist der Schall sehr kurz und man meint, bald nach dem Aufschlagen absolute Stille zu vernehmen. Wir können aber das Experiment modificiren. Ein Gehülfe schlägt mit dem Hammer auf den Tisch, während wir mit den Fingern beide Gehörgänge zudrücken. Oeffnen wir die Gehörgänge 0.5 – 1.0 Secunden nach dem Aufschlagen, so hören wir den Schlag neu entstehn. Wir können nach dem Aufschlagen einigemale die Gehörgänge rasch öffnen und schliessen und hören bei jedem Oeffnen einen neuen Schlag, der natürlich desto schwächer ausfällt, je später das Oeffnen nach dem Aufschlagen erfolgt. Dies erklärt sich aus dem im Zimmer fortbestehenden allmälig abnehmenden Schallvorgang, der nur von dem nicht ermüdeten Organe bemerkt wird oder wenn das Organ kurze Zeit Gelegenheit hatte, sich zu erholen.

Teleologisch ist es ganz begreiflich, warum alle Vorgänge des Gehörsinns als eines Zeitsinnes sehr rasch ablaufen müssen, wenngleich die physiologische Erklärung schwierig genug sein mag. Eigentliche Nachbilder des Schalles kann man daher direct gar nicht

---

*) M a c h, Einleitung in die Helmholtz'sche Musiktheorie. Graz 1866. S. 29.

only through the fusion of intermittent acoustic stimuli. The recovery of the excited organ most likely occurs very rapidly. Of course, one cannot speak of a negative sound after-image. There do not exist two sound sensations that could cancel each other.

The most varied aftereffects following excitation can be observed in the vision. The visual sense is a spatial sense and all processes run very slowly there. One such after-image phenomenon that, at least from a superficial view, seems closely connected to our main argument will be discussed below in more detail.

4.

If one draws an Archimedes' spiral on a disk, and rotates it counterclockwise about an axis normal to the disk and passing through the center of the spiral, the spiral seems to get wider. If after some time one suddenly stops the disk, then the disk or any other object one looks at seems to shrink further and further. One observes the opposite if one rotates the disk such that the spiral shrinks, and then stops it. This experiment came from Plateau. Similar observations, without such an apparatus, were previouly made by Purkyně and later by Oppel.[6.6] Oppel extended the experiments with a special apparatus.

Fig. 9.

Plateau and Oppel already assumed that the motion aftereffects of images

beobachten. Die Existenz derselben wird nur wahrscheinlich durch das Verschmelzen intermittirender Schallreize. Die Erholung des erregten Organs findet muthmasslich sehr rasch statt. Selbstverständlich kann von einem negativen Schallnachbilde nicht die Rede sein. Es gibt nicht zwei Schallempfindungen, von denen die eine die andere aufheben könnte.

Die mannigfaltigsten Folgezustände einer Erregung lassen sich im Gebiete des Gesichtssinnes beobachten. Der Gesichtssinn ist ein Raumsinn und alle Vorgänge laufen hier sehr langsam ab. Eine solche Nachbilderscheinung, welche zu unserm Hauptgegenstande bei oberflächlicher Betrachtung in naher Beziehuug zu stehn scheint, wollen wir hier eingehender besprechen.

4.

Malt man auf eine Scheibe eine Archimedes'sche Spirale und dreht dieselbe um eine durch den Ausgangspunkt der Spirale senkrecht gegen die Scheibe gehende Axe langsam dem Sinne des Uhrzeigers entgegen, so scheinen sich die Windungen der Spirale zu erweitern. Hält man nun nach einiger Zeit die Scheibe plötzlich an, so scheint die Scheibe oder irgend ein Gegenstand, den man anblickt, fort und fort zu schrumpfen. Das Gegentheil tritt ein, wenn man die Scheibe so dreht, dass die Spirale schrumpft und dann an-

Fig. 9.

hält. Das beschriebene Experiment rührt von Plateau her. Aehnliche Beobachtungen sind ohne Apparat schon früher von Purkyně und später von Oppel gemacht worden. Oppel hat die Versuche mit einem besondern Apparat fortgesetzt.

Plateau und Oppel haben schon angenommen, dass die Be-

result from processes in the retina. To explain them by eye movements, as Helmholtz[6,7] later attempted, could be ruled out on the basis of evidence from details of older experiments. Since eye movements can only shift the visual field as a whole, the following is already sufficient proof against the eye movement hypothesis. The illusory enlarging spiral causes a shrinking of objects towards a certain central point after the spiral stops, and the Oppel motion after-image moves faster in the center than in the periphery. Further, the assertion by Helmholtz that the phenomena do not occur under strict fixation of a target does not stand up under closer scrutiny. Dr. Dvořak,*) who repeated the experiments in my laboratory, changed Plateau's experiment to yield a striking result. On top of a spiral on a large white disk one places a smaller concentric one whose spiral winds in the opposite sense, and within it a third smaller one, with a movement sense like the first. A small black circle is placed at the common center on top of all of this. A few black strings may be stretched in front of this composite disk. When the disk is rotated one can readily fixate on its center, since any gaze deviation is immediately recognizable by the bright borders of after-images of the center and of the black strings. If one then looks at a lined white screen, the dark after-image of the disk separates into three rings, some of which shrink and some of which enlarge. Within this after-image the bright after-images of the center and of the strings appear absolutely stationary. It is worth noting that the illusory movements in the after-image only grasp faint small points and blemishes [on the screen] but never the clearly seen points and lines. These visual motion aftereffects are just as much local phenomena of the retina as are light and color after-images and, like those, they occur during steady fixation.

---

*) On the after-images of stimulus changes. Proc. Vienna Acad., vol. 61.

wegungsnachbilder durch Vorgänge in der Retina bedingt sind. Die Erklärung derselben durch Augenbewegungen, welche Helmholtz später versucht hat, lässt sich schon durch die Details der älteren Versuche ad absurdum führen. Denn wenn eine scheinbar sich erweiternde gedrehte Spirale ein scheinbares Schrumpfen der nachher betrachteten Gegenstände gegen ein bestimmtes Centrum hin bewirkt, wenn das Oppel'sche Bewegungsnachbild in der Mitte rascher strömt als am Rande, so sind dies Beweise genug gegen die Augenbewegung; denn Augenbewegungen können nur eine gleichmässige Verschiebung des ganzen Gesichtsfehldes bewirken. Auch die Behauptung von Helmholtz, dass die Erscheinungen bei strenger Fixation eines Punktes nicht eintreten, erweist sich bei näherer Prüfung als unhaltbar. Herr Dr. Dvořak*), der in meinem Laboratorium diese Experimente wiederholte, hat den Plateau'schen Versuch in eine sehr eclatante Form gebracht. Man lege auf eine grosse weisse Scheibe mit einer Spirale eine kleinere concentrische mit einer entgegengesetzt laufenden Spirale, auf diese etwa noch eine dritte, noch kleinere, mit einer der ersten gleichlaufenden Spirale, und auf das gemeinschaftliche Centrum aller Spirelen einen kleinen schwarzen Kreis. Vor der so zusammengesetzten Scheibe mögen noch einige schwarze Fäden gespannt sein. Während nun die Scheibe gedreht wird, kann man das Centrum ganz scharf fixiren, indem sich jede Blickschwankung sofort durch die hellen Nachbildränder des Centrums und der schwarzen Fäden verräth. Sieht man dann nach einem weissen linirten Schirme, so erscheint auf demselben das dunkle Nachbild der Scheibe in drei theils schrumpfende, theils schwellende Ringe getheilt, und in diesem Nachbilde ganz fest und ruhig die hellen Nachbilder des Centrums und der Fäden. Hiebei ist zu bemerken, dass die scheinbare Bewegung im Nachbilde immer nur schwächere Pünktchen und Fleckchen ergreift, nie aber deutlich gesehene Punkte und Linien. Diese optischen Bewegungsnachbilder sind also eben so locale Erscheinungen der Netzhaut als die Licht- und Farbennachbilder und treten wie diese bei ruhiger Fixation auf.

---

*) Ueber die Nachbilder von Reizveränderungen. Sitzgsber. d. Wiener Akad. Bd. 61.

I have mentioned this phenomenon, which can be interpreted to be an after-image of a stimulus change, because Plateau considers this a case of his general law. In a letter to me he expressed the opinion that his law also covers the sensations I observed during passive rotation. We will see that the latter proves not to be the case.

5.

It is not uninteresting to note that yet another stimulus change gives rise to an after-image as well. In the course of a long series of experiments, Dr. Dvořak*) found another after-image associated with a change of light intensity. The experimental result was negative for all other changes where after-images were suspected. If one increases the light intensity in a room rather quickly from a value i to another level $i + \Delta i$, then suddenly drops to i, then increases it to $i + \Delta i$ again, and so forth, and then suddenly leaves the light intensity constant, it clearly appears to get dimmer and dimmer. The reversal of the experiment is self-evident.

The experiment can easily be performed as follows: the hole in a window shutter in a darkened room is covered with a hat shaped device, which points into the room. Its top contains a slit covered with frosted glass. The slit is positioned radially and in front of a moveable opaque disk whose circumference forms a spiral. Now one places another frosted glass in front which receives stronger or weaker diffuse light, depending on the position of the slit, and one observes it. It is clear that, depending on the direction of rotation of the disk, the slit is always either gradually opened and suddenly closed, or gradually closed and suddenly opened. In the first case, the glass seems to get darker and darker when one stops the rotation, in the second case, it seems to get brighter and brighter.

The speed was 2 to 3 revolutions per second, the trial time on average 1 minute, and the slit was 1.5 inches long, 2 lines wide; the diameter

---

*) Cited above.

Ich bin auf dieses Phänomen, welches sich als das Nachbild einer Reizveränderung auffassen lässt, hier eingegangen, weil Plateau dasselbe als einen Fall seines allgemeinen Gesetzes ansieht und in einem Schreiben an mich auch die Ansicht ausgesprochen hat, dass die von mir bei passiven Bewegungen beobachteten Empfindungen ebenfalls unter sein Gesetz fallen. Wir werden sehn, dass Letzteres nicht zutrifft.

5.

Es ist nicht uninteressant zu bemerken, dass noch eine andere Reizveränderung ebenfalls ein Nachbild gibt. Bei seinen zahlreichen Versuchen fand nämlich Dr. Dvořak*) noch ein Nachbild der Lichtintensitätsänderung. Für alle übrigen Veränderungen, bei welchen Nachbilder vermuthet wurden, war das Versuchsresultat negativ. Lässt man die Lichtintensität in einem Zimmer von einem gewissen Werth i ziemlich schnell auf einen andern $i + \Delta i$ wachsen, dann plötzlich auf i fallen und wieder auf $i + \Delta i$ steigen, und lässt dann nach oftmaliger Wiederholung des Processes die Lichtintensität plötzlich constant, so scheint dieselbe deutlich fort und fort kleiner zu werden, Die Umkehrung des Versuches ist selbstverständlich.

Das Experiment lässt sich einfach so ausführen: Das Fensterladenloch eines verdunkelten Zimmers wird mit einem hutförmigen, in das Zimmer ragenden Aufsatz verschlossen, welcher am Deckel eine mit mattem Glas bedeckte Spalte trägt. Die Spalte steht radial zu einer vor derselben drehbaren undurchsichtigen Scheibe, deren Umfang durch einen Spiralgang gebildet wird. Nun setzt man noch eine matte Glastafel vor, die je nach der Stelle der Spirale stärker oder schwächer diffus beleuchtet wird. und die man beobachtet. Es ist klar, dass je nach dem Drehungssinn der Scheibe die Spalte immer allmälig geöffnet und plötzlich geschlossen oder allmälig geschlossen und plötzlich geöffnet wird. Im ersten Falle scheint sich die Glastafel, wenn man mit der Drehung aufhört, fort und fort zu verdunkeln, im zweiten fort und fort zu erhellen.

Die Zahl der Umdrehungen war 2 bis 3 in der Secunde, die Versuchszeit im Durchschnitt 1 Minute; die Spalte war 1.5 Zoll lang,

---

*) a. a. O.

of the disk was 2 feet, the height of the spiral 1.5 inch.

If during the experiment one observes the diffraction image of a polished pin head brought close to the eye, then corresponding to changes in pupil diameter one sees how the image suddenly contracts a little with every sudden brightening, and how it suddenly enlarges slightly with every sudden darkening. After the rotation is stopped, one notices no further changes in the diffraction image.

In order to eliminate the influence of the pupil size on the phenomenon, and to investigate whether simultaneous brightening and darkening in different parts of the visual field can produce the aftereffects, the experiment was performed as follows:

Fig. 10.

The top of the hat-like box contains two slits, which are positioned along the same radius on a large glass disk that can be rotated in front [of the box] (Fig. 10). The disk is painted black and only two ring shaped spaces are left transparent. Their outer boundary is circular, their inner boundary is part of a spiral. The spirals run in opposite directions on the two rings. During rotation as one slit is slowly shut, the other one is slowly opened.

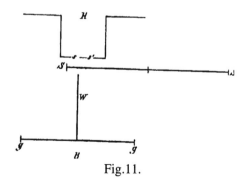

Fig.11.

Fig. 11 shows the experimental arrangement. The hat is indicated by *H*, the slit with frosted glass by *ss'*, the disk by *SS*. The light of the two slits is separated by the opaque sheet *W* and distributed on both halves of a matte board *gg*, which is seen by the observer *B*. Here one sees during the rotation

2 Linien breit; der Durchmesser der Scheibe war 2 Fuss, die Gang-
höhe der Spirale 1.5 Zoll.

Beobachtet man während des Versuches das Zerstreuungsbild
eines nahe an das Auge gebrachten glänzenden Stecknadelknopfes,
so sieht man dasselbe entsprechend den Veränderungen der Pupillenweite bei jeder plötzlichen Erhellung schwach zusammenzucken,
bei jeder plötzlichen Verdunklung sich schwach vergrössern. Nach
Beendigung der Drehung kann man nichts mehr am Zerstreuungs-
bilde bemerken.

Um den Einfluss der Pupillenweite bei der Erscheinung möglichst zu eliminiren und zu untersuchen, ob gleichzeitige Erhellung
und Verdunklung in verschiedenen Theilen des Gesichtsfeldes als
Nacherscheinung eintreten könne, wurde der Versuch folgendermaassen angestellt:

Fig. 10.

Der Hutdeckel erhält zwei Spalten,
welche in demselben Radius einer grossen
vor denselben drehbaren Glasscheibe (Fig.
10) liegen. Die Scheibe ist schwarz bemalt und nur zwei ringfömige Räume an
derselben sind durchsichtig gelassen. Diese
sind nach aussen durch einen Kreis, nach
innen durch einen Spiralgang begrenzt.
Die Spiralen laufen bei beiden Ringen entgegengesetzt. Wird nun bei der Drehung
die eine Spalte allmälig geschlossen, so wird die andere ebenso allmälig geöffnet.

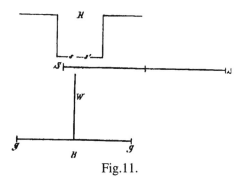

Fig.11.

Die Fig. 11 gibt die An-
ordnung des Versuches. Der
Hut ist durch A; die Spalten mit matten Gläsern durch
ss', die Scheibe durch SS
angedeutet. Das Licht der
beiden Spalten wird durch
die undurchsichtige Zwi-
schenwand W gesondert und
auf die beiden Hälften einer
matten Tafel gg vertheilt,

how one half always gets lighter while the other gets darker. When one suddenly which is seen by the observer *B*. Here one sees during the rotation how one half always gets lighter while the other gets darker. When one suddenly stops the disk, the reverse immediately becomes evident. The half that previously got brighter now gets darker, and the other one visibly gets lighter. This also shows that these brightening and darkening after-images, as they could aptly be called, are local retinal phenomena and cannot be caused by pupillary changes alone.

Upon further consideration of the diffraction experiment one sees that for each opening of the slit the pupil quickly contracts a bit. Although the spirals were chosen such that the total intensity of the light that penetrates through the two slits remains constant, still every new light distribution seems to have its effect on the iris. The brightening and darkening after-images seem to be slightly stronger with experiments using one slit compared to two. It therefore remains open to what extent changes of pupil width contribute to these phenomena.

6.

In both of the two last cases investigated by Plateau and Dvořak, the after-image can indeed be called the negative opposite image of the original phenomenon. However, the time course of the following events is not oscillatory. Earlier I put forward a hypothesis that I see no reason to abandon yet. According to this hypothesis one has to bring into play as many different physical processes during the course of a sensation as there are psychologically differentiable qualities of that sensation*). For similar sensations one should assume similar processes, for dissimilar sensations dissimilar processes.

Accordingly we must consider that with the movement of a retinal image a special process is triggered which is not present at rest. During

---

*) Mach, On the effect of the spatial distribution of a light stimulus on the retina. Proc. Vienna Acad. vol. 52.

welche vom Beobachter *B* betrachtet wird. Hier sieht man während der Drehung immer die eine Hälfte sich erhellen, die andere sich verdunkeln. Hält man die Scheibe plötzlich an, so tritt sofort augenfällig das Umgekehrte ein; die Hälfte, welche sich zuvor erhellte, verdunkelt sich, die andere erhellt sich zusehends. Also auch diese Erhellungs- und Verdunkelungsnachbilder, wie man sie passend nennen könnte, sind locale Netzhauterscheinungen und können nicht von Pupillenänderungen allein herrühren.

Wenn man bei diesem Versuch wieder das Zerstreuungsbild zu Rathe zieht, so findet man, dass bei jedesmaliger Oeffnung der Spalte die Pupille etwas zusammenzuckt. Obgleich nun die Spiralen so gewählt sind, dass die Intensität des Gesammtlichtes, welches durch beide Spalten eindringt, constant bleibt, so scheint doch jede neue Lichtvertheilung als Reiz auf die Iris zu wirken. Das Erhellungs- und Verdunklungsnachbild schien bei Versuchen mit einer Spalte etwas stärker aufzutreten, als bei zwei Spalten. Es muss demnach dahin gestellt bleiben, ob und welchen Antheil Aenderungen der Pupillenweite bei der Erscheinung haben.

6.

In den beiden letzten von Plateau und Dvořak untersuchten Fällen lässt sich nun das Nachbild in der That als negativer Gegensatz der anfänglichen Erregung bezeichnen. Aber oscillatorisch ist der Verlauf der Folgezustände nicht. Schon früher habe ich eine Hypothese ausgesprochen, welche zu verlassen mir bisher unnöthig erschienen ist. Nach dieser Hypothese hat man bei einem Empfindungsvorgang so viele verschiedene physikalische Processe anzunehmen, als man Empfindungsqualitäten psychologisch in demselben zu unterscheiden vermag*). Für ähnliche Empfindungen wird man ähnliche Processe, für unähnliche Empfindungen unähnliche Processe annehmen müssen.

Dem entsprechend werden wir daran denken müssen, dass mit der Bewegung eines Netzhautbildes ein besonderer Process erregt

---

*) Mach, über die Wirkung der räumlichen Vertheilung des Lichtreizes auf die Netzhaut. Sitzgsber. d Wiener Akademie. Bd. 52.

opposing movements very similar processes are triggered in similar organs. These opposing processes neutralize each other, however, so that with the occurrence of one the other gets eliminated and that with exhaustion of one the other surfaces. This is not a theory, but the physical manifestation of what can be observed psychologically. Completely analogous statements can be made about after-images from changes of light intensity.

7.

For the time being we leave open the question of whether special organs for movement detection exist and where one should search for them. However, we can compare the movement sensations themselves with other areas of sensation and can stress the following points. Movement sensation is elicited by certain body parts sensitive to acceleration differences. To the extent that acceleration can produce nervous activity, it acts as a stimulus for motion sensation. Several of the previously mentioned experiments lead to the conclusion that movement sensation continues even after the acceleration disappears. Furthermore, the experiments teach us that with continued acceleration (i.e., continued stimulation) the sensation decays.[6.8] Plateau's law, however[6.9], does not apply here. With the termination of the stimulus no negative phase of motion sensation occurs. Following the cessation of the acceleration the motion sensation slowly disappears, without turning in the opposite sense. Acceleration in the opposite direction is required to produce an opposite direction of motion sensation. Oppositely directed accelerations produce very similar processes of sensation that are mediated by similar organs. These processes relate to each other in such an opposition that if both occur simultaneously, they can neutralize each other, as shown in experiment 3, p. 26. This can hardly be considered as other than the existence of very similar but oppositely directed organs that are stimulated by oppositely directed accelerations.

wird, der bei der Ruhe nicht vorhanden ist, und dass bei entgegengesetzten Bewegungen ganz ähnliche Processe in ähnlichen Organen erregt werden, welche sich aber gegenseitig in der Art ausschliessen, dass mit dem Eintreten des einen der andere erlöschen muss, und mit der Erschöpfung des einen der andere eintritt. Es ist dies keine Theorie, sondern ein physikalischer Ausdruck des psychologisch Beobachteten. Ganz Analoges lässt, sich über das Nachbild der Lichtintensitätsänderung sagen.

7.

Indem wir es zunächst ganz in Frage lassen, ob es besondere Organe der Bewegnngsempfindung gibt und wo dieselben zu suchen seien, können wir, die Bewegungsempfindungen selbst mit andern Empfindungsgebieten vergleichend, leicht folgende Punkte hervorheben. Bewegungsempfindungen werden durch Beschleunigungsdifferenzen gewisser Körpertheile erregt. Die Beschleunigung wirkt also, sofern sie Nervenarbeit auslösen kann, als Reiz für die Bewegungsempfindungen. Wie aus mehreren vorher angeführten Versuchen hervorgeht, dauert die Bewegungsempfindung noch fort, wenn die Beschleunigung schon verschwunden ist. Ferner lehren die Versuche, dass bei fortdauernder Beschleunigung (also fortdauerndem Reize) Erschöpfuug der Beweguugsempfindung eintritt. Das Plateau'sche Gesetz gilt also [aber] hier nicht. Beim Erlöschen des Reizes zeigt sich keine negative Phase der Bewegungsempfindung. Mit dem Aufhören der Beschleunigung verschwindet, allmälig die Bewegungsempfindung, ohne in das Gegentheil umzuschlagen. Vielmehr ist zur Erzeugung der entgegengesetzten Bewegungsempfindung auch die entgegengesetzte Beschleunigung nöthig. Entgegengesetzte Beschleunigungen erregen demnach einander ganz ähnliche Empfindungsprocesse, die also auch durch ähnliche Organe vermittelt sind. Diese Processe stehn aber in einem derartigen Gegensatz, dass beide gleichzeitig eintretend einander aufheben können, wie der Versuch 3. S. 26 lehrt. Es ist dies kaum anders denkbar als dadurch, dass es verschiedene ein-

Further Investigations of the Phenomena.

1.

The previous material provisionally identified the labyrinth of the ear as the most likely principal organ for movement sensation. Here, we shall once again investigate and critically evaluate the different possible sources for movement sensation. As we have already mentioned, the acceleration of two body parts relative to each other can induce sensations that can be used to indicate motion. We will therefore study whether the sensations providing us information about motion of our body result

a) from connective tissue and bones,
b) from the skin,
c) from the muscles,
d) from blood,
e) from the eyes,
f) from the brain, or finally
g) from a separate organ in the head.

2.

Let us begin with the first possibility. If the connective tissue and the bones contain sensitive nerve fibers, then relative accelerations of parts of these tissues can give rise to local pressure sensation. However, it is obvious that such sensations are of very doubtful usefulness as indicators for locomotion. Let us consider two such body parts A and B. If A is accelerated toward B by an outside force, then a sensation E arises from the pressure and counterpressure between A and B. The same sensation occurs, however, if B is again the same as for an acceleration of B away from A, i.e., again identical for oppositely directed locomotion. In a word, connective tissue and bones can possibly sense the strains and pressures that occur during

ander sehr ähnliche aber entgegenwirkende Organe sind, welche durch entgegengesetzte Beschleunigungen gereizt werden.

Weitere Untersuchung der Erscheinungen.

1.

In dem Vorhergehenden wurde ganz vorläufig das Ohrlabyrinth als wahrscheinliches Haupt-Organ der Bewegungsempfindungen bezeichnet. Hier sollen nun die verschiedenen möglichen Quellen der Bewegungsempfindungen noch einmal untersucht und kritisch durchgemustert werden. Wie wir schon bemerkt haben, kann die Relativbeschleunigung je zweier Körpertheile gegeneinander Anlass zu Empfindungen geben, welche als Merkmale der Bewegung benutzt werden konnten. Wir wollen also untersuchen, ob die Empfindungen, durch welche wir über die Bewegungen unseres Körpers Aufschluss erhalten, herrühren

a) von dem Bindegewebe und den Knochen,

b) von der Haut,

c) von den Muskeln,

d) vom Blute,

e) von den Augen,

f) vom Hirn oder endlich

g) von einem eigenen Organe im Kopfe.

2.

Betrachten wir zunächst die erste Möglichkeit. Wenn das Bindegewebe und die Knochen sensible Nervenfasern enthalten, so können Relativbeschleunigungen der Theile dieser Gewebe gegeneinander immer Anlass zu localen Druckempfindungen geben. Man sieht aber leicht, dass diese Empfindungen in Bezug auf ihre Verwerthbarkeit als Merkmale der Locomotion sehr zweifelhafter Natur sein müssen. Betrachten wir zwei solche Körpertheile A und B. Erhält A durch eine äussere Kraft eine Beschleunigung gegen B hin, so entsteht durch den Druck und Gegendruck von A und B eine Empfindung E. Dieselbe Empfindung tritt aber auch auf, wenn B

locomotion, however, these strains and pressures can reveal nothing about the direction of the acceleration and motion. Since the experiment clearly shows that we can undoubtedly sense the direction of acceleration, the hypothesis that the sensitive nerves in the connective tissue and bones detect motion perception is untenable.

3.

All passive movements of our body result from accelerations that first affect the skin. Active movements also inevitably create tension and compression of the body against the sensitive skin covering the body surface. One might therefore suppose that movement sensations are skin sensations. Although no one pursuing these experiments would attach too much importance to skin sensations, one must nevertheless admit that they cannot be fully excluded by these experiments. However, we can drastically modify the skin sensations without inducing any illusion of passive body movement. It is therefore quite unlikely that they play an important role in movement sensation.

Experiment 1. We take an angled, bent pedestal (Fig. 12) where the part *fg hi* is very flat and only *ab cd* is open [actually, *f* and *g* belong higher on the drawing]. We place a tight-fitting piston *S* in this opening. By rapidly pulling up this piston a considerable thinning of the air and a considerable reduction of air pressure occurs in the box. We now pour plaster on the surface *fg hi* and step on it with bare feet. Once the cast has hardened, an

gegen *A* beschleunigt ist, wobei also der Sinn der Locomotion gerade entgegengesetzt sein wird. Eine andere Empfindung *F* kann auftreten, wenn *A* von *B* weg beschleunigt wird, also Zug und Gegenzug zwischen *A* und *B* entsteht, der aber wieder derselbe ist bei der Beschleunigung des *B* von *A* weg, also wieder beim entgegengesetzten Sinn der Locomotion. Mit einem Wort, Bindegewebe und Knochen können möglicher Weise die bei Locomotionen auftretenden Zerrungen und Pressungen empfinden, welche Zerrungen und Pressungen aber keinen Aufschluss über den Sinn der Beschleunigung und der Locomotion geben. Da wir nun, wie der Versuch lehrt, den Sinn der Beschleunigung unzweifelhaft empfinden, so ist die Hypothese unhaltbar, dass die betreffenden Bewegungsempfindungen durch die sensiblen Nerven des Bindegewebes und der Knochen geliefert werden.

3.

Alle passiven Bewegungen, die unser Körper erhält, rühren von Beschleunigungen her, welche zunächst die Haut ergreifen, und auch bei den activen Bewegungen ist ein Stützen und Stemmen des Körpers auf mit empfindlicher Haut versehene Oberflächentheile unvermeidlich. Der Gedanke liegt also nahe, dass die Bewegungsempfindurgen Hautempfindungen seien. Wenn auch wohl niemand, welcher die hier beschriebenen Versuche ausführt, unter dem Eindruck derselben den Hautempfindungen die grösste Wichtigkeit zuschreiben wird, so muss man doch zugeben, dass sie bei den Experimenten nicht vollständig eliminirt werden können. Wir können aber die Hautempfindungen sehr gewaltig modificiren, ohne eine Täuschung über die Art der passiven Körperbewegung herbeizuführen, und es wird hiedurch sehr unwahrscheinlich, dass sie als Bewegungsempfindungen eine wichtige Rolle spielen.

Versuch 1. Wir nehmen eine winkelig geknickte Kiste (Fig. 12), an welcher der Theil *fg hi* sehr flach ist und die nur bei *ab cd* offen ist. In diese Oeffnung setzen wir einen gutpassenden kistenförmigen Stempel *S*. Beim raschen Herausziehn dieses Stempels tritt in der Kiste eine bedeutende Luftverdünnung und eine bedeutende Verminderung des Luftdruckes ein. Wir begiessen nun

area of the plaster on which the feet stand is broken away, so that when one steps on the box again the soles of the feet themselves form part of the box wall and help make the interior of the box airtight. Then the whole box is bolted to the floor and the piston pulled up forcefully. Normally, the soles of the feet support only the entire weight of the body. With the pulling of the piston, however, the soles support an additional weight [force] result from the external air pressure acting against the reduced air

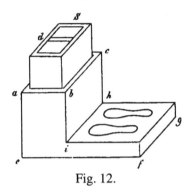

Fig. 12.

pressure in the box. One immediately feels a strong pressure on the feet and it feels like the floor under the feet is rising slightly. One does not, however, experience the sensation that the whole body rises, which would be the critical finding. If one makes the <u>hole only under one foot</u>[7.1], the floor seems to rise more beneath that foot.

The air pressure against the whole [upper] surface area of the body remains constant, and is also not sensed. What one feels in this experiment is only an overpressure against the soles of the feet. If skin sensations were identical to movement sensations, this experiment would have to induce the illusion of our <u>body rising</u>.[7.2]

Experiment 2. If we lie horizontally supine inside a box which starts to rotate in a clockwise direction as seen from above, we feel the accelerating pressure of the box walls on the right arm, the right shoulder, and the left calf. This pressure becomes much stronger if the bottom of the box on which we rest is moveable independently from the walls. The pressure reverses if we make the side walls of the box movable and use stones to make them heavier. At the start of the clockwise movement, we then experience pressure against the left shoulder and the right leg. Despite this major modification of skin sensation, we are, while enclosed in the box, never in doubt about the sense of rotation, and can readily separate skin and

die Fläche *fg hi* mit Gyps und treten mit den blossen Fusssohlen dar auf. Ist der Abdruck erhärtet, so wird die von den Fusssohlen berührte Fläche durchbrochen, so zwar, dass wenn man sich wieder auf die Kiste stellt, die Fusssohlen selbst einen Theil der Kistenwand bilden und den Innenraum der Kiste luftdicht abschliessen helfen. Alsdann wird der ganze Apparat am Boden festgeschraubt und der Stempel kräftig aufgezogen.

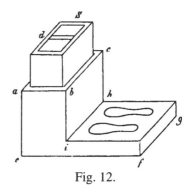

Fig. 12.

Die Fusssohlen haben für gewöhnlich bloss die ganze Last des Körpers zu tragen, beim Aufziehen des Stempels aber noch das Uebergewicht des äussern Luftdrucks gegen den auf die Sohlenfläche entfallenden Luftdruck im Innern der Kiste. Man fühlt auch sofort ein mächtiges Andrücken der Fusssohlen und es hat den Anschein, als ob der Boden sich unter den Füssen etwas heben würde. Die Empfindung, dass der ganze Körper steigt, auf die es eben ankäme, hat man aber nicht. Macht man bloss unter der einen Sohle eine Oeffnung, so scheint sich der Boden mehr unter dieser zu heben.

Der Luftdruck an der ganzen Oberfläche des Körpers ändert sich nicht, wird auch nicht empfunden. Was man bei dem Versuch empfindet, ist bloss ein Ueberdruck auf die Fusssohlen. Dieser müsste uns ein Steigen unseres Körpers vorspiegeln, wenn die Hautempfindungen identisch mit den Bewegungsempfindungen wären.

Versuch 2. Wenn wir horizontal in einer Kiste auf dem Rücken liegen, welche von oben gesehn sich im Sinne des Uhrzeigers zu drehen beginnt, so fühlen wir den beschleunigenden Druck der Kistenwände an dem rechten Arm und der rechten Schulter und an der linken Wade. Dieser Druck wird viel stärker, wenn der Boden der Kiste, auf welchem wir liegen, für sich beweglich ist. Der Druck kehrt sich um, wenn wir im Gegentheil die Seitenwände der Kiste für sich beweglich machen und mit Steinen beschweren. Wir empfinden dann den Druck bei der beginnenden Uhrzeigerbewegung an der linken Schulter und am rechten Bein. Trotz dieser gewaltigen

movement sensations if they are in conflict. We will return to this and similar experiments while investigating muscle sensations.

Let us call an element of the skin surface df, X the <u>force</u>[7.3] acting on the unit surface in the direction x, and dt the unit time. Any given displacement of our body in a direction x is determined only by the value of the integral $\int\int X \, df . dt$ in different cases. It is certainly correct that we can change and modify the distribution of the pressure forces against the skin. However, we cannot change the value of that integral without actually changing the magnitude of our body's movement. One could get the idea that in some sort of subconscious manner we could experience the value of this integral and use that to characterize the movement. But even the ardent supporters of subconscious conclusions will find difficulties with this interpretation.

Considerable changes in skin sensation can occur without altering the value of the integral. Furthermore, not every change of force will correspond to the change in sensation in a simple manner. If the value of the above integral were actually to be perceived one would have to assume that the pressure forces X led to the same sensation, whether they were applied to the skin elements normally or tangentially. The elements of the integral and the corresponding elements of sensation certainly do not parallel each other.

In the air pressure experiment mentioned above the value of the integral remains constant since the body continues to remain at rest. The skin sensation underwent a considerable change leading to the illusion discussed earlier without, however, leading to any false sensation of body movement.

That which applies for translational forces here also holds true for the turning sensations of the rotation experiments in the box. The sum of the rotational moments applied to the skin must, of course, always remain

Modificationen der Hautempfindung sind wir, in der Kiste eingeschlossen, doch nie im Zweifel über den Drehungssinn und sondern
die Haut- und Bewegungsempfindungen sehr wohl von einander,
wenn sie in Widerspruch gerathen. Wir kommen auf dieses und
ähnliche Experimente noch bei Untersuchung der Muskelempfindungen zurück.

Nennen wir ein Element der Hautoberfläche df, die auf die
Flächeneinheit nach der Richtung x wirkende Kraft X und dt das
Zeitelement, so muss, damit unserem Körper in verschiedenen Fällen
nach der Richtung x dieselbe Bewegungsgrösse ertheilt werde, das
Integral $\iint X \, df.dt$ in allen diesen Fällen denselben Werth haben.

Nun ist es allerdings richtig, dass wir die Vertheilung der Druckkräfte auf die Haut beliebig abzuändern und zu modificiren, dass wir
aber den Werth dieses Integrales nicht zu ändern vermögen, ohne auch
wirklich die Bewegungsgrösse unseres Körpers abzuändern. Man
könnte also auf den Einfall kommen, dass wir durch irgend einen
unbewussten Schluss den Werth dieses Integrals erfahren und als
Kennzeichen der Bewegung benützen. Aber auch die Liebhaber der
unbewussten Schlüsse werden in dieser Auffassung Schwierigkeiten
finden.

Es werden beträchtliche Veränderungen in der Hautempfindung
eintreten können, ohne dass sich der Werth des Integrals ändert.
Ferner bringt nicht jede Veränderung der Kräfte eine entsprechende
zu ihr in einfacher Beziehung stehende Aenderung der Empfindungen hervor. Man müsste, damit der Werth des obigen Integrals
auch empfunden werde, annehmen, dass die Druckkräfte X dieselbe
Empfindung auslösen, ob sie normal oder tangential auf die Haut-
elemente wirken. Die Elemente des Integrals und die entsprechenden
Elemente der Empfindung gehen durchaus nicht parallel.

Bei dem obigen Luftdruckversuch ändert sich der Werth des
Integrals nicht. Der Körper bleibt ja nach wie vor in Ruhe. Die
Hautempfindung hat aber eine wesentliche Aenderung erfahren,
welche zu der erwähnten Täuschung Anlass gibt, ohne jedoch eine
Bewegung des Körpers vorzuspiegeln.

Dasselbe, was hier für die Fortschrittskräfte gilt, kann man bei
dem Drehversuch in der Kiste von den Drehungsmomenten sagen.
Die Summe der auf die Haut applicirten Drehungsmomente muss

constant for a constant angular acceleration of the body. The skin sensations can vary widely without affecting the perception of the movement. Because our knowledge of pressure forces comes exclusively from the induced sensation, and a simple relationship between forces and sensation cannot be given, one would find the assumption that the skin[7.4] provides the relevant movement sensations very unlikely.

4.

The existence of a special muscle sense has been suggested but also disputed. One cannot deny that one senses muscle pain and fatigue as well as the extent of muscular effort, and therefore sensory elements probably reside in muscles. Here we can be satisfied with the fact that muscle effort is perceived, and it is immaterial whether the origin of this sensation lies in the muscle itself or in the central organ [central nervous system] or in the skin after all. The last seems unlikely since skinning[7.5] of an extremity does not seem to lead to a loss of muscle sense. We therefore accept the muscle sense as something given and observable, without delving into its explanation.

One can now consider that this muscle sense provides knowledge to distinguish between active and passive movements. If the relative position of the body parts are not maintained by the muscles when the trunk is actively or passively moved, one can imagine that those parts that are movable against the trunk would lag behind. In that case, however, muscle forces $m(\varphi' - \varphi)$, or torques $T(\varphi' - \varphi)$ will be necessary, which one will perceive. If one performs an active movement during a passive one, then as noted earlier [page 21], the effort is $m[\psi + (\varphi' - \varphi)]$, and $T[\psi + (\varphi' - \varphi)]$ respectively. The sensation of that effort cannot only make the passive movement noticeable but, because it is unusual, it can also cause movements to be executed differently than intended. Purkyně claims that one's handwriting changes if one writes on the platform of a merry-go-round. I was unable to detect anything like that for such small movement excursions. However, the contrast between the intended and executed movement is

natürlich bei gleicher Winkelbeschleunigung des Körpers immer dieselbe bleiben. Die Hautempfindungen können dabei sehr mannig-
faltig variiren, ohne die Bewegungsvorstellung zu alteriren. Da wir
nun von den Druckkräften nur durch die erregten Empfindungen
Kenntniss haben, ein einfacher Zusammenhang zwischen den Kräften
und Empfindungen aber nicht angegeben werden kann, so wird
man die Annahme, dass die Haut die maassgebenden Bewegungsempfindungen liefert, sehr unwahrscheinlich
finden.

4.

Dass es ein besonderes Muskelgefühl gibt, ist behauptet und
auch bestritten worden. Geläugnet kann nicht werden, dass man
Schmerz und Ermüdung der Muskel, so wie den Grad der Anstrengung derselben empfindet, und dass daher sensible Elemente in den
Muskeln wahrscheinlich sind. Wir können uns hier mit der Thatsache der Empfindung der Muskelanstrengung begnügen, und es ist
uns gleichgültig, ob die Quelle dieser Empfindung in den Muskeln
selbst oder in den Centralorganen oder gar in der Haut liegt. Letzteres wird unwahrscheinlich durch den Umstand, dass bei Enthäutung
der Extremitäten das Muskelgefühl nicht verloren zu gehen scheint.
Wir nehmen also das Muskelgefühl als etwas Gegebenes, Beobacht-
bares, ohne uns um die Erklärung desselben zu kümmern.

Man kann nun daran denken, dass dieses Muskelgefühl die
Kenntniss der activen und passiven Bewegungen vermittelt. Wird
der Rumpf in active oder passive Bewegung gesetzt, so kann man
sich vorstellen, dass die gegen den Rumpf beweglichen Theile zurückbleiben würden, wenn die relative Lage der Körpertheile nicht
durch die Muskeln erhalten würde. Hiezu wird aber die Muskel-
anstrengung $m(\varphi' - \varphi)$ beziehungsweise $T(\varphi' - \varphi)$ nöthig sein, und
diese wird man empfinden. Führt man während der passiven Be-
wegung noch eine active aus, so gehört hiezu nach den früheren
Bemerkungen die Anstrengung $m[\psi + (\varphi' - \varphi)]$ beziehungsweise
$T[\psi + (\varphi' - \varphi)]$. Die Empfindung derselben kann nicht nur die passive Bewegung bemerklich machen, sondern auch weil sie eine un-
gewohnte ist, Ursache werden, dass andere als die beabsichtigten
Bewegungen ausgeführt werden. Purkyně führt an, dass die Schrift-

evident for larger motions attempted during passive rotation.

It is unreasonable to assume that the parts that are movable against the trunk are maintained in their position by some special muscular effort with some special innervation. For slow movements the mere stiffness and tension of the muscle with the help of friction is sufficient to maintain its position. With more rapid disturbances the muscular apparatus shows a kind of elasticity that one can easily visualize by a simple experiment. If one strikes an extended finger of one hand with the other, the finger yields against the stroke and, like a spring, resumes its original extended position. The muscular apparatus probably behaves this way in general. No special innervation[7.6] is necessary to maintain position or to regain it. Furthermore, no special perception will occur. If one lies horizontally on a disk that starts rotating rapidly enough, the body parts are indeed displaced. The muscles have not maintained their position.

If movement sensations originate from muscular efforts, these movement sensations would have to vary greatly, depending on the extent to which a passive movement is executed in a more or less supported position. I cannot observe this. We will now, nevertheless, perform several experiments concerning muscle sensation.

Experiment 1. When pouring mercury out of a vessel, one has the sensation of a powerful elevation of the hand holding the vessel. By careful observation one can actually see the hand rising a little. One senses the reduction in muscle effort necessary to support the vessel and, because one cannot reduce it quickly enough to keep the vessel in place, the hand rises a little. Every such reduction of the resistance against a muscle seems to be sensed by the active muscle as a decrease of resistance and as a movement in a direction opposite to the resistance.

züge beim Schreiben auf der Scheibe des Ringelspiels sich ändern. Ich habe bei so kleinen Bewegungsexcursionen nichts Derartiges wahrnehmen können. Wohl aber ist der Contrast zwischen der beabsichtigten und der ausgeführten Bewegung auffallend bei ausgiebigeren Bewegungen, die man während der passiven Rotation versucht.

Es ist aber kaum anzunehmen, dass die gegen den Rumpf beweglichen Theile durch eine besondere Muskelanstrengung, durch eine besondere Innervation erhalten werden. Bei langsameren Bewegungen genügt die blosse Steifigkeit und Spannung der Muskel mit Hilfe der Reibung, um diese Lage zu erhalten. Bei rascheren Störungen zeigt der Muskelapparat eine Art Elasticität, die man sich durch ein einfaches Experiment zur Anschauung bringen kann. Schlägt man gegen den ausgestreckten Finger einer Hand mit der andern Hand, so weicht der Finger sich biegend dem Schlage und nimmt wie eine Feder seine ursprüngliche gestreckte Lage wieder ein. So dürfte sich überhaupt der Muskelapparat verhalten. Es sind also hier keine besondern Innervationen nöthig, um die Lage zu erhalten oder herzustellen. Demnach werden auch keine besondern Empfindungen auftreten. Liegt man horizontal auf einer Scheibe, welche hinreichend plötzlich in Rotation versetzt wird, so kommen die Körpertheile in der That aus ihrer Lage. Die Muskel haben dieselbe nicht erhalten.

Rühren Bewegungsempfindungen von Muskelanstrengungen her, so müssen diese Bewegungsempfindungen sehr verschieden sein, je nachdem man eine passive Bewegung in mehr oder weniger unterstützter Lage ausführt. Dies kann ich nicht beobachten. Wir wollen nun dennoch mehrere Versuche über die Muskelempfindungen ausführen.

Versuch 1. Beim Ausgiessen von Quecksilber aus einem Gefäss hat man die Empfindung der gewaltsamen Hebung derjenigen Hand, welche das Gefäss hält Man kann auch beim aufmerksamen Zusehen wahrnehmen, dass sich die Hand wirklich etwas hebt. Man empfindet die Verminderung der Muskelanstrengung, welche zur Erhaltung des Gefässes nöthig ist, und indem man dieselbe nicht ebenso rasch zu vermindern vermag, als es eben zur Erhaltung des Gefässes erfordert wird, hebt sich die Hand etwas. Jede solche

Experiment 2. I produced two tin vessels (Fig. 13) with comfortable handles, each with a capacity of about four kilograms of water. They have a nozzle on the bottom, over which is placed a short rubber tube of about 15^{mm} inner diameter that is closed with a spring clip. The first experiment performed using this simple apparatus was the following. One takes one vessel in each hand. On command both spring clips are opened simultaneously, and the water pours into reservoirs placed below. One notices a clear elevation of the arms, especially towards the end of the experiment, since then the largest percentage of the remaining weight flows out per unit time.

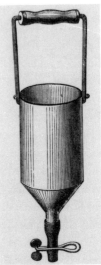

Experiment 3. A cushioned wooden cap is placed on the head, tight enough so that the head must turn with any rotation of the cap. Hooks are placed near

Fig. 13.

the right and the left ears of the cap. A string connected to the right hook leads horizontally forward and then over a pulley. One of the vessels described above hangs from the string. A second string running horizontally backwards from the left hook and over a second pulley carries the second vessel. The two vessels together exert an arbitrarily chosen rotational moment on the head, which is counterbalanced by muscles. If one then lets the water flow out of the vessels, one feels a rotation of the head opposite to the direction it would be rotated by the weights. Remarkably, all such rotations are sensed very strongly, but their excursions are hardly noticeable.

Experiment 4. One positions a wooden bar across the back on both shoulders arranged similarly to the experiment above, so that on the right

Verminderung eines Widerstandes gegen einen Muskel scheint von dem activen Muskel als ein Weichen des Widerstandes und ein Fortbewegen in der dem Widerstande entgegengesetzten Richtung empfunden zu werden.

Versuch 2. Ich habe mir zwei Blechgefässe (Fig. 13), jedes für etwa vier Kilogramm Wasser, mit bequemen Handhaben anfertigen lassen. Dieselben haben am Boden ein Ansatzrohr, über welches ein kurzes Kautschukrohr von 15^{mm} Durchmesser im Lichten gezogen und durch einen Quetschhahn verschlossen wird. Mit diesem einfachen Apparat wurde zunächst folgender Versuch angestellt. Man fasst mit jeder Hand ein Gefäss. Auf Commando werden beide Quetschhähne gleichzeitig geöffnet, und das Wasser fliesst in untergestellte Behälter ab. Man verspürt hiebei namentlich gegen das Ende des Versuches, weil da in der Zeiteinheit der grösste Bruchtheil des noch vorhandenen Gewichtes abfliesst, eine deutliche Erhebung der Arme.

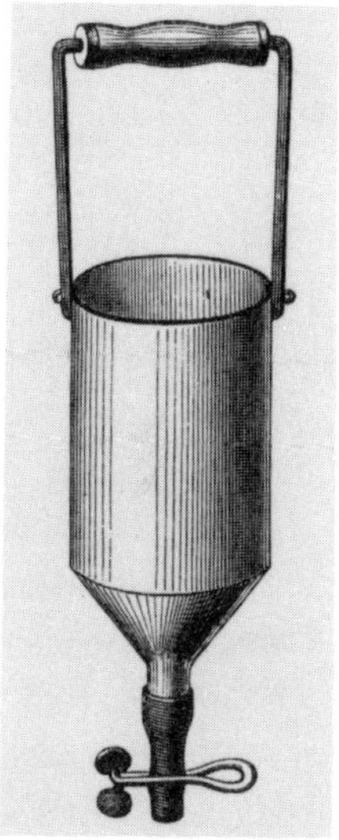

Fig. 13.

Versuch 3. Auf den Kopf wird eine innen gepolsterte Holzkappe gesetzt, jedoch so fest, dass man den Kopf nothwendig mit drehen muss, wenn man die Kappe dreht. In der Nähe des rechten und linken Ohres befindet sich an der Kappe ein Haken. Am rechten Haken greift eine Schnur ein, welche horizontal nach vorne über eine Rolle geht und an der eines der beschriebenen Gefässe hängt. Eine zweite Schnur geht vom linken Haken horizontal nach hinten über eine zweite Rolle und trägt das zweite Gefäss. Beide Gefässe zusammen ertheilen dem Kopfe ein übrigens beliebig variables Drehungsmoment, welches durch die Muskel im Gleichgewicht gehalten wird. Lässt man nun das Wasser aus den Gefässen abfliessen, so fühlt man eine Drehung des Kopfes entgegen demjenigen Sinne, in welchem er durch die Gewichte gedreht würde. Bemerkenswerth ist, dass alle diese Drehungen sehr stark empfunden, aber nur unmerklich ausgeführt werden.

Versuch 4. Man befestigt quer über den Rücken an beiden Schultern eine Holzleiste und lässt in der eben angegebenen Weise

side one vessel pulls forward, and on the left side the other one pulls backward. Here again one feels rotation against the resistance during the outflow of the water.

Experiment 5. One vessel is attached to the left end of the same bar and pulls downward; the other one is connected to a string that leads from the right end of the bar and goes vertically upward over a pulley. One feels oneself sinking to the right when the water flows out.

Experiment 6. Vessels hang from both the right and left end of the bar. During the outflow one feels that one stretches and is elevated from the floor.

In none of the six experiments did any phenomena occur which were comparable to the peculiar spatial perceptions and optical manifestations described earlier.

In the experiments just described a strange observation is that one does not relate the perception to movement of the entire body, but rather to a change of the positions of the affected parts relative to each other.

In experiments 2 and 6 one feels elevation of the arms, stretching of the body, and rising up from the floor. On this occasion a reduction of body weight actually occurs. Such a reduction of the body weight, however, as observed with moving subjects, causes the perception that the body sinks.[7.7]

In experiment 3 a turning moment on the head tends to rotate it in a counterclockwise manner as seen from above. The muscles compensate for this rotational moment. If the turning moment is suddenly reduced, one feels that the head is rotated clockwise; if it is rapidly increased, one senses a counterclockwise rotation. If the trunk were angularly accelerated clockwise, the head would receive an opposite relative angular acceleration with respect to the trunk. The muscles that attempt to counter this stretching moment in order to keep the head in its position, but are unprepared for this increasing rotational moment, would command a head rotation[7.8] in the counterclockwise direction, according to the experiments described above. Based on observations of rotated subjects, however, rotation of the whole body is sensed in a clockwise direction.

rechts das eine Gefäss nach vorn, links das andere nach hinten ziehn. Auch hier meint man sich beim Abfliessen des Wassers dem Widerstande entgegen zu drehn.

Versuch 5. An dem linken Ende derselben Leiste hängt das eine Gefäss direct und zieht nach unten, das andere befindet sich an einer Schnur, welche von dem rechten Ende der Leiste vertical nach oben über eine Rolle abgeht. Man meint beim Abfliessen des Wassers nach rechts umzusinken.

Versuch 6. An beiden Enden der Leiste rechts und links hängt eines der Gefässe. Beim Abfliessen meint man sich zu strecken und sich aus dem Boden zu erheben.

Bei keinem der sechs Versuche trat eine Erscheinung ein, die sich mit den früher beschriebenen eigenthümlichen Raumvorstellungen und optischen Phänomenen in Vergleich bringen liesse.

In den beschriebenen Versuchen herrscht nun die Eigenthümlichkeit vor, dass man die entstehenden Empfindungen nicht auf eine Bewegung des ganzen Körpers, sondern auf eine Aenderung der relativen Stellung der eben afficirten Theile zu einander bezieht.

Bei den Versuchen 2 und 6 fühlt man Erhebung der Arme, Streckung des Körpers, Emporwachsen aus dem Boden. Hiebei geht nun eigentlich eine Verminderung des Körpergewichtes vor sich. Eine solche Verminderung des Körpergewichtes bedingt aber nach den Beobachtungen an bewegten Menschen die Empfindung des Versinkens des Körpers.

In Versuch 3 greift am Kopf ein Drehungsmoment an, welches denselben von oben gesehn verkehrt wie den Uhrzeiger zu drehen strebt. Diesem Drehungsmoment halten die Muskel das Gleichgewicht. Wird das Drehungsmoment rasch vermindert, so meint man, der Kopf werde wie der Uhrzeiger gedreht; wird es rasch vermehrt, so empfindet man eine Kopfdrehung dem Uhrzeiger entgegen. Gesetzt nun, der Rumpf würde eine Winkelbeschleunigung im Sinne des Uhrzeigers erhalten, so erhielte der Kopf gegen den Rumpf eine umgekehrte relative Winkelbeschleunigung. Die Muskel, welche dieses Streckungsmoment zu vernichten streben, den Kopf in seiner Lage zu halten suchen, aber auf dieses anwachsende Drehungsmoment nicht gefasst sind, würden nach den oben mitgetheilten Versuchen eine Kopfdrehung dem Uhrzeiger ent-

Although the principles used in my <u>1865 publication</u>[7.9] are certainly correct, their application to muscle sensations as an explanation for movement sensation seems insufficient since we are accustomed to see muscle sensations partly unrelated and partly contradictory to the expected movement indication. Other information seems necessary to provide a positive perception of locomotion. Teleologically this is indeed understandable. Muscle sensations must account for the outcome of mechanical work, including that which is unrelated to locomotion. This result is primarily determined by changes of the relative positions of the body parts. It has nothing to do with locomotion. Therefore, it seems that a separate organ is necessary for the perception of locomotion.

I will illustrate by a further example how different results are produced if one applies the same mechanical principles to different body parts. The observer who rotates in a clockwise direction (seen from above) is, so to speak, a <u>Fessel flywheel machine</u>[7.10], and all such phenomena have to apply to him, although they will be sensed differently in different parts. If he makes a nodding head movement, the whole head will be slightly twisted counterclockwise, as seen from behind. At the same time the head is thrown a little to the left. Both movements can be sensed through the muscles. In addition the observer feels that the whole room in front of him, including himself, rotates in a clockwise direction, which is exactly opposite to his sensed head rotation.

5.

For one and the same passive movement, one can artificially vary skin sensation. The same experiment can also be performed with muscle sensations. Since no illusions about the kind of movement arise here, one can convince oneself of the minor role played by muscles in the detection of movement.

gegen signalisiren. Nach den Beobachtungen an gedrehten
Menschen empfindet man aber eine Drehung des ganzen
Körpers im Sinne des Uhrzeigers.

Obgleich also die in meiner Arbeit von 1865 benützten Grundsätze gewiss richtig sind, so scheint die Anwendung derselben auf
die Muskelgefühle zur Erklärung der Bewegungsempfindungen doch
insofern nicht auszureichen, als wir in den Muskelgefühlen theils
andere, theils auch den geforderten widersprechende Bewegungs-
anzeigen zu sehen gewohnt sind. Zur sichern Erkenntniss der Locomotion scheint noch ein anderes Mittel gefordert. Teleologisch ist dies
auch begreiflich. Die Muskelempfindungen haben zu berichten über
den Erfolg mechanischer Arbeiten, auch solcher, welche mit der Locomotion nichts zu schaffen haben. Dieser Erfolg ist hauptsächlich bestimmt durch die Veränderungen in der Relativstellung der Körpertheile. Er hat mit der Locomotion nichts zu schaffen. Es scheint
also ein eigenes Organ zur Perception der Locomotion nöthig.

Noch durch ein Beispiel will ich erläutern, wie verschiedene
Ergebnisse man erhält, wenn man dieselben mechanischen Grundsätze auf verschiedene Körpertheile anwendet. Der Beobachter,
welcher im Sinne des Uhrzeigers (von oben gesehn) rotirt, ist so zu
sagen eine Fessel'sche Schwungmaschine, und alle Erscheinungen
derselben müssen an ihm auftreten, wenngleich sie an verschiedenen Theilen in verschiedener Weise empfunden werden. Nickt
er bejahend mit dem Kopfe, so wird hiebei dieser ganze Kopf etwas
verdreht, und zwar von hinten gesehn verkehrt wie der Uhrzeiger.
Gleichzeitig wird der Kopf etwas nach links hinübergeworfen. Beide
Bewegungen kann man durch die Muskel merken. Ausserdem meint
aber der Beobachter, dass sich der ganze vor ihm befindliche Raum
sammt ihm selbst im Sinne des Uhrzeigers dreht, also gerade entgegengesetzt, als er seinen Kopf gedreht fühlt.

5.

Man kann bei einer und derselben passiven Bewegung die
Hautempfindung künstlich variiren. Das gleiche Experiment kann
man nun mit den Muskelempfindungen anstellen, und da hiedurch
Täuschungen über die Art der Bewegung nicht entstehn, so kann
man sich durch dieses Mittel von der untergeordneten Rolle über-

Experiment 1. If we start a passive clockwise rotation while seated on a chair, the head and the trunk both experience an oppositely directed relative angular acceleration. Assume for the moment that the muscles maintain the position of the body parts relative to each other, and that the corresponding muscle sensations appear consciously as movement sensations in a clockwise direction. The perception of movement might be expected to change over a wide range if we vary those muscle sensations artificially. This can easily be done as follows. Inside the frame of the rotation apparatus described earlier we install two large heavy disks that may easily be rotated about vertical axes. The observer is seated on a chair inside the paper box and two strings that go around these disks are fastened to his head and shoulders such that for each acceleration of the apparatus, these body parts receive an angular acceleration in keeping with the principle of the conservation of areas. One can arrange the strings such that the angular acceleration is either in the direction of the natural acceleration of the body parts, or opposite. We will sketch the situations A and B for the head schematically; they are the same for the shoulders. The disk is indicated by S, the head, connected to strings through a tight fitting headband, by K, and a pulley, behind the observer's head by R. For the arrangement A, the head of the observer behaves as if it has a much larger moment of inertia. For the arrangement B, the moment of inertia can be compensated or even become negative. In case A the muscle sensations must be stronger than usual, in B they must be reduced or totally eliminated, or can even act in the opposite sense.

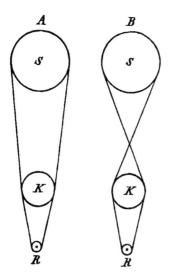

Fig. 14.

zeugen, welche die Muskel bei Erkenntniss der Locomotionen spielen.

Versuch 1. Beginnen wir auf einem Stuhle sitzend eine passive Drehung im Sinne des Uhrzeigers, so erhält der Kopf und der Oberkörper eine entgegengesetzte relative Winkelbeschleunigung. Nehmen wir fur einen Augenblick an, dass die Muskel die relative Ruhe der Körpertheile gegeneinander erhalten, und dass die betreffenden Muskelgefühle uns als Bewegungsempfindungen der Uhrzeigerdrehung zum Bewusstsein kommen. Die Empfindung der Bewegung muss sich dann bedeutend ändern lassen, wenn wir jene Muskelgefühle künstlich variiren. Dies kann leicht auf folgende Weise geschehn. In dem Rahmen des beschriebenen Rotationsapparates bringen wir zwei grosse schwere Scheiben auf verticalen Axen leicht drehbar an. Der Beobachter setzt sich in den Papierkasten auf den Stuhl, und zwei Schnüre, welche um jene Scheiben gelegt sind, werden an dessen Kopf und Schultern befestigt, so dass diese Körper bei jeder Beschleunigung des Apparates durch die Scheiben, welche das Flächenprincip zu erfüllen streben, eine Winkelbeschleunigung erhalten. Man kann die Schnur so legen, dass die Winkelbeschleunigung der natürlichen, welche die Körpertheile für sich haben, gleich oder entgegen gerichtet ist. Für den Kopf wollen wir diese beiden Anordnungen A und B schematisch darstellen; sie sind dieselben für die Schultern. Die Scheibe ist durch S, der Kopf, welcher eine fest angezogene Gurte trägt, an der die Schnüre angreifen, ist durch K, und eine Rolle hinter dem Kopf des Beobachters ist durch R angedeutet. Bei der Anordnung A verhält sich der Kopf des Beobachters, wie wenn sein Trägheitsmoment enorm vermehrt wäre. Bei der Anordnung B kann das Trägheitsmoment compensirt oder

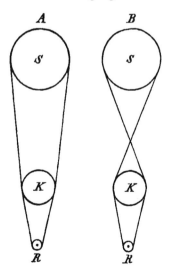

Fig. 14.

The experiment demonstrates that the observer enclosed in the paper box cannot be misled about the direction of motion. Although he feels the head and shoulders strongly rotated in the accustomed or unaccustomed sense, he always knows which way the whole apparatus turns. As one can easily notice with the arrangements A and B, one could always receive an indication of clockwise rotation by skin sensation evoked by the head belts, if the muscles indicate the opposite rotation, or vise versa. However, this presumes that skin and muscles are relevant at all.

Experiment 2. The observer stretches out horizontally on a 2^m long board that is suspended with two strings in the frame of our rotation apparatus. The points where the strings are attached to hooks bolted to the board must, of course, be located well above the observer, so that the board will not tip. A rather closely fitting box without any bottom is now placed over the observer. The box hangs freely suspended, just like the board. Both the board and the box, which may be weighted down with stones if necessary, can be fixed by special bolts.

The arrangement in which the board with the observer can be moved freely but with the box fixed [to the frame] we will call A, whereas the one in which the box is free but the board with the observer stays fixed we call B. If now the whole rotation apparatus is set in motion clockwise (as seen from above), then for arrangement A, the body of the observer attempts to fulfill the principle of conservation of areas.[7.11] But he will be prevented from doing so by the box, which is firmly connected to the frame and gives the body a clockwise rotational moment. This rotational moment can be sensed through the skin, or also through the muscles if he resists. With the arrangement B, on the other hand, the box follows the principle of conservation of areas and attempts to remain behind in a counterclockwise direction, but is prevented from doing so by the observer's body, which lies on the board fixed to the frame. The rotational moment sensed by the skin or the muscles is now in exactly the opposite direction. The free space between

sogar negativ sein. Bei A müssen also die Muskelempfinduugen stärker ausfallen als gewöhnlich, bei B müssen sie vermindert oder ganz vernichtet sein oder können sogar den entgegengesetzten Sinn haben.

Das Experiment lehrt nun, dass alles dies den im Papierkasten eingeschlossenen Beobachter über den Bewegungssinn nicht täuschen kann. Er fühlt Kopf und Schultern mächtig im gewohnten oder ungewohnten Sinn gedreht, weiss aber immer, wie sich der ganze Apparat bewegt. Wie man leicht bemerken kann, würde man bei den Anordnungen A und B immer durch die Hautempfindung der am Kopfe angreifenden Gurte eine Anzeige für die Uhrzeigerdrehung erhalten, wenn die Muskel die entgegengesetzte Drehung signalisiren und umgekehrt, vorausgesetzt, dass Haut und Muskel überhaupt maassgebend wären.

Versuch 2. In den Rahmen unseres Rotationsapparates hängen wir an zwei Seilen ein 2^m langes Brett, auf welches sich der Beobachter horizontal ausstreckt. Die Anknüpfungspunkte der Seile am Brette müssen natürlich an fest mit dem Brette verbundenen Haken ziemlich hoch über dem Beobachter sich befinden, damit das Brett nicht umkippt. Ueber den Beobachter wird nun eine ziemlich gut anschliessende Kiste ohne Boden gestülpt und in gleicher Weise frei aufgehängt wie das Brett. Sowohl das Brett als auch die nöthigenfalls mit Steinen beschwerte Kiste können durch besondere Riegel festgestellt und unbeweglich gemacht werden.

Die Disposition, bei welcher das Brett mit dem Beobachter frei beweglich, die Kiste aber festgestellt ist, nennen wir A, jene, bei welcher die Kiste frei, das Brett mit dem Beobachter aber fest ist, B. Wird nun der ganze Rotationsapparat in Bewegung gesetzt (von oben gesehn) im Sinne des Uhrzeigers, so strebt bei der Disposition A der Körper des Beobachters das Flächenprincip zu erfüllen, wird aber durch die fest mit dem Rahmen verbundene Kiste daran verhindert, welche ihm ein Drehungsmoment im Sinne des Uhrzeigers ertheilt. Dieses Drehungsmoment kann er durch die Haut, oder wenn er sich dagegen sträubt, auch durch die Muskel empfinden. Bei der Disposition B sucht umgekehrt die Kiste das Flächenprincip erfüllend, dem Uhrzeiger entgegen, zurückzubleiben, wird aber daran durch den Körper des Beobachters verhindert, welcher auf dem im Rahmen festgestellten Brett liegt. Das durch die

the box and the observer must be very small for this experiment, so that the suspended parts cannot oscillate, because then the pressure forces would become periodic, and the outcome of the experiment would be difficult to observe.

In this experiment the observer senses a large pressure in one direction or the other, although he is certainly never in doubt about the direction in which he is being rotated. For example, he might feel himself pressed to the right and rotated to the left. Without doubt the sum of all the rotational moments acting on the body surface determines the angular acceleration of the observer's body. However, the sensations associated with these rotational moments can vary considerably, and need not show a simple relation to the sum of the moments at all. For sensation it is certainly not irrelevant whether a moment of rotation is exerted by shearing forces on the skin of the back on which one lies, or by pressure on the sides of the body, or even by the sum or difference of these two signals. However, since such variations in pressure sensation create no illusions about the direction and magnitude of the movement, one is then entitled to conclude that they make no substantial contribution to movement sensations. Thus this experimental series makes it improbable that the skin and muscle sensation, which cannot always be separated easily, participate in movement sensation.

6.

As already mentioned, it seems that for the unequivocal detection of locomotion a special element is needed, one that is not already dedicated to another, essentially different role. Elements that allow us to judge the position of body parts relative to each other are useless for the characterization of locomotion. Blood, along with other body fluids, could seem to be appropriate to indicate elements of locomotion. The fluid content of the body pushes backwards with every forward acceleration of the body and upwards with every downwards acceleration of the body. This

Haut oder die Muskel empfundene Drehungsmoment ist nun gerade entgegengesetzt. Der Spielraum der Kiste und des Beobachters muss bei diesem Versuch sehr klein sein, damit die aufgehängten Theile nicht in Schwingungen gerathen können, wobei die Druckkräfte periodisch würden, und der Erfolg des Versuches schwer zu beobachten wäre.

Bei diesem Versuch empfindet nun der Beobachter einen mächtigen Druck in dem einen oder andern Sinne, während er doch nie darüber im Zweifel ist, in welchem Sinne er gedreht wird. Er fühlt sich z. B. nach rechts gedrückt und nach links gedreht. Ohne Zweifel muss, wenn der Körper des Beobachters dieselbe Drehbeschleunigung annimmt, auch die Summe aller auf seine Körperoberfläche wirkenden Drehungsmomente immer dieselbe sein. Allein die Empfindung dieser Drehungsmomente kann sehr variiren und steht überhaupt in keinem einfachen Zusammenhang mit der Summe der Drehungsmomente. Für die Empfindung kann es doch nicht einerlei sein, ob ein Drehungsmoment durch Zerrung der Rückenhaut, auf welcher man liegt, oder durch Druck auf die Seitenflächen des Körpers oder gar durch die Summe oder Differenz beider Affectionen auf den Körper ausgeübt wird. Bringen aber solche Variationen der Druckempfindung keine Täuschungen über den Sinn und das Ausmaass der Bewegung hervor, so ist man wohl zu der Annahme berechtigt, dass sie nicht zu den maassgebenden Bewegungsempfindungen gehören. Also auch diese Versuchsreihe macht die Rolle der Haut- und Muskelgefühle, welche sich nicht immer gut trennen lassen, als Bewegungsempfindungen unwahrscheinlich.

6.

Wie schon bemerkt wurde, scheint für die unzweideutige Erkenntniss der Locomotion ein besonderes Merkmal gefordert, dem nicht schon eine andere wesentlich verschiedene Rolle angewiesen ist. Merkmale, nach welchen wir die Relativstellung der Körpertheile beurtheilen, sind zur Bezeichnung der Locomotion unbrauchbar. Die Blutmasse, überhaupt die im Körper enthaltenen Flüssigkeiten könnten nun ganz geeignet scheinen, Merkmale der Locomotion zu liefern. Der flüssige Körperinhalt drängt rückwärts

locomotion acceleration is always associated with the weight [gravito-inertial] acceleration of the blood. In free fall of the body, for example, the blood weight must completely disappear. The distribution of the blood in the body is very variable, however, and is determined by changes in activity and changed nutritional demands of the organs. On the other hand, the blood distribution will be influenced by changes in blood pressure occurring during strong movements. Evidently, this makes the estimation of locomotion parameters already quite complicated. Finally, however, and this is the main point, one cannot state that a direct perception of blood pressure and blood distribution even exists. For this purpose, sensory nerves would presumably have to lie parallel to vasomotor nerves.

Everyone recognizes the influence of acceleration on blood mass [distribution]. The veins empty in an elevated hand. However, one can fill them quite extensively if one swings the outstretched raised hand, or swings the whole body about a vertical axis with the arm extended. Because the activity of all organs depends critically on the blood supply, a change in blood mass during acceleration, that is, its redistribution in the body, must bear strong secondary consequences. These consequences should be more noticeable as the regulation of blood distribution to compensate for the mass acceleration gets slower and weaker. Purkyně already speaks of an effect of centrifugal forces on blood mass. He also considers a therapeutic application[7.12] of this fact – possibly a fertile thought!

I have to leave the final decision to the physiologists as to whether acceleration of blood can be used as an indicator of locomotion, whether a part of movement sensation, directly or indirectly, stems from the acceleration of blood. The experiments that will be described make it unlikely that direct generation of movement sensation comes by way of the blood. I was brought to this view through a communication of my friend J. Popper.[7.13]

Let us assume a rectangular vertical frame R, which carries a horizontal

bei jeder Beschleunigung, die der Körper vorwärts erhält, aufwärts bei jeder Beschleunigung abwärts. Diese Locomotionsbeschleunigung muss sich mit der Schwerebeschleunigung des Blutes fortwährend zusammensetzen. Im freien Fall des Körpers müsste z. B. das Blutgewicht ganz verschwinden. Die Vertheilung des Blutes im Körper ist aber sehr variabel und wird durch die wechselnde Thätigkeit und das. wechselnde Ernährungsbedürfniss der Organe bestimmt. Umgekehrt muss die Abänderung des Blutdruckes bei heftigen Bewegungen auf die Blutvertheilung Einfluss üben. Damit werden die Merkmale der Locomotion schon recht complicirt. Endlich aber, und das scheint die Hauptsache, kann man nicht behaupten, dass man eine unmittelbare Empfindung des Blutdruckes und der Blutvertheilung hat. Zu diesem Zwecke müssten wohl sensible Nerven den vasomotorischen parallel gehn.

Den Einfluss von Beschleunigungen auf die Blutmasse kennt Jedermann. In der erhobenen Hand leeren sich die Venen. Man kann sie aber ziemlich stark füllen, wenn man die ausgestreckte gehobene Hand oder den ganzen Körper bei ausgestrecktem Arm rasch um eine Verticalaxe schwingt. Da nun die Thätigkeit aller Organe wesentlich von der Blutzufuhr abhängt, so muss die Affection der Blutmasse durch Beschleunigungen, die Veränderung ihrer Vertheilung im Körper, mächtige secundäre Folgen haben. Diese Folgen müssen desto auffallender werden, je weniger der die Blutvertheilung regulirende Apparat schnell und kräftig genug wirken kann, um die Massenbeschleunigungen zu compensiren. Schon Purkyně spricht von einer Affection der Blutmasse durch die Centrifugalkraft. Er denkt auch an eine therapeutische Verwerthung dieser Thatsache – vielleicht ein fruchtbarer Gedanke!

Ich muss die endgültige Entscheidung darüber, ob die Blutbeschleunigung als Merkmal der Locomotion verwerthbar ist, ob ein Theil der Bewegungsgefühle direct oder indirect von der Blutbeschleunigung herrührt, den Physiologen überlassen. Hier mögen nur Experimente folgen, welche die directe Erregung von Bewegungsempfindungen durch das Blut unwahrscheinlich machen. Ich werde zu denselben durch eine Mittheilung meines Freundes J. Popper veranlasst.

Stellen wir uns einen rechteckigen verticalen Rahmen R vor,

board *B* suspended by two vertical bars *ss*. A wooden watertight trough *T* placed below this board can be moved up and down by a winch, guided by railings just as used in theaters to lower the stage. The observer, wearing

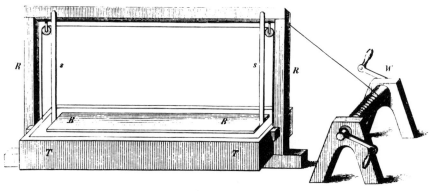

Fig. 15.

light clothes, lies down on the horizontal board, and the trough filled with lukewarm water is now slowly elevated and lowered. While the observer remains absolutely stationary the water level is alternately raised above his body and lowered under it. If the temperature of the water is properly chosen, one hardly feels this. The skin sensation is further reduced by the water soaked clothes which cling to the body.

In this experiment one senses the alternating decrease and increase of body weight as a pressure change on the supporting board which, according to Archimedes' principle, occurs with the rising and sinking of the water level. Neither with my eyes open or closed, however, was I able to observe the sensation of body movement. This latter sensation would have to occur, however, if the pressure on the lower area of the skin were to be perceived as movement. If one performs the experiment such that the body is supported by the force of one's own muscles, the negative result argues against muscle sensation as a source for movement sensation. Finally the assumption that the blood mass itself could contribute to the perception of locomotion is also

welcher an zwei verticalen Stäben *ss* ein horizontales Brett *B* trägt. Unter diesem Brett befindet sich ein hölzerner wasserdichter Trog *T,* der mit Hilfe einer Winde, wie sie bei Theaterversenkungen ge-

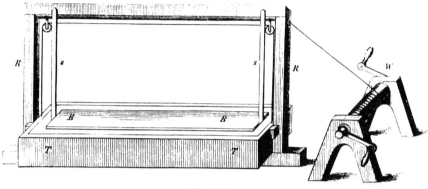

Fig. 15.

braucht wird, in einer Führung auf und ab bewegt werden kann. Der Beobachter legt sich in leichten Kleidern auf das horizontale Brett und der mit lauem Wasser gefüllte Trog wird nun langsam gehoben und gesenkt. Während der Beobachter in absoluter Ruhe ist, wird das Wasserniveau abwechselnd über seinen Körper gehoben und unter denselben gesenkt. Wenn die Temperatur des Wassers gut getroffen ist, fühlt man dasselbe fast gar nicht. Die Hautempfindung wird noch durch das fortwährende Anliegen der mit Wasser angesaugten Kleider vermindert.

Man empfindet nun bei dem Versuche die abwechselnde Verkleinerung und Vergrösserung des Körpergewichtes, welche vermöge des Archimedes'schen Principes beim Heben und Senken des Wasserniveau's eintritt, als veränderten Druck auf die Unterlage. Weder bei offenen noch bei geschlossenen Augen habe ich aber die Empfindung einer Bewegung des Körpers beobachten können. Letztere Empfindung müsste aber eintreten, wenn der Druck auf die untere Hautfläche als Bewegung empfunden werden könnte. Macht man den Versuch so, dass man den Körper durch die eigene Muskelkraft stützt, so spricht der negative Erfolg gegen die Herkunft der Be-

set aside, since <u>blood loses weight</u>[7.14] during submersion in water.

If even one of these arguments were useful as an explanation of movement sensations, one should feel a sinking of the body during raising of the water level, i.e., a reduction of body weight, and a sensation of body elevation during lowering of the water level.

The experiment would be even more striking if it were performed in an otherwise similar manner in a vertical position. I was not yet able to accomplish this. However, it is already unlikely at this stage that acceleration of the blood directly leads to a movement sensation.

7.

Among the phenomena that one perceives during and after passive or active rotation of the body, those coming from the eyes play an important role. Purkyně says about this *):

"If one carefully concentrates on his visual field when starting to rotate, the seen objects at first remain relatively at rest as the eye movements permit compensation for the continuously changing spatial relations caused by the body rotation. Soon, however, the eye muscles start to <u>fatigue</u>[7.15] and become stiff, partly because of the recurrent stereotypical movements, partly because of a peculiar spastic excitation originating in the brain. The eyes no longer follow the rotation of the body to the same extent, but only intermittently, such that the objects sometimes seem to be moving and sometimes appear stationary. Finally, this struggle also ceases and the unconsciously stationary eye moves steadily with the rest of the body, while the visible objects rotate faster and faster in the opposite direction. However these objects certainly do not move steadily, since from time to time the eye becomes weakly attached to single objects, and thereby brings about a reduction of the illusory movement, but not a complete cessation.

---

*) Medical Yearbooks of the Austrian State. vol. 6 part 2. year. 1820. p. 93.

wegungsempfindungen aus Muskelempfindungen. Endlich ist auch die Annahme widerlegt, dass die Blutmasse zur Perception der Locomotion diene. Denn das Blut verliert beim Eintauchen ins Wasser an Gewicht.

Man müsste, wenn nur eines dieser Momente zur Erklärung der Bewegungsempfindungen brauchbar sein sollte, bei Erhebung des Wasserniveau's, also Verminderung des Körpergewichtes, ein Versinken, bei Senkung des Wasserniveaus ein Erheben des Körpers fühlen.

Der Versuch wäre noch schlagender, wenn er, in übrigens gleicher Form, bei verticaler Lage ausgeführt würde. So habe ich ihn noch nicht anstellen können. Doch ist es auch jetzt schon unwahrscheinlich, dass die Beschleunigung der Blutmasse direct eine Bewegungsempfindung liefert.

7.

Unter den Erscheinuugen, welche man während und nach der passiven oder activen Drehung des Körpers wahrnimmt, spielen die von den Augen herrührenden eine wichtige Rolle. Purkyně sagt hierüber *):

»Wenn man beim Anfange des Sichherumdrehens genau auf sein Gesichtsfeld achtet, so beharren die Gegenstände erst in relativer Ruhe, indem das Auge durch seine Bewegungen die sich vermöge der Drehung des Körpers immerfort ändernden Raumverhältnisse derselben ausgleicht. Bald aber fangen die Augenmuskeln an zu ermüden und starr zu werden, theils wegen der wiederholten gleichartigen Bewegungen, theils von einem eigenthümlichen krampferregenden Einflusse des Gehirns; sie folgen nicht mehr in gleichem Maasse der Drehung des Körpers, sondern nur absatzweise, indem die Gegenstände jetzt bewegt, jetzt ruhend erscheinen; endlich hört auch dieser Kampf auf, und das unwillkührlich fixirte Auge bewegt sich gleichmässig mit dem übrigen Körper, indess die sichtbaren Gegenstände nach entgegengesetzter Richtung immer schneller und schneller umlaufen, jedoch nicht durchaus gleichmässig, indem das

---

*) Medizinische Jahrbücher des österreichischen Staates. Bd. 6 Stück 2. Jahrg. 1820. S. 93.

Incidentally, the outlines of the objects seem to flow together, points elongate to lines and lines spread to surfaces, just as one observes with moving torchlights. The colors become faded, and one may predict that during very rapid artificially induced body rotation about its own axis, the elliptical surface of the visual field would change to a gray cylindrical surface. If one fixates the eyes from the very beginning of the rotation, e.g., by staring at a finger held near the eye, this replaces the previously mentioned spastic fixation [nystagmus] of the eye. Visible objects immediately rotate along with the body and their movements again become irregular only if one again releases the fixating eye."

"Once the relative movement of the visual objects caused by the rotation has reached steady state and one comes to a stop, a vertigo [illusory] movement occurs that is, under close examination, decidedly different from that discussed above, which is evoked by changes of spatial relations. Whereas movements in the first case are in the same direction and always present a new visual field, in the second case, although movements are still in the same direction, they repeatedly present one and the same visual field. Therefore, in the latter case the same objects are always brought back to their original position following repeated deviation in the same direction. The basis of vertigo movement thus is an oscillation, which one can detect by lightly touching one's own eye and by observation of someone else's eye."

"Upon stopping, one can immediately arrest even a strong vertigo movement by fixating a finger held near the eye. This phenomenon, as well as the movement of the eyeballs, which can be seen and touched, leads to the trail to understanding the conditions of visual vertigo."

"Visual vertigo[7.16] is first of all caused by a struggle between unconscious involuntary muscle reactions and voluntary conscious ones in the opposite direction. The former, as an unconscious movement, is placed

Auge noch von Zeit zu Zeit an einzelnen schwach haften bleibt, und dadurch eine Retardirung der Scheinbewegung, aber keinen völligen Stillstand hervorbringt. Uebrigens scheinen die Umrisse der Gestalten in einander zu verfliessen, die Punkte ziehn sich in Linien, die Linien in Flächen nach den Gesetzen des geschwungenen Feuerbrandes. Die Farben werden matter, und es ist vorauszusehn, dass bei einer sehr schnellen, künstlich hervorgebrachten Drehung des Körpers um seine Axe die elliptische Fläche des Gesichtsfeldes sich in eine graue Cylinderfläche umwandeln müsste. Wenn man beim Sichdrehen gleich Anfangs die Augen fixirt, z. B. durch starres Anblicken eines nahe dem Auge gehaltenen Fingers, so vertritt dies die oben bemerkte krampfhafte Fixirung des Auges, und die sichtbaren Gegenstände drehen sich sogleich der Bewegung des Körpers gemäss, und ihre Bewegung wird nur dann wieder ungleichmässig, wenn man das fixirte Auge wieder frei lässt.«

»Wenn die relative Bewegung der Gesichtsgegenstände durch die Drehung hinlänglich eingeschwungen ist, und man hält stille, so zeigt sich nun die Schwindelbewegung, die bei näherer Betrachtung von der vorigen relativen durch Aenderung der Ortsverhältnisse hervorgebrachten wesentlich verschieden ist. Wenn jene nach einer und derselben Richtung ist, und immer ein neues Gesichtsfeld darbietet, so ist diese zwar auch nach derselben Richtung, aber bei einem und demselben Gesichtsfelde; es müssen also hier dieselben Gegenstände immerfort an ihren Ort wieder zurückkehren, werden aber immerfort nach derselben Richtung abgeleitet. Der Schwindelbewegung liegt also eine Oscillation zum Grunde, wie man dies schon am eigenen Auge durch leises Betasten, und an fremden augenscheinlich bemerken kann.«

»Man kann beim Stillestehn die Schwindelbewegung, wenn sie noch so heftig wäre, sogleich aufhalten, wenn man den nahe vorgehaltenen Finger mit den Augen fixirt. Dieser Umstand, so wie auch die tastbare und sichtbare Bewegung der Augäpfel bietet einen Weg dar, den Bedingungen des Augenschwindels auf die Spur zu kommen. «

»Der Augenschwindel ist nämlich zunächst bedingt durch einen Kampf bewusstloser unwillkührlicher Muskelactionen und willkührlicher bewusster nach der entgegengesetzten Richtung. Die erstere

on the objects, and they seem to be moving, although only the eyeball moved. This may be regarded as similar to the passive movement of the eyeball, which is induced by pressing sideward with the finger, whereby objects also seem to undergo apparent motion. For example, if one attentively fixates a point and gently presses the finger against the eye[7.17] from any side, the image of the point moves out of the area of clear vision in the opposite direction; however, since attention was not voluntarily removed from the image it is seen by peripheral vision, and since attention follows the moving image its movement will also be perceived."

Purkyně then mentions phenomena that are caused by the Plateau motion afterimage and tries to explain them also by involuntary ("acquired by temporal adaptation") eye movements. We have already seen that these phenomena are certainly different in nature. Not only are these restricted to that part of the retina affected by the movement, but furthermore the movement aftereffect never occurs for objects that had been seen clearly. Rather, a shadow of movement passes over the objects.

Breuer added important new experiments to Purkyně's observations on eye movements and it seems to me that he significantly improved upon Purkyně's theory. During one active rotation, he found about ten rapid jerks[7.18] of the eyeball in the rotation direction, which can be felt by a light touch. During a head turn, the eyeball therefore at first lags behind and then suddenly moves forward.[7.19] According to Breuer these eye movements cannot be caused by a fixation attempt, because it can be observed with closed eyes or in the blind as well. It is also not caused by the inertia of the eyeball, because then this activity would be finished after the first powerful return jerk. Breuer thinks that these "compensatory" eye movements are triggered by reflexes. Concerning the influence of these eye movements on visual perception Breuer states the following:

"If one does turn the head not too fast, the objects appear stationary, as is

wird als eine bewusstlose auf die Gegenstände übertragen, und es scheinen diese in Bewegung, indess nur der Augapfel sich bewegte; sie ist der passiven Bewegung des Augapfels, in die er durch einen Seitendruck mit dem Finger versetzt wird, und wobei die Gegenstände ebenfalls scheinbar bewegt erscheinen, ganz gleich zu achten. Wenn man nämlich z. B- einen Punkt mit Aufmerksamkeit fixirt, und drückt sanft mit dem Finger das Auge von irgend einer Seite, so rückt das Bild des Punktes aus dem Orte des deutlichen Sehens nach der entgegengesetzten Richtung; da aber die Aufmerksamkeit von dem Bilde willkührlich nicht abgewendet wurde, so wird es durch indirectes Sehen festgehalten, und, weil dem bewegten Bilde die Aufmerksamkeit folgt, so wird auch seine Bewegung bemerkt.«

Purkyně erwähnt hierauf Erscheinungen, welche auf das Plateau'sche Bewegungsnachbild zurückzuführen sind, und versucht dieselben ebenfalls durch unwillkührliche (»durch temporelle Eingewöhnung erworbene«) Augenbewegungen zu erklären. Wir haben schon gesehn, dass diese Erscheinuugen durchaus anderer Natur sind. Nicht nur sind dieselben auf die durch die Bewegung afficirte Netzhautstelle beschränkt, sondern es tritt auch nie die Bewegung deutlich gesehener Objecte als Nacherscheinung auf. Es zieht vielmehr ein Schatten von Bewegung über die Objecte hin.

Breuer hat den Purkyně'schen Beobachtungen über die Augenbewegung neue wichtige Versuche hinzugefügt und wie mir scheint, die Purkyně'sche Theorie wesentlich verbessert. Er findet bei einer activen Umdrehung etwa ein zehnmaliges Rucken des Bulbus in der Drehrichtung, welches sich durch leises Betasten constatiren lässt. Der Bulbus bleibt also bei Kopfdrehungen anfangs zurück und rückt dann plötzlich nach. Von einem Fixirbestreben kann diese Augenbewegung nach Breuer nicht herrühren, denn sie lässt sich in gleicher Weise bei geschlossenen Augen und an Blinden beobachten. Auch von der Trägheit des Augapfels rührt sie nicht her, denn sonst müsste dieses Spiel mit dem ersten gewaltsamen Nachrücken ein Ende haben. Breuer hält diese »compensirenden« Augenbewegungen für reflectorisch ausgelöst. Ueber den Einfluss dieser Augenbewegungen auf die Gesichtswahrnehmungen sagt Breuer Folgendes:

»Dreht man nämlich den Kopf nicht zu rasch, so sieht man

commonly known. However, if one stabilizes the (open) eyes by finger pressure, and thereby blocks or at least reduces such compensatory movement, these objects seem to move in a direction opposite to that of the head rotation. With the eyes closed and the head rotating back and forth I also see afterimages that remain motionless in place; however, if I produce a pressure figure by finger pressure, which of course immobilizes the eyeball as well, and turn my head, the figure follows all movements of the head in space. (Afterimages disappear too rapidly for me to observe them with ordinary finger pressure immobilization)."

"With very rapid head rotation the objects seem to move. This appears to me as proof that the compensating movement is an active one. With rapid movements the compensating muscle action cannot develop.[7.20] The rotational moment tending to keep the eye in place, however, should then have its greatest effect, such that both eyeballs would remain stationary, but especially in this case, they rotate with the head."

"If I attach a vertically lined paper to a Helmholtz's bite board and rotate the head, e.g., to the right, while the teeth hold the board, the lines make an illusory movement in the same direction, whereas the immediately visible afterimages dissociate from them and seem to remain in place."

Breuer develops the following theory of eye vertigo based on these experiments. During clockwise rotation of the head the eyes lag behind, not because of inertia and not by any attempt to fixate, but rather by a reflex-activated muscle action (from the semicircular canals). They make a slow movement opposite to the head movement, interrupted at times by jerky movements in the direction of the head movement. Because the eyes lag behind, the images maintain their position on the retina and visual objects retain their position in space, while we receive absolutely no visual impression from the jerky return movements. After prolonged clockwise movement the eyeball movements cease. The eyes keep their position relative to the head and the objects seem to move in the opposite direction because of the shift of their image on the retina. With a sudden stop of the clockwise rotation the movements of the eyeball reappear. But now the eye

bekanntlich die Objecte ruhig; sie scheinen aber die der Kopfdrehung entgegengesetzte Bewegung zu machen, wenn man die
(offenen) Augen durch den aufgedrückten Finger fixirt und dadurch
jene compensirende Bewegung hindert oder doch vermindert. Eben
so sehe ich Nachbilder bei geschlossenen Augen und hin und her
sich drehendem Kopfe, unverrückt ihren Ort bewahren; erzeuge
ich mir dagegen durch Fingerdruck die Druckfigur, wobei natürlich
der Bulbus fixirt wird, und drehe den Kopf, so macht die Figur
alle Bewegungen desselben im Raume mit. (Nachbilder schwinden
mir bei ordentlich fixirendem Fingerdruck zu rasch, als dass ich
dies an ihnen beobachten könnte).«

»Auch das scheint mir ein Beweis dafür, dass die compensirende
Bewegung eine active ist, dass bei sehr rascher Kopfdrehung die
Objecte sich zu bewegen scheinen; bei rascher Bewegung kann sich
nun zwar die compensirende Muskelaction nicht entwickeln, jenes
zurückhaltende Drehungsmoment aber müsste dabei am stärksten
wirksam sein; es müssten die Bulbi also dabei ruhig bleiben, aber
eben da drehen sie sich mit dem Kopfe.«

»Habe ich auf das Helmholtz'sche Visirbrettchen ein vertical
linirtes Papier befestigt und drehe nun, während die Zähne das
Brettchen festhalten, den Kopf z. B. nach rechts, so machen die
Linien eine scheinbare Bewegung eben dahin, während die momen-
tan sichtbaren Nachbilder derselben, sich von ihnen abspaltend, am
Platze zu bleiben scheinen. «

Breuer schöpft nun aus den Versuchen folgende Theorie des
Augenschwindels. Bei der Uhrzeigerdrehung des Kopfes bleiben
die Augen nicht vermöge der Trägheit und nicht durch ein Fixirbestreben, sondern durch eine reflectorisch (von den Bogengängen
aus) erregte Muskelaction theilweise zurück, führen also eine langsame Gegenbewegung aus, worauf sie dann immer ruckweise der
Bewegung des Kopfes folgen. In Folge dieses Zurückbleibens behalten die Bilder ihren Ort auf der Netzhaut und die Gesichtsobjecte
ihre Stelle im Raume, während wir von den ruckweisen Bewegungen überhaupt keinen Gesichtseindruck. erhalten. Nach längerer
Dauer der Uhrzeigerbewegung hören nun die Bulbusbewegungen
auf, die Augen bewahren ihre relative Lage gegen den Kopf und
die Gegenstände scheinen nun vermöge der Verschiebung ihrer Bilder

rotates slowly in a clockwise direction and returns in jerks in the reverse direction. This movement can neither be explained by an attempt to fixate nor by temporal adaptation, as Purkyně suggests. However, according to Breuer it can be explained if one assumes that at the end of rotation the same reflex mechanism becomes active, and follows the same rules as at the beginning of the rotation. As a result of the retinal image slip objects appear to rotate slowly in the opposite direction due to the slow clockwise movement of the eyeball, whereas we again receive <u>no visual impression</u>[7.21] during rapid gaze shifts. Upon stopping, therefore, all objects seem to rotate in a counterclockwise direction, without changing their position. This exactly describes the phenomenon of visual vertigo. To me the form of the theory given by Breuer seems to be the only acceptable one that explains visual vertigo on the basis of eye movements.

8.

When I issued my first communication on the subject of this publication, I had a different view of visual vertigo. The description that Helmholtz gave about visual vertigo did not seem to correspond to the facts and his explanation seemed farfetched. I had already learned from our own earlier investigations how much one was tempted to explain by means of eye movements, which was totally unrelated, e.g., Plateau's motion afterimage. Therefore, I assumed from the very beginning that eye movements do not cause visual vertigo. I was strengthened in this opinion by experiments. In the paper box of my rotation apparatus there was a black cross on a white background that reveals any gaze shift by a sharp afterimage. If I fixated this cross during the rotation experiments I obtained no afterimage from gaze movements, but still observed the previously described phenomena of visual vertigo. In my experience visual vertigo does not disappear with fixation. However, Dr. Breuer wrote to me that with fixation visual vertigo becomes weaker and the nystagmic movements then become very small and can be difficult to observe. I think that I sometimes, but not always, noticed a decrease of visual vertigo upon repetition of

auf der Netzhaut die entgegengesetzte Drehung auszuführen. Beim plötzlichen Anhalten der Uhrzeigerdrehung tritt die Bulbusbewegung wieder auf. Nun dreht sich aber der Bulbus langsam im Sinne des Uhrzeigers und kehrt ruckweise in umgekehrter Richtung zurück. Diese Bewegung lässt sich nun weder durch ein Fixirbestreben noch durch temporelle Eingewöhnung erklären, wie Purkyně will. Sie erklärt sich aber nach Breuer, wenn man beim Anhalten der Rotation denselben Reflexmechanismus und nach denselben Regeln als wirksam annimmt, wie beim Beginn der Drehung. Die Gegenstände scheinen nun bei der langsamen Bewegung des Bulbus im Sinne des Uhrzeigers vermöge der Verschiebung der Netzhautbilder eine langsame Gegendrehung auszuführen, während wir von der raschen Blickbewegung wieder keinen Gesichtseindruck erhalten. Daher scheinen beim Stehenbleiben alle Gegenstände sich dem Uhrzeiger entgegen zu drehn, ohne doch von der Stelle zu kommen, worin eben das Phänomen des Augenschwindels besteht. Erst in der Form, welche die Erklärung des Augenschwindels aus Augenbewegungen durch Breuer erhalten hat, scheint mir die Theorie annehmbar.

8.

Als ich meine erste Mittheilung über den Gegenstand dieser Schrift publicirte, hatte ich über den Augenschwindel eine andere Ansicht. Die Beschreibung, welche Helmholtz von dem Augenschwindel gegeben, schien mir den Thatsachen nicht zu entsprechen und seine Erklärung sehr gezwungen. Es war mir ferner schon aus frühern eigenen Untersuchungen bekannt, wie viel man aus Augenbewegungen zu erklären pflegte, was damit in gar keinem Zusammenhange steht, so z. B. das Plateau'sche Bewegungsnachbild. Ich hatte also von Anfang an die Vermuthung, dass der Augenschwindel nicht durch Augenbewegungen bedingt sei. In dieser Meinung wurde ich durch den Versuch bestärkt. In dem Papierkasten meines Rotationsapparates befand. sich auf weissem Grunde ein schwarzes Kreuz, welches jede Blickschwankung durch ein scharfes Nachbild verräth. Wenn ich nun bei den Rotationsversuchen dies Kreuz fixirte, so erhielt ich keine von Blickschwankungen herrührenden Nachbilder und beobachtete gleichwohl die schon beschriebenen Phänomene des Augenschwindels. Ich kann

experiments. I will therefore not claim that Breuer's theory is incorrect. I will explain below why it appears to me that it is not sufficient alone and without modification.

I observed the nystagmic movements[7.22] described by Breuer in the following manner. I produce a strong and lasting afterimage with a piece of flaming magnesium wire. If I now rotate actively about the body axis in a clockwise direction (as seen from above), I notice how the afterimage sticks on an object and then suddenly catches up. The eyes rotate slowly counterclockwise and jerk back clockwise. This movement gradually weakens until finally the afterimage rotates with the observer. After stopping, the afterimage[7.23] moves slowly across the objects clockwise, interrupted by jerklike movements in the opposite direction. The objects then move counterclockwise.

If I rotate passively in the paper box with eyes closed the nystagmic movements are strong, as can be made evident by touching the eyeballs. With eyes open, however, the displacements of the afterimage are much smaller than with free active rotation. If I produce a pressure figure while rotating with eyes fixated, then that figure rotates with me, just as Breuer also observed. However, in this case I clearly feel the attempts of the eyeball to rotate.

It would be very natural to assume that every movement of an image on the retina that is not brought about by a conscious eye movement is seen as an object movement in space. Even from the last experiments it became evident that stationary images on the retina also can appear to move. How else could one understand that the pressure figure of the immobilized eye

also nach meinen Erfahrungen nicht behaupten, dass beim Fixiren der Augenschwindel wegfällt. Freilich hat mich Herr Dr. B r e u e r brieflich aufmerksam gemacht, dass beim Fixiren der Augenschwindel doch schwächer wird und dass die nystagmatischen Bewegungen dann sehr klein und schwer beobachtbar werden können. Eine Abschwächung des Augenschwindels glaubte ich bei wiederholten Versuchen zuweilen wahrzunehmen, doch nicht immer. Ich will ' also nicht behaupten, dass die B r e u e r'sche Theorie unrichtig ist. Warum sie mir allein und ohne Modification nicht auszureichen scheint, will ich dann gleich angeben.

Die von B r e u e r beschriebenen nystagmatischen Bewegungen habe ich auf folgende Weise beobachtet. Mit einem Stückchen brennenden Magnesiumdraht erzeuge ich mir ein starkes und dauerndes Nachbild. Drehe ich mich nun activ (von oben gesehn) im Sinne des Uhrzeigers um die Körperaxe, so sehe ich wie das Nachbild immer auf einem Gegenstande hängen bleibt und dann plötzlich nachrückt. Die Augen drehn sich langsam dem Uhrzeiger entgegen und ruckweise dem Uhrzeiger nach. Diese Bewegung wird allmälig schwächer, bis endlich das Nachbild sich einfach mit dreht. Beim Stehnbleiben geht das Nachbild im Sinne des Uhrzeigers langsam über die Gegenstände hin und macht darauf immer ruckweise die verkehrte Bewegung. Die Gegenstände laufen hiebei der Uhrzeigerbewegung entgegen.

Führe ich nun im Papierkasten eine passive Drehung aus, so sind bei verschlossenen Augen die nystagmatischen Bewegungen heftig, wie man sich durch Betasten der Bulbi überzeugen kann. Bei offenen Augen fallen aber die Verschiebungen des Nachbildes im Kasten bei passiver Drehung sehr viel geringer aus als bei freier activer Drehung. Erzeuge ich, die Augen fixirend, bei passiver Drehung eine Druckfigur, so dreht sich diese wie bei B r e u e r auch bei mir mit. Ich fühle aber hiebei deutlich die Anstrengungen des Bulbus zu Bewegungen.

Es wäre nun sehr natürlich anzunehmen, dass jede Bewegung eines Bildes auf der Netzhaut, welche nicht durch eine bewusste Augenbewegung hervorgebracht ist, als Bewegung des Objectes im Raume gesehn wird. Allein aus den letzten Versuchen geht doch hervor, dass auch auf der Netzhaut ruhende Bilder bewegt erscheinen können.

rotates with oneself as Breuer had also observed? Might the innervation of the eye muscles alone induce the image of movement? Without trying to give an explanation, I believe I can best characterize these relations by postulating that visual space is related to a second space that we construct from our movement perceptions.

The following experiment makes it clear that nystagmic movements of the eyes during rotation have nothing to do with inertia. We sit in the paper box that can be easily rotated inside the fixed frame of our rotation apparatus. If one performs a rapid rotation, approximately one revolution every two seconds, and after the rotation had become constant directs one's gaze quickly up and down, of course, without moving the head, then one notices absolutely no movement on the walls of the box. However, in this situation the eye behaves like a Fessel flywheel machine, and therefore involuntary eye movements and their associated movement sensation would necessarily occur if inertia and relatively frictionless movement of the eyes were playing a role.

9.

Apparently, visual sensations can be modified by motion sensation. Conversely, motion sensation can also be initiated by visual means. This can be demonstrated by some experiments.

Experiment 1. Consider a horizontal circular wooden disk of $1\frac{1}{2}^m$ diameter placed above the head of the observer that can be rotated about an axis that passes near the observer's head. Around the rim of that disk we attach a $\frac{1}{2}^m$ broad paper stripe covered on the inside with equidistant vertical lines. The observer therefore sits inside a hollow, turning, lined cylinder. When this drum alone is kept turning for a few minutes one quickly notices himself turning in the direction opposite to the drum[7.24], along with all those things that are not hidden by the drum. However, on occasion one feels quickly returned to rest with the drum still turning, and the two subjective states alternate frequently. After several repetitions of this experiment it

Wie anders soll man es verstehn, wenn die Druckfigur des fixirten Auges sich mitdreht, wie ja auch B r e u e r beobachtet hat. Sollten die blossen Innervationen der Augenmuskel diese Bewegungsvorstellung zu Stande bringen? Ohne damit eine Erklärung geben zu wollen, glaube ich dies Verhältniss doch am besten so zu bezeichnen, dass ich sage, d e r o p t i s c h e R a u m w i r d a u f e i n e n z w e i t e n R a u m b e z o g e n, d e n w i r a u s u n s e r n B e w e g u n g s e m p f i n d u n g e n c o n s t r u i r e n.

Dass die nystagmatischen Bewegungen des Auges bei Drehungen des Körpers nichts mit der Trägheit zu schaffen haben, wird noch durch folgenden Versuch klar. Wir setzen uns in den Papierkasten, welcher für sich in dem festgestellten Rahmen unseres Rotationsapparates leicht gedreht werden kann. Führt man nun die Umdrehung recht rasch aus, etwa eine in zwei Secunden, und blickt man mit den Augen, nachdem die Drehung gleichförmig geworden, rasch auf und ab, natürlich ohne den Kopf zu bewegen, so bemerkt man an den Wänden des Kastens gar keine Bewegung. Hiebei verhält sich aber das Auge wie eine F e s s e l 'sche Schwungmaschine und es müssten folglich nicht beabsichtigte Augendrehungen und damit subjective Bewegungserscheinungen eintreten, wenn die Trägheit und leichte Beweglichkeit der Augen eben eine Rolle spielen würde.

9.

Wie es scheint, können also optische Empfindungen durch Bewegungsempfindungen modificirt werden. Umgekehrt werden aber auch Bewegungsempfindungen auf optischem Wege angeregt. Dies soll durch einige Experimente demonstrirt werden.

V e r s u c h 1. Wir denken uns über dem Kopfe des Beobachters eine horizontale kreisformige Holzscheibe von $1\frac{1}{2}^{m}$ Durchmesser auf einer nahe am Kopfe des Beobachters vorbeigehenden verticalen Axe drehbar. Um den Rand dieser Scheibe legen wir einen $\frac{1}{2}^{m}$ breiten Papierstreifen, welcher auf der Innenseite mit aequidistanten verticalen Linien überzogen wird. Der Beobachter sitzt also im Innern eines hohlen drehbaren linirten Cylinders. Wird diese Trommel allein einige Minuten lang in Drehung gehalten, so meint man bald, sich mit jenen Gegenständen, welche die Trommel nicht verdeckt, verkehrt zu drehen, bald glaubt man wieder zu ruhen, während die Trommel

seemed to me as if the peripheral part of the visual field were placed in motion most easily. This is good evidence that the phenomenon can be explained by Breuer's principle. However, the subjective process is hardly limited to the optical part. I, at least, cannot suppress some movement sensation.

Many various forms of this phenomenon are well known, as when one stands on a bridge above flowing water, or if one observes several <u>moving railway</u>[7.25] cars from a motionless one, whereby the visual field is partitioned into several parts, each having different motion.

Experiment 2. We place the paper drum over the observer in the rotation apparatus as previously described. Now we start the apparatus turning with a slight acceleration, such that friction will keep the drum stationary relative to the apparatus. When we stop the apparatus suddenly the drum retains its final velocity (according to the principle of conservation of areas), which it only gradually loses. The observer perceives the drum as stationary and feels himself turning equally fast in the opposite direction. If the apparatus is initially turned clockwise, as seen from above, then after the stop the observer feels himself and all stationary objects turning oppositely, in a counterclockwise direction, whereas the drum, which actually is turning clockwise, <u>appears to be stationary</u>.[7.26] Therefore, an apparent motion can be compensated by a true motion. If the drum loses its velocity too rapidly it takes on the apparent motion of the observer. It would therefore be possible to measure the intensity and persistence of the apparent motion. It is noteworthy that the <u>powerful nausea</u>[7.27] which normally accompanies a stop after a strong rotation does not occur in this experiment.

Experiment 3. The compensation of a subjective movement by an objective one is a special case of the combination of the two kinds of movements, an example of which might be the following. If in the course of the preceding experiment the head is nodded down to the chest during the rotation, and after the stopping of the apparatus is brought up again, then the

sich dreht und beide subjective Zustände wechseln oft. Es schien mir bei mehrmaliger Wiederholung des Versuches, als ob der nicht fixirte Theil des Gesichtsfeldes am leichtesten in Bewegung gerathen würde. Dies würde einen guten Anhaltspunkt für die Erklärung der Erscheinung nach dem Breuer'schen Princip geben. Doch wird der subjective Vorgang kaum mit dem optischen Theil erschöpft sein. Ich kann mich wenigstens eines Bewegungsgefühls nicht erwehren.

In der manigfaltigsten Form sieht man solche Erscheinungen bekanntlich, wenn man auf einer Brücke über fliessendem Wasser steht oder von einem ruhenden Eisenbahnzuge aus mehrere bewegte beobachtet, wobei also das Gesichtsfeld in mehrere Theile von verschiedener Bewegung abgetheilt ist.

Versuch 2. Bringen wir die Papiertrommel in der angegebenen Weise über dem Beobachter im Rotationsapparat an. Setzen wir nun den Apparat mit geringer Beschleunigung in Drehung, so bleibt die Trommel wegen der Reibung in relativer Ruhe gegen den Apparat. Halten wir dann den Apparat plötzlich an, so behält die Trommel (nach dem Princip der Erhaltung der Flächen) ihre Endwinkelgeschwindigkeit bei, welche sie nur allmälig verliert. Der Beobachter hält dann die bewegte Trommel für ruhend und sich selbst fühlt er in desto schnellerer Gegendrehung. Drehte sich der Apparat anfänglich, von oben gesehn, wie der Uhrzeiger, so meint der Beobachter nach dem Anhalten sich und alle festen Gegenstände verkehrt wie den Uhrzeiger gedreht zu sehn, mit Ausnahme der Trommel, welche sich wirklich wie der Uhrzeiger dreht und die zu ruhen scheint. Die scheinbare Bewegung lässt sich also durch eine wirkliche compensiren. Verliert die Trommel ihre Geschwindigkeit zu rasch, so nimmt sie dann an der Scheinbewegung des Beobachters Theil. Es liesse sich so die Intensität und Nachdauer der Scheinbewegung messen. Merkwürdig ist, dass bei diesem Versuch das gewaltige Ekelgefühl, welches der Sistirung starker Rotationen zu folgen pflegt, nicht eintritt.

Versuch 3. Die Compensation einer subjectiven Bewegung durch eine objective ist ein specieller Fall der Zusammensetzung beider Bewegungsarten, von welcher hier ein Beispiel folgen mag. Wenn man im vorigen Versuch bei der Drehung den Kopf auf die Brust vorneigt und nach dem Anhalten des Apparates

87

true rotation of the drum about a vertical axis is combined with the subjective rotation about a horizontal axis to yield a rotation about a tilted axis. The results of experiments 2 and 3 may also be explained, fully or in part, by the Purkyně-Breuer principle.

10.

We cannot ignore Hitzig's[7.28] related observations of eye movements during galvanic stimulation of the head. The following description is in his words.

"The induced movements in healthy persons are nearly always being binocular and are best compared to a pattern called nystagmus. Particularly with weak current intensities, one can always clearly differentiate a fast jerky movement to one side and a slower one to the other side. In some individuals, at a certain stimulus intensity, the iris moves like a fisherman's float, which slowly drifts in a river until it is suddenly snatched back in the opposite direction by the line. With increasing current intensity the rhythm becomes faster and faster, until finally the direction of the short jerky movement dominates and, with very large currents, the eyeball will be driven to small oscillations in the corner of the eye[7.29]."

"The direction of the individual movements – and this is one of the most interesting aspects of the whole question – depends upon the choice of the inducing current, such that the faster jerky movements, which we will consider alone to keep things simple at first, always occur in the direction of the positive current, and the slower ones in the opposite direction. If therefore the anode is placed on the right, and the cathode on the left mastoid, the jerk goes to the left, and with strong currents both eyeballs will be kept in the left corners."

Evidently these eye movements are the same as those observed during rotational vertigo, and one can assume that the movement perception[7.30] induced by the current triggers them. Breuer explains these movements also by stimulation of the semicircular canals. Assuming that it was also demonstrated that the semicircular canals are end organs for the perception

aufrichtet, so combinirt sich die wirkliche Axendrehung der Trommel um eine Verticalaxe mit der subjectiven des Beobachters um eine Horizontalaxe zu einer Drehung um eine schief liegende Axe. Auch die Ergebnisse der Versuche 2 und 3 lassen sich vielleicht ganz oder theilweise nach dem Purkyně - Breuer'schen Princip erklären.

10.

Wir können hier nicht die von Hitzig beim Galvanisiren des Kopfes beobachteten Augenbewegungen übergehn. Die Beschreibung möge mit seinen eigenen Worten folgen.

»Ihrem Character nach sind die so an Gesunden hervorgebrachten Bewegungen fast immer associirte und lassen sich am Besten mit der Nystagmus genannten Affection vergleichen. Nur unterscheidet man hier immer deutlich, namentlich bei geringeren Stromintensitäten, eine schnell ruckartig ausgeführte Bewegung nach der einen Seite und eine langsamere nach der andern Seite. Bei manchen Individuen gleicht unter einer bestimmten Reizgrösse die Iris dem Schwimmer eines Angelfischers, der langsam auf einem Flusse dahintreibt, bis er plötzlich an der Leine in entgegengesetzter Richtung zurück ge-rissen wird. Bei zunehmender Stromintensität wird der Rhythmus schneller und schneller, bis endlich die Richtung der kurzen rucken-den Bewegung dominirt und der Bulbus bei sehr starken Strömen nur noch leise oscillirend im Augenwinkel festgehalten wird.«

»Die Richtuug der einzelnen Bewegungen – und dies ist einer der interessantesten Punkte der ganzen Frage – hängt derart von der Wahl der Einströmungsstellen ab, dass die schnellere ruckende Bewegung, die wir der Einfachheit wegen zunächst allein berück-sichtigen werden, immer in der Richtung des positiven Stromes er-folgt, die langsamere in der entgegengesetzten Richtung. Wenn sich also die Anode in der rechten und die Kathode in der linken Fossa mastoidea befindet, so erfolgt der Ruck nach links, und bei starken Strömen werden beide Bulbi in den linken Winkeln festgehalten.«

Wie man sieht, sind diese Augenbewegungen dieselben wie die beim Drehschwindel beobachteten, und es liegt die Vermuthung nahe, dass die durch den Strom hervorgebrachte Bewegungsempfindung sie auslöst. Breuer führt auch diese Bewegungen auf Reizung der Bogengänge zurück. Gesetzt aber es wäre auch bewiesen, dass die

of movement, then undoubtedly all phenomena which can be stimulated by the end organs can also result from direct stimulation of the central organ. Hitzig thinks such stimulation of the central organ is responsible for the eye movements and believes that it can at least not be excluded by means of this experiment. On the other hand, Flourens' experiment yielded very similar eye movements, providing further support for Breuer's viewpoint. In my opinion the question cannot yet be decided.

11.

I have already pointed out the difficulties encountered if one tries to explain all optical illusory movements [illusions] by means of eye movements. Further insight into these difficulties comes from the following observations which <u>Rollmann</u>[7.31] and I*) made at about the same time, and which were probably often seen in other forms much earlier. If one looks at the wire figure of a <u>cube</u>[7.32] monocularly, as <u>Plateau</u>[7.33] did to demonstrate the equilibrium states of fluids, then corner a can either be seen as convex or concave. This is because the monocular retinal image is a perspective picture that can be interpreted in different ways. If one imagines the edge ab closer than cd at the intersection s, then a seems to be convex, in the opposite case a appears as a concave corner. If a in reality is e.g., a convex corner, while one sees it as concave by inverting the image, which is especially easily achieved with a black wire figure in front of a light background,

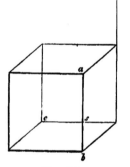

Fig. 16.

*) Mach, Observations about monocular stereopsis. Proc. Vienna Acad. vol. 58.

Bogengänge End organe für die Bewegungsempfindung sind, so könnte doch nicht bezweifelt werden, dass alle Vorgänge, welche von den End organen aus erregt werden können, auch durch directe Affection des Centralorgans entstehen können. Eine solche Affection des Cen-. tralorgans hält nun Hitzig für die Ursache der Augenbewegungen und sie lässt sich bei der Form des Versuches wenigstens nicht ausschliessen. Anderseits werden beim Flourens'schen Versuch ganz ähnliche Augenbewegungen erregt und die Breuer'sche Ansicht erhält dadurch eine gute Stütze. Die Frage kann, wie ich glaube, noch nicht entschieden werden.

11.

Ich habe schon auf die Schwierigkeit hingewiesen, der man bei dem Versuch begegnet, alle optischen Scheinbewegungen aus Augenbewegungen zu erklären. Einen Einblick in diese Schwierigkeit wird man noch durch folgende Beobachtungen erhalten, welche in der hier beschriebenen Form fast gleichzeitig von Rollmann und mir*), in anderen Formen aber wohl wiederholt schon viel früher gemacht worden sind. Wenn man das Kantengerüst eines Würfels aus Draht, wie es Plateau zur Darstellung seiner Gleichgewichtsgestalten der Flüssigkeiten benutzt, monocular betrachtet, so steht es frei, die Ecke *a* als eine convexe oder concave zu sehen. Denn das monoculare Netzhautbild ist ein perspectivisches Bild, welches eine verschiedene Auslegung zulässt. Stellt man sich an dem Durchschnittspunkt *s* die Kante *ab* näher vor als *cd*, so erscheint *a* als convexe, im umgekehrten Fall als concave Ecke. Wenn nun *a* in Wirklichkeit z. B. eine convexe Ecke ist, während man es als eine concave sieht, wenn man also das Bild invertirt, was besonders leicht bei einem schwarzen Drahtgerüst vor einem hellen Grund gelingt, so sieht man jede

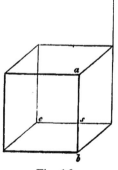

Fig. 16.

then one perceives every rotation of the wire figure as going in the reverse direction. This requires no special explanation.

However, the following phenomenon is surprising. One places a Plateau grid on the table, looks at it monocularly and moves the head sideways, say from right to left. The grid seems to be rigid and motionless. If we invert the grid then it seems to rotate with every movement of the head. The phenomenon is even more pleasing if one connects two grids with handles and puts them on the table with one of them inverted. The inverted part rotates with the observer's head, while the other one remains still.

This now seems to be very noteworthy, since it demonstrates the precise relationship between visual perception and even the smallest movement of the body, particularly the head. Both images change on the retina, but only one of these changes is seen as movement. Eye movements play no part[7.34] in this phenomenon. Not only do the eyes have no reason to rotate perceptibly with such a movement of the head, but one can also convince oneself by moving the eyeball with the finger, in which case the whole monocular image is shifted, but the object, of course, does not appear to rotate. The different parts of the retinal image undergo an unequal shift and, when interpreting this shift, the head movements and not the eye movements are taken into account.

With the head and the eyeball both motionless, one perceives the shift of the retinal image in a certain direction as a space displacement in the opposite direction that becomes larger the more distant the object is placed. However, if at the same time the head is moved in the direction of the image motion, the fraction of the head movement subtracted from the object shift is smaller the further away the object is placed. A retinal image can therefore undergo a certain change without resulting in the perception of object motion. It again appears almost as though a special perceptual space[7.35] exists that is different from the visual space. This also gets support from the fact that one can easily imagine a visual scene or an object behind one's own back, where the visual field never reaches.

Drehung, die mit dem Drahtkörper vorgenommen wird, verkehrt vor sich gehn. Dies bedarf keiner besondern Erklärung.

Ueberraschend ist aber folgende Erscheinung. Man lege ein Plateau'sches Netz auf den Tisch, betrachte es monocular und bewege den Kopf hin und her, etwa von rechts nach links. Das Netz erscheint fest und unbeweglich. Invertiren wir aber das Netz, so scheint es bei jeder Bewegung des Kopfes eine Drehung auszuführen. Noch hübscher ist die Erscheinung, wenn man zwei Netze mit den Stielen verbindet, auf den Tisch legt und das eine invertirt. Der invertirte Theil dreht sich mit dem Kopfe des Beobachters, der andere liegt ruhig.

Dies scheint nun desshalb sehr merkwürdig, weil es zeigt, welcher exacte Connex der Gesichtswahrnehmungen mit den leisesten Bewegungen des Körpers, namentlich des Kopfes besteht. Beide Bilder auf der Netzhaut verändern sich, doch wird nur die eine Veränderung als Bewegung gesehn Die Augenbewegungen haben an der Erscheinuug keinen Antheil. Nicht nur, dass die Augen bei der beschriebenen Bewegung des Kopfes keinen Anlass haben, eine merkliche Drehung auszuführen, man kann sich auch durch Bewegung des Bulbus mit dem Finger überzeugen, dass bei dieser das ganze monoculare Bild verschoben, der Gegenstand aber, wie natürlich, nicht gedreht erscheint. Die verschiedenen Theile des Netzhautbildes erfahren hier eine ungleiche Verschiebung und bei Auslegung dieser Verschiebung wird nicht die Augen-, sondern die Kopfbewegung in Rechnung gezogen.

Bei ruhigem Kopfe und ruhigem Bulbus sieht man die Verschiebung des Netzhautbildes in einem bestimmten Sinne als eine desto grössere entgegengesetzte Raumverschiebuug, in je grössere Ferne man das Object versetzt. Wird aber gleichzeitig der Kopf im Sinne der Bildverschiebung bewegt, so wird ein desto kleinerer Bruchtheil der Kopfverschiebung von der Objectverschiebung in Abzug gebracht, in je grössere Ferne man das Object setzt. Ein Netzhautbild kann also gewisse Veränderungen erfahren, ohne dass daraus die Wahrnehmung eines veränderten Objectes hervorgeht. Es sieht fast wieder so aus, als ob es einen besonderen Vorstellungsraum gäbe, der von dem Sehraum verschieden ist. Darauf würde auch der Umstand deuten, dass man sich leicht ein Spectrum oder sonst

In summary, we see that the phenomena of visual vertigo can be explained by eye movements, which are induced by reflexes stemming from <u>movement sensations</u>.[7.36] In addition, however, single phenomena remain to which this explanation does not seem to apply.

12.

Since, as we will soon see, the source of movement sensations, particularly those that cause rotational vertigo, can be shown to lie in the head, one is easily led to the explanation of trying to deduce all such sensations from direct stimulation of the brain during movement. Purkyně has already attempted to do this, and he says:

"During rotations of the body about the longitudinal axis, the movement of the brain, because it is soft, will necessarily tend to momentarily lag behind the movement of the cranium, just as we notice with a fluid whose container is rotated. The fluid elements remain stationary relative to outer [inertial] space until they are gradually carried along with the moving walls of the vessel by friction. The same cannot occur in the brain because its adhesion is too large. However, it is a non-rigid, and somewhat deformable mass and thereby displays properties of fluids. Therefore one must conclude that to the extent that the brain is more or less vigorously moved from the outside, its parts also shift relative to each other, and are pushed and pulled, however, without any discontinuity. Such pulls might then produce effects similar to real mechanical lesions, differing from those only in degree. On the basis of animal experiments, I would be inclined to believe that [mechanical] influences on the cerebellum and the brachium cause these effects, whereas the resulting stupor would have to be ascribed to the cerebral hemispheres."

ein Object hinter dem eigenen Rücken vorstellen kann, wohin doch das Gesichtsfeld nie reicht.

Alles zusammengefasst sehen wir, dass die Phänomene des Augenschwindels sich durch Augenbewegungen erklären lassen, welche reflectorisch durch die Bewegungsempfindungen ausgelöst werden. Daneben bleiben aber einzelne Erscheinungen übrig, auf welche jene Erklärung nicht anwendbar scheint.

12.

Da man, wie wir gleich sehn werden, nachweisen kann, dass die Quelle der Bewegungsempfindungen, namentlich derjenigen, welche den Drehschwindel bedingen, im Kopfe liegt, so verfällt man leicht auf den Versuch, alle diese Empfindungen aus einer directen Affection des Gehirns bei Bewegungen abzuleiten. Diesen Versuch hat schon Purkyně gemacht, indem er sagt: »Bei den Drehungen des Körpers um die Längenaxe muss nothwendig das Gehim, da es weich ist, in seiner Bewegung gegen die Bewegung der Hirnschale momentan zurückzubleiben die Tendenz haben, so wie wir dies an einer Flüssigkeit bemerken, deren Gefäss gedreht wird. Die Theile der Flüssigkeit behaupten ihre Lage gegen den äussern Raum so lange, bis sie vermöge der Adhäsion an den Wänden des Gefässes in dessen Bewegung allmälig mit fortgerissen werden. Dasselbe kann nun im Gehirn nicht stattfinden, da seine Cohäsion dafür zu gross ist; da, es jedoch eine weiche, in sich bis auf einen gewissen Grad verschiebbare Masse darstellt, die in dieser Hinsicht an den Eigenschaften des Flüssigen participirt, so muss man nothwendig annehmen, dass in dem Grade, als es von aussen in mehr oder weniger heftige Beweguug versetzt wird, auch seine Theile gegeneinander, wenn gleich ohne Continuitätsveränderung in gleichem Grade verschoben und gedehnt werden. Solche Zerrungen mögen dann ähnliche Wirkungen hervorbringen, wie wirkliche mechanische Läsionen, die dann von dieser nur graduell verschieden sein werden. Ich wäre geneigt zu glauben, dass den Versuchen an Thieren gemäss auch hier zunächst die Affectionen des kleinen Gehirns und der Schenkel des grossen jene Wirkungen

This proposition is further expanded elsewhere:

"For our purpose it is sufficient to propose that during stronger prolonged movements accompanied by one or several return movements the brain is subject to a change in its cohesion, so that its parts, if not actually in movement, nevertheless have a tendency to move and are pushed one way or the other. This motion tendency outlasts the real movement by some time and is the cause of vertigo. Therefore, the direction of vertiginous movement depends on the direction in which the centrifugal force has affected the brain. The cohesion as well as the movement tendency is a more or less profound interaction of mass particles. The brain thereby senses its own mechanical state: the brain is capable of sensing; therefore, the increased mechanical distortion makes its way into consciousness as a sensation. For common body positions and movements the influence of weight is continuously sensed in each body part, and this sensation serves to regulate movement and to establish the equilibrium of the body. If, however, the weight of any [body] part, especially of the head as the most noble one that is served by all instincts of the rest of the body, is applied in a different direction, then the instinct of the whole muscular system uprises and produces involuntary movements aimed at restoring the normal position of the head and eliminating its apparent motion. Therefore, all movements associated with tactile vertigo are caused by illusory movements of brain vertigo. That the illusory motion seems real follows directly from the lack of consciousness of tactile motion sensors. Because they have no subjective basis in consciousness, they are transformed and become objective."

According to Purkyně the brain directly senses its <u>mechanical strain</u>[7.37] for just what it is, namely its movement tendency. This view is somewhat primitive, reminiscent of the position of the Ionic School, which, by the way, we encounter repeatedly with the genial Purkyně in the field of sensory

hervorbringen, indessen die dabei stattfindende Betäubung dem grossen Gehirne zuzeignen wäre.«

An einer andern Stelle wird diese Vorstellung noch weiter ausgeführt:

»Für unsern Zweck ist es hinlänglich, sich vorzustellen, dass das Gehirn bei anhaltenden heftigeren Bewegungen nach einer oder mehreren zurückkehrenden Richtungen eine Abänderung in seiner Cohäsion erleide, dass seine Theile, wenn nicht in eine wirkliche Bewegung, doch in eine Bewegungstendenz nach einer oder mehreren Seiten versetzt werden, welche Bewegungstendenz nach vorhergegangener realer Bewegung noch einige Zeit nachbleibt und Ursache des Schwindels ist. Daher die Richtung der Schwindelbewegung von der Richtung abhängt, in welcher die Schwungkraft auf das Gehirn eingewirkt hat. Die Cohäsion sowohl, als die Bewegungstendenz, ist eine mehr oder weniger innige Wechselberührung der Massentheilchen; durch sie kommt das Gehirn in Selbstberührung. Dieses ist aber ein Empfindendes: die vermehrte Selbstberührung wird also als Empfindung ins Bewusstsein treten. Zwar wird in den gewöhnlichen Stellungen und Bewegungen des Körpers die Einwirkung der Schwere in jedem Theile immer fort empfunden, und diese Empfindung dient als Regulativ der Bewegungen und der Erhaltuug des Gleichgewichts des Körpers; wenn aber die Schwere in irgend einem Theile, vorzüglich im Kopfe, als dem edelsten, dem aller Instinct des übrigen Körpers dient, eine andere Richtung erhält, so kömmt der Instinkt des ganzen Muskelsystems in Aufruhr, und bringt bewusstlose Bewegungen hervor, die die normale Lage des Kopfes herstellen, seine scheinbare Bewegung zum Stillstande bringen sollen. Daher die jedesmalige Bewegung des Tastschwindels durch die Scheinbewegung des Hirnschwindels bedingt ist. Dass die Scheinbewegung als objectiv erscheint, kommt bloss von der Unwillkührlichkeit der Tastbewegungen, die, weil sich für dieselben keine subjectiven Gründe im Bewusstsein finden lassen, ins Objective übertragen werden.«

Nach Purkyně empfindet also das Hirn die Bewegungstendenz, welche es erhält, als eben diese Bewegungstendenz. Diese Ansicht hat etwas Primitives an den Standpunkt der Jonischen Schule erinnerndes, wie wir es übrigens im Gebiete der Sinnes-

physiology. It is based on confusing the sensation with the means by which it is induced. The two are, as far as we can see, fundamentally different. Another question is whether or not there are organs of movement sensation in the brain that are specifically activated by head movements. This question remains open.

The interpretation suggested by Purkyně, more or less modified and generally in a very vague form, has been kept alive to this day. It is supported by Hitzig, who brought sound new scientific experimental evidence to bear on it, and was accepted by <u>Wundt</u>*).[7.38]

Even if one agrees that all sensations and conceptions that can originate in end organs can also be initiated in the central organ, which I will certainly not dispute, one is not satisfied by Wundt's explanation that vertigo is caused by a »functional disturbance of the cerebellum«. One encounters further difficulties in imagining a certain mechanical conception of the process that Wundt**) describes with the words: "As a consequence of rotational movement the centrifugal forces on the cerebellum will create a pressure towards the side of the movement."... Considering the complicated anatomy of the cerebellum it must be considered extremely unlikely that angular acceleration about any given axis passing through the cerebellum will always produce the sensation of rotation about that same axis. How peculiarly would the cerebellum have had to be constructed if it were to react to mechanical pressure in that way.

Aside from the inconclusiveness of the experiments that have been performed in this direction, it seems improbable to me that pressure on the brain during commonly occurring linear and angular accelerations plays a role and induces movement sensation. The angular acceleration of the brain nevertheless remains the same, whether the head is on or off the rotation axis in my apparatus. Only the pressure that depends on the centrifugal

---

*) Physiological Psychology. Leipzig 1874. p. 208.
**) As referenced, p. 211.

physiologie bei dem genialen Purkyně wiederholt antreffen. Sie beruht auf einer Verwechslung der Empfindung mit den Mitteln, durch welche die Empfindung erregt wird. Beide sind ja, so viel wir sehen, grundverschieden. Eine andere Frage ist die, ob nicht Organe der Bewegungsempfindung im Hirn durch die Kopfbewegungen selbst zu der ihnen eigenthümlichen Thätigkeit erregt werden. Diese Frage kann discutirt werden.

Die durch Purkyně angeregte Auffassuug hat sich in mehr oder weniger modificirter, meist in sehr unbestimmter Form bis auf den heutigen Tag erhalten. Sie ist in sehr wissenschaftlicher Weise von Hitzig vertreten, der neue experimentelle Stützen für dieselbe beigebracht, und von Wundt*) acceptirt worden.

Wenn man auch zugibt, dass alle Empfindungen und Vorstellungen, welche von den Endorganen aus auslösbar sind, auch im Centralorgan entstehn können, was ich gar nicht in Abrede stellen will, so wird man doch die Wundt'sche Erklärung, nach welcher der Drehschwindel durch eine »Functionsstörung des Kleinhirns« bedingt ist, nicht für befriedigend halten. Man wird ferner Mühe haben, sich eine bestimmte mechanische Vorstellung von dem Vorgange zu machen, den Wundt**) mit den Worten meint: »In Folge der Drehbewegung wird das Cerebellum durch die Centrifugalkraft einen Druck nach der Seite der Drehung erfahren«.... Bei dem complicirten Bau des Kleinhirns muss es als sehr unwahrscheinlich angesehn werden, dass die Winkelbeschleunigung um irgend eine durch das Kleinhirn gehende Axe immer die Empfindung einer Drehung um dieselbe Axe auslöst. Wie eigenthümlich müsste das Kleinhirn gebaut sein, wenn es auf mechanischen Druck so reagiren sollte.

Dass Druck auf das Gehirn bei den gewöhnlich vorkommenden Progressiv- und Winkelbeschleunigungen eine Rolle spielt und die Bewegungsempfindungen auslöst, ist mir, abgesehn von der Unbestimmtheit der Ableitungen, die man in dieser Richtung versucht hat, durch die Experimente unwahrscheinlich. Die Winkelbeschleunigung des Hirns bleibt allerdings dieselbe, ob sich in meinem Ap-

---

*) Physiologische Psychologie. Leipzig 1874. S. 208.
**) a. a. O. S. 211.

acceleration changes significantly if one steps away from the axis. However, I did not notice any significant change of the phenomenon either in this situation or with other head position changes. Direct mechanical effects on the brain might have caused certain slight sensations of pressure in the head, which were very vague and difficult to describe, however, they are hardly responsible for the specific perception of space that is activated by motion.

Hitzig*) has shown by several experiments that direct stimulation of the brain can cause fits of vertigo. Specific examples follow.

"If one opens the back of a rabbit's skull and removes the small lateral lobe of the cerebellum that corresponds to the flocculus in humans, a small cavity appears. It is formed by the tentorium that is bony in the rabbit, and into which the stalk of the flocculus intrudes. If one now puts a few fragments of ice into that cavity, or carefully injects cold water into it, the animal suddenly rights itself, makes shaking head movements and occasionally body movements, similar to those seen after the rotation experiments, and then falls to the contralateral side, while both eyes are displaced to the lesioned side and show strong nystagmus. For a while it remains lying on its side, then it suddenly jumps up and sits there quietly, just as if nothing had happened to it." ...

"Finally I impart on a long series of lesion, cutting and excision experiments on the cerebellum. The overall result substantially clarifies the state of affairs. If deep cuts severed one hemisphere such that its connections with the middle and hind brachium must have been broken for the most part, the same picture emerged as with strong galvanic stimulation; the animals rotated frantically toward the lesioned side."

Without doubt one can induce vertigo by general or localized changes of

---

*) Investigations about the Brain. Berlin 1874. p. 264.

parate der Kopf in oder ausser der Rotationsaxe befindet. Allein der von der Centrifugalbeschleunigung abhängige Druck ändert sich wesentlich, wenn man aus der Axe tritt. Mir ist aber hiebei, so wie bei andern Lageänderungen des Kopfes keine wesentliche Aenderung der Erscheinung aufgefallen. Gewisse mir wenigstens sehr unbestimmt erscheinende und schwer zu beschreibende Druckgefühle im Kopfe mögen sich aus der directen Affection des Gehirns ableiten lassen, schwerlich aber die eigenthümlichen Raumvorstellungen, die durch die Bewegung erregt werden.

Hitzig*) hat durch mehrere Versuche dargethan, dass directe Erregung des Hirns Schwindelzufälle hervorbringen kann. Wir wollen Beispiele anführen.

»Wenn man dem Kaninchen das Hinterhaupt eröffnet und ihm den kleinen Seitenlappen des Kleinhirns exstirpirt, der der Flocke beim Menschen entspricht, so bleibt eine, von dem beim Kaninchen verknöcherten Tentorium gebildete Höhle zurück, in die der Flockenstiel hineinragt. Bringt man nun in diese Höhle einige Fragmente Eis, oder spritzt vorsichtig kaltes Wasser hinein, so richtet sich das Thier plötzlich auf, macht ähnliche wackelnde Bewegungen mit dem Kopfe, manchmal auch mit dem Körper wie nach den Drehversuchen, und stürzt dann auf die entgegengesetzte Seite, während sich nun beide Augen unter heftigem Nystagmus in die Winkel der verletzten Seite stellen. Eine Weile bleibt es so liegen, dann springt es plötzlich wieder auf, und sitzt ruhig da, als wenn ihm nichts geschehen wäre.«....

»Endlich nehme ich noch eine lange Reihe von Verletzungs-, Durchschneidungs- und Exstirpationsversuchen am Kleinhirn vor, deren Gesammtresultat zur fernern Aufklärung des Sachverhaltes wesentlich beiträgt. Wenn tiefgehende Schnitte die eine Hemisphäre der Art trennten, dass ihre Verbindungen mit dem mittleren und hinteren Schenkel zum grösseren Theile unterbrochen sein mussten, so entstand dasselbe Bild, wie bei starker galvanischer Reizung, die Thiere rotirten mit rasender Vehemenz nach der verletzten Seite.«

Ohne Zweifel lassen sich also durch allgemeine oder locale Aen-

---

*) Untersuchungen über das Gehirn. Berlin 1874. S. 264.

the functional state of the cerebellum. This is not surprising. For the cerebellum to sense every angular acceleration about an axis as rotation about that same axis, however, is so unlikely a priori that it cannot be assumed without very compelling evidence. This is about as likely as the expectation that a picture would be seen clearly if a lens projected it onto the surface of the brain.

The participation of direct brain stimulation in the phenomena of vertigo cannot entirely be ruled out. However, even if it were accepted, it could not explain the details of the phenomena.

13.

The location of the movement sensation organ in the head had already been apparent on the basis of Purkyně's earlier experiments involving the change of head position following cessation of rotation. However, one can object to the conclusion drawn from that experiment. One can consider, e.g., that during body acceleration the neck muscles have to maintain head position, and that even after a changed head position the respective muscle sensations continue in a modified way. Even though this opinion might seem farfetched, the possibility exists. Therefore, it seemed appropriate to examine the influence of head position on the phenomena of vertigo more precisely.

Experiment 1. We sit in the rotation apparatus, with the head fixed by means of a four-sided wooden cap placed in a railing between two boards. By this arrangement the head can only be nodded forward or backward, but is otherwise prevented from rotation or tilt to the side, while the back is resting securely against a support. If the apparatus is now placed in rotation and the head is pitched, the same subjective phenomena occur during or after rotation with the same strength as described in the earlier experiments, as if the head were free to move.

Experiment 2. We place a four-sided wooden cap on the head, with the two side parts extending almost to the shoulders on the right and left sides. Then we sit on a chair with the back firmly supported and make

derung der Zustände des Kleinhirns Schwindelerscheinungen hervorbringen. Dies hat nichts Befremdendes. Dass aber das Kleinhirn jede Winkelbeschleunigung um eine Axe als Drehung um dieselbe Axe empfinden soll, ist a priori so
unwahrscheinlich, dass es ohne sehr stichhaltige Beweise nicht angenommen werden kann. Mit gleichem
Rechte könnte man erwarten, das mit einer Linse auf die Hirn-
oberfläche entworfene Bild werde deutlich gesehen.

Die Mitbetheiligung der directen Hirnerregung bei den Erscheinungen des Drehschwindels lässt sich zwar nicht absolut ausschliessen, kann aber auch, wenn sie angenommen wird, die Details
der Erscheinungen nicht erklären.

13.

Der Sitz des Organs der Bewegungsempfindungen im Kopfe
wird eigentlich schon durch den Versuch Purkyně's mit der Veränderung der Kopfstellung nach Unterbrechung der Rotation dargethan. Man kann aber gegen die Beweiskraft des Versuches Einwendungen erheben. Man kann z. B. daran denken, dass bei Beschleunigungen des Körpers die Halsmuskel den Kopf in der Lage
zu erhalten haben und dass auch nach veränderter Kopfstellung die
betreffenden Muskelempfindungen in modificirter Weise fortbestehn.
Ist diese Auffassung auch eine gezwungene, so kann sie doch auftreten. Desshalb schien es mir gut, den Einfluss der Kopfstellung
auf die Schwindelphänomene noch genauer zu constatiren.

Versuch 1. Wir setzen uns in den Rotationsapparat, befestigen auf dem Kopfe eine vierkantige Holzkappe und bringen diese
in eine Führung aus zwei Brettern, so dass der Kopf bei fest an-
gelehntem Rücken nur vor- und rückwärts geneigt, aber weder sonst
wie gedreht noch seitwärts geführt werden kann. Setzt man nun
den Apparat in Rotation, so treten bei Kopfneigungen sowohl während
der Rotation als nach Beendigung derselben die bei den früheren
Versuchen beschriebenen subjectiven Erscheinungen eben so deutlich
auf, wie wenn der Kopf frei beweglich wäre.

Versuch 2. Wir setzen auf den Kopf eine vierkantige Holzkappe, an welcher die beiden Seitentheile rechts und links fast bis
zu den Schultern herabreichen. Dann setzen wir uns auf einen Stuhl

forward and backward <u>pitching movements</u>[7.39] of the head. With the help of a pencil attached to the side part such that it does not draw a curve during head movements, but only marks a small point on the wood, one easily finds the intersection of the rotation axis of the head and the boards of the wooden cap. The cap is now bolted to the frame r above (Fig. 7), and the chair is allowed to move freely about the axis of the head in the frame r. The observer now sits on the chair and pushes his head firmly into the cap. An assistant who is seated opposite in the frame R places the chair into oscillation [about a horizontal axis] by means of a bar attached to the chair. If one now puts the rotation apparatus in motion and simultaneously strongly oscillates the chair with the body of the observer while the head of the observer remains fixed, then one does not notice the phenomena that usually occur with movements of the head. Furthermore, such phenomena do not even occur after stopping the rotation, with the chair oscillating. In this and the previous experiment the movements of body relative to head are the same. The muscles must perform the same function in both cases. Since nothing at all noticeable happens with only the body moving, whereas with only head movement all the known phenomena appear, it seems to me proof that the seat of the organ of movement sensation is in the head.

Head position determines not only the axis of subjective rotation, but also the form of eye and tactile vertigo, as mentioned earlier. Both are explained by Purkyně by means of unconscious continuation of eye and muscle movements acquired by temporary adaptation. Since these movements are not generated consciously they are ascribed to object movement. An object whose image is shifted on the retina of the unconsciously moving eye is perceived as moving. If we unconsciously pull an object towards us with our arms, we feel that it is falling against us. Indeed, one knows that objects seem to be moving when the <u>eyeball is moved by an outside force</u>.[7.40] Furthermore, aside from a tingling skin sensation, one can produce phenomena similar to tactile vertigo, for example, by supporting one's hand against a table and then inducing current in the arm. I have performed this latter experiment. The hypothesis of

den Rücken fest anlehnend und führen Neigungen des Kopfes nach vorn und rückwärts aus. Mit Hilfe eines Bleistiftes, den man an die Seitentheile so setzt, dass derselbe keinen Bogen bei den Kopfneigungen schreibt, sondern auf dem Holz nur eine kleine Stelle markirt, findet man leicht die Durchschnittspunkte der Drehungsaxe des Kopfes mit den Wänden der Holzkappe. Nun wird die Kappe oben am Rahmen r (Fig. 7) befestigt und der Stuhl um die Axe des Kopfes im Rahmen s beweglich gemacht. Der Beobachter setzt sich nun auf den Stuhl und keilt den Kopf fest in die Kappe ein. An dem Stuhle wird ein Stab angebracht, durch welchen ein Gehilfe, der gegenüber auf einem Sitze in dem Rahmen R Platz nimmt, den Stuhl in Schwingungen versetzt. Versetzt man nun den Rotationsapparat in Bewegung und zugleich den Stuhl mit dem Körper des Beobachters in heftige Schwingungen, während der Kopf des Beobachters ruhig bleibt, so bemerkt man nichts von den Erscheinungen, welche bei Bewegung des Kopfes aufzutreten pflegen. Ebensowenig treten nach dem Anhalten der Rotation beim Schwingen des Stuhles solche Erscheinungen auf. Bei diesem und dem vorigen Versuch sind die Relativbewegungen von Kopf und Rumpf dieselben. Die Muskel haben also in beiden Fällen dieselbe Verrichtung. Wenn nun dennoch bei blosser Bewegung des Körpers gar nichts Auffallendes eintritt, bei blosser Bewegung des Kopfes aber alle bekannten Erscheinungen sich zeigen, so scheint mir damit der Sitz des Organs der Bewegungsempfindungen im Kopfe nachgewiesen.

Von der Kopfstellung hängt nicht nur die Lage der Axe der subjectiven Drehung, sondern, wie schon erwähnt wurde, auch die Art des Augen- und Tastschwindels ab. Beide werden von Purkyne erklärt durch unbewusste Fortsetzungen der durch temporelle Eingewöhnung erworbenen Augen- und Muskelbewegungen, deren Effect, weil er eben unbewusst hervorgebracht ist, den Objecten zugeschrieben wird. Ein Object, dessen Bild sich auf der Netzhaut des unbewusst bewegten Auges verschiebt, wird für bewegt gehalten. Von einem Gegenstande, den wir unbewusst mit unsern Armen heranziehen, meinen wir, dass er gegen uns heran stürzt. In der That weiss man, dass bei Bewegung des Augapfels durch eine fremde äussere Kraft die Gegenstände bewegt erscheinen, und man kann, abgesehn von den prickelnden Hautempfindungen, dem Tastschwindel

unconscious movements thus explains the phenomena quite well.

However, Breuer's hypothesis that unconscious movements reflect the response of an organ of movement sensation, which Purkyně also considered at one time, has clear advantages over the assumption of temporary adaptation. Difficulties already occurred in explaining the details of eye movements by temporary adaptation. The reversal in direction of rapid eye movements at the beginning and the end of rotation gives a decisive advantage to Breuer's viewpoint.

The dependence of the illusory movement on head position during eye vertigo is not obvious, as long as the eyes remain essentially fixed relative to the head.[7.41] On the other hand it is more difficult to understand why the nature of tactile vertigo is affected by head position. If the objects we touch, or which we lean on, change from horizontal to vertical rotation as soon as we change head position, then one can surely not assign this to temporary adaptation of muscle movement. The rapid and reliable alteration of the reflex mechanism with each change of head position is really quite remarkable. If, following the stopping of the rotation, one makes rapid and large repositioning movements of the supporting extremities, while keeping the head still, then at least I experience no change in the rotational movement of any objects I am holding. If this theory is to be considered complete, it must finally also explain how involuntary muscle activities depend on the position of the limbs, as well as on head position.

durchaus ähnliche Erscheinungen hervorrufen, wenn man etwa die Hand auf den Tisch stützt und nun den Arm mit Inductionsströmen behandelt. Letzteres Experiment habe ich angestellt. Die Annahme von unbewussten Bewegungen ist also sehr gut zur Erklärung der Erscheinungen brauchbar.

Die Breuer'sche Annahme aber einer reflectorischen Auslösung der unbewussten Bewegungen von einem Organ der Bewegungsempfindungen aus, die auch Purkyně vorübergehend macht, empfiehlt sich sehr gegenüber der Annahme der temporellen Eingewöhnung. Der Schwierigkeit ist schon gedacht worden, die Augenbewegungen im Detail durch temporelle Eingewöhnung zu erklären. Die Umkehrung des Sinnes der ruckenden Augenbewegung beim Beginn und Ende der Rotation gibt der Breuer'schen Ansicht den entschiedenen Vorzug.

Dass die Scheinbewegung des Augenschwindels von der Kopfstellung abhängt, kann nicht auffallen, indem die Augen im Ganzen ihre Relativstellung gegen den Kopf beibehalten. Dagegen ist es schwerer verständlich, wie so die Art des Tastschwindels von der Kopfstellung afficirt wird. Wenn die angefassten Gegenstände, oder die Gegenstände auf die wir uns stützen, aus einer horizontalen in eine verticale Drehung gerathen, sobald. wir die Lage des Kopfes ändern, so kann gewiss nicht von einer Wirkung der temporellen Eingewöhnuog der Muskelbewegung die Rede sein. Allein auch der Reflexmechanismus ist sehr merkwürdig, der von der Aenderung der Kopfstellung so rasch und sicher Notiz nimmt. Wenn man nach dem Anhalten der Rotation bei unveränderter Kopfstellung die Lage der Extremitäten, auf die man sich stützt, sehr rasch und sehr bedeutend abändert, so gelingt es mir wenigstens nicht, hiebei eine Aenderung der Drehbewegung an den erfassten Gegenständen wahrzunehmen. Soll also diese Theorie auf Vollendung Anspruch machen, so muss einmal auch aufgeklärt werden, wie so die unbewussten Muskelerregungen nicht nur von der Kopfstellung, sondern auch von der Stellung der Glieder abhängen.

Theory of the Phenomena.

1

The theory of movement sensation at which I arrived and that I described in a preliminary and brief form was previously proposed by two other researchers. To emphasize the degree of agreement as well as the differences between myself and Breuer and Brown, I will reprint their preliminary communications, in addition to some excerpts from the first communication that I sent to the Vienna Academy.

Session of the Vienna Academy of Nov. 6, 1873.

The c. m. [corresponding member] Prof. E. Mach from Prague submits a report: "Physical experiments concerning the sense of equilibrium in human".

From numerous experiments that Mach conducted on himself, he reaches the conclusion that a single point of view can explain all of the following phenomena: Flourens' rotation phenomena, the maintenance of equilibrium and posture, the common phenomena of rotational vertigo, Goltz' phenomena, and several optical movement sensations. One must simply assume that the nerves of the ampullae of the semicircular canals of the ear labyrinth respond to each stimulation with a sensation of rotation. (This stimulation is normally caused by a rotational moment acting on the contents of the semicircular canal.)

Session of the R.I. [Royal Imperial] Society of Physicians in Vienna of November 14, 1873.

Dr. J. Breuer requests permission to publish already in this week's bulletin the most important results of a paper "On the semicircular canals of the labyrinth", about which he hopes to be able to report in detail next Friday. Because Prof. E. Mach presented a communication on the same subject at the last session of the Academy, and, as far as can be seen from the short publication, arrived at the same results as Dr. B., it seems desirable to him to state the independence of his work by the most rapid publication possible. For this reason he requests permission to present the following preliminary communication.

Stimulated by the publication of Goltz concerning the semicircular canals (Arch. Physiol. 1870), I started to spend considerable time on this subject. I am convinced that Goltz presented not only a new but also fully correct assumption that the semicircular canals are "sensory organs for the balance of the head and indirectly of the whole body". I believe that this is the solution to the riddle, but also that the concept of the function of the sensory apparatus given by Goltz is hardly defensible. In the following I try, as briefly as possible, to develop another concept and to evaluate it as to its correctness and fruitfulness.

In a system of three nearly orthogonal fluid-filled tubular rings, as in the case of the semicircular canals, currents are produced by every rotational movement of the

Theorie der Erscheinungen.

1

Die Theorie der Bewegungsempfindungen zu der ich gelangt bin, und die vorläufig und kurz schon angegeben wurde, ist ausserdem noch von zwei andern Forschern aufgestellt worden. Um den Grad der Uebereinstimmung und zugleich die Differenzpunkte ersichtlich zu machen, will ich ausser dem Auszuge, den ich meiner ersten an die Wiener Akademie gesandten Arbeit beigegeben habe, auch die vorläufigen Anzeigen von B r e u e r und B r o w n hieher setzen.

Sitzung der Wiener Akademie vom 6. Nov. 1873.

Das c. M. Herr Prof. E. Mach in Prag übersendet eine Abhandlung: »Physikalische Versuche über den Gleichgewichtssinn des Menschen«.

Aus zahlreichen Versuchen, die Mach an sich selbst angestellt hat, zieht er den Schluss, dass man die Flourens'schen Dreherscheinungen, die Orientirung des Gleichgewichtes und der Bewegung, die gewöhnlichen Erscheinungen des Drehschwindels, die Goltz'schen Phänomene und einige optische Bewegungserscheinungen aus einem Gesichtspunkt begreifen kann, wenn man annimmt, dass die Nerven der Ampullen der Bogengänge des Ohrlabyrinthes jeden Reiz (welcher gewöhnlich durch ein Drehungsmoment an dem Inhalt des Bogenganges ausgeübt wird) mit einer Drehempfindung beantworten.

Sitzung d. k. k. Gesellschaft der Aerzte in Wien vom 14. November 1873.

Dr. J. Breuer bittet, die wichtigsten Resultate einer Arbeit: »Ueber die Bogengänge des Labyrinths«, über welche er nächsten Freitag ausführlich zu referiren hofft, schon im diesswöchentlichen Anzeiger veröffentlichen zu dürfen.. Da Herr Prof. E. Mach in der letzten Akademiesitzung eine Abhandlung über denselben Gegenstand vorgelegt hat, und, soweit aus einer kurzen publicirten Note zu ersehen, zu denselben Resultaten gelangt wie Dr. B., erscheint es diesem wünschenswerth, durch möglichst rasche Publication die Unabhängigkeit seiner Arbeit zu constatiren. Er erbittet aus diesem Grnnde [Grunde] die Erlaubniss zu der folgenden vorläufigen Mittheilung:

Angeregt durch die Abhandlung von Goltz über die Bogengänge (Arch. f. Physiol. 1870.) beschäftige ich mich seit geraumer Zeit mit diesem Gegenstande. Ich halte mich überzeugt, dass Goltz einen nicht bloss neuen, sondern auch durchaus richtigen Satz aufgestellt hat mit der Annahme, die Bogengänge seien »Sinnesorgane für das Gleichgewicht des Kopfes und mittelbar des ganzen Körpers«. Ich glaube, dass dies das Lösungswort des Räthsels ist, aber auch, dass die Vorstellung von den Functionen des Apparates, wie sie Goltz gibt, kaum haltbar sei. Ich versuche im Folgenden in möglichster Kürze eine andere Vorstellung zu entwickeln und auf ihre Richtigkeit und Fruchtbarkeit zu prüfen.

In einem Systeme von drei annähernd aufeinander senkrecht stehenden, mit Flüssigkeit erfüllten Röhrenringen, wie es die Bogengänge darstellen, entstehen

whole system (the head), in a direction opposite to the head rotation. The strength of the current in each single canal depends on the plane of head rotation and the extent of the rotation. There is a very tight relationship between head rotation and the flow of the enclosed fluid. If the currents were perceived, they could convey very accurate information about every turn of the head. As possible organs for perception we can consider the "acoustic hairs", which perpendicularly project into the lumen of the canal at a location where the canal broadens and is flattened. In this place they are as sensitive as possible to fluid currents, and additionally they are connected to the nerve whose end organ they form. In a completely analogous manner the very similar end organs of the nerves in "lateral line organs"[8.1] of fish seem to function for perception of currents, resistance, etc, in the surrounding water (F. E. Schulze).

If we want to relate these two facts to Goltz' theory, we have to assume that the ampullar nerve detects each endolymph current. This in turn leads to the perception of head rotation in the plane of the respective canal in a direction opposite to that of the current. However, the perceptions of the 6 ampullae of both labyrinths are combined to form a global sensation.

In support of this assumption we have:

1. The considerations which, taken together, lead Goltz to claim that the semicircular canals are balance organs.

2. This assumption explains the phenomena following a lesion of the membranous canals. Thereby endolymph always flows out or blood flows in, or both, and consequently relatively violent currents of the endolymph must occur. According to our assumption these currents generate the sensation of head rotation in the plane of the lesioned canal (even if such intense currents subsequently overflow to other canals and thereby complicate the phenomena). From observations in humans we know that stronger rotations of the body (falling) are countered by powerful reflexive compensatory movements. Animals will therefore move the head and the body in the plane of the lesioned canal, up to the point of somersaulting. If the movement impetus is more moderate it will continuously be cancelled by correcting influences, but then the movement restarts immediately, thus producing the pendular movement of the head and the nystagmus of the eyes that Flourens' experiments demonstrated. For those phenomena, which persist after healing of the wounds, we have to be content with the explanations by Goltz without being able to go into further details, since research on this aspect of the organ has just begun and precise post-mortem findings are absent.

To the known experiments I must add the observation that currents of endolymph, produced by sticking a needle into a canal and withdrawing it, lead to movements in opposite directions (chicken, pigeons). The same happened when a cannula was successfully inserted in a canal, even if only for a short time, and endolymph was moved inward and outward.

3. They explain the phenomena of electrical vertigo, as recently once again

bei jeder drehenden Bewegung des ganzen Systems (des Kopfes) Strömungen der Flüssigkeit, in einer der Kopfdrehung entgegengesetzten Richtung; das Ausmaass der Strömung in jedem einzelnen Gang hängt von der Ebene ab, in welcher der Kopf gedreht wird, und von dem Ausmaass der Drehung. Zwischen der Kopfdrehung und den Strömungen der eingeschlossenen Flüssigkeit bestehen ganz feste Beziehungen; würden die Strömungen percipirt, so könnten sie also über jede Drehung des Kopfes genaue Auskunft geben. Als mögliche Perceptions-Apparate stellen sich unserer Betrachtung die »Hörhaare« dar, welche an einer verbreiterten und abgeplatteten Stelle des Canals senkrecht in das Lumen hineinragen, also möglichst empfindlich gegen Flüssigkeitsströmungen placirt sind und anderseits mit Nerven zusammen hängen, deren End organ sie bilden. In ganz analoger Weise scheinen die ganz analogen Nervenendapparate in den »Seitenorganen« der Fische zur Perception der Strömungen, Widerstände u. s. w. des umgebenden Wassers zu dienen. (F. E. S c h u l z e.)

Wollen wir diese beiden Thatsachen im Sinne der G o l t z 'schen Theorie verwerthen, so müssen wir a n n e h m e n , dass jede Strömung der Endolymphe, durch die Ampullennerven percipirt, die Vorstellung einer Drehung des Kopfes in der Ebene des betreffenden Bogenganges und in der jener Stömung entgegengesetzten Richtung hervorrufe, dass aber die Perceptionen der 6 Ampullen beider Labyrinthe sich zu einer Gesammtvorstellung combiniren.

Für diese Annahme sprechen:

1. Die Erwägungen, welche nach G o l t z überhaupt dafür sprechen, dass die Bogengänge Organe des Gleichgewichtssinnes sind.

2. Erklärt diese Annahme die Erscheinungen bei Verletzung der häutigen Canäle. Dabei tritt immer Endolymphe aus oder Blut ein, oder es geschieht beides, und es müssen dadurch relativ gewaltige Strömungen der Endolymphe entstehen. Diese Strömungen erzeugen unserer Annahme nach die Vorstellung von Drehung des Kopfes in der Ebene des verletzten Canals (wenn auch solches intensive Strömen secundär in andere Bogengänge übergreifen und dadurch die Erscheinungen compliciren wird). Aus Erfahrung am Menschen wissen wir, dass stärkere Drehungen des Körpers (Umstürzen) durch energische compensirende Bewegungen reflectorisch beantwortet werden. Die Thiere werden also Kopf und Körper i n d e r E b e n e d e s v e r l e t z t e n B o g e n g a n g e s bewegen, bis zum Ueberkugeln. Sind die Bewegungsantriebe mässiger,. so werden sie immer wieder durch corrigirende Einflüsse aufgehoben, um sogleich wieder zu beginnen und jenes Pendeln des Kopfes, und den Nystagmus der Augen herzustellen, welche die F l o u r e n s 'schen Versuche zeigen. Für jene Erscheinungen, die nach Verheilung der Wunden fortbestehn, müssen wir uns, da die Betrachtuug des Apparates in diesem Sinne ja eben erst begonnen wird, und bei dem Mangel an genauen Sectionsbefunden, mit der G o l t z 'schen Erklärung begnügen, ohne ins Detail gehen zu können.

Den bekannten Versuchen habe ich hinzuzüfugen, dass Strömungen der Endolymphe, erzeugt durch Einstechen und Ausziehen einer Nadelspitze in einen Canal, mit entgegengesetzten Bewegungen beantwortet werden (Hühner, Tauben); ebenso wenn es, allerdings nur für kurze Zeit, gelang, eine Canüle in einen Gang zu fixiren, das Ansaugen und Wiedereintreiben der Endolymphe durch diese.

3. Erklären sich die in letzter Zeit von H i t z i g wieder genau beobachteten

observed in detail by <u>Hitzig</u>.[8.2] The current affects precisely the end organs of the equilibrium sense. The fact that they [the phenomena] become stronger as the electrode approaches the mastoid process already makes it likely that an organ located within the petrous bone is stimulated. I inserted a fine wire electrode in the semicircular canals and saw intense head movements when galvanic currents passed through it.

4. Our assumption has a necessary consequence. With continuous rotation of the head (of course, together with the body), the initial oppositely directed movement of the endolymph is slowed down by friction against the walls. If the head is then suddenly stopped, the endolymph, because of its inertia, must continue to flow in the sense of the head rotation; therefore, the sensation must appear that the head and the body rotate back in a direction opposite to that of the former rotation and in the plane of those canals that lie closest to the previous rotation plane. If I turn around while standing upright, the horizontal canals primarily lie in the plane of rotation for the usual head position. The plane of rotation lies closest to the <u>sagittal canals</u>[8.3] when the head is tilted to one or the other shoulder, and closest to the frontal canals for the head pitched upward or downward. The illusory aftermotion of the body will occur for one case in the horizontal, and for the other cases in the sagittal or frontal direction. This is the law that Purkyně formulated and generalized from experiments of his own as well as those of the elder <u>Darwin</u>.[8.4] "Consider a cut through the head (taken as a sphere) as defining the axis about which the first movement took place. The vertigo movements with any subsequent head position always occur about that axis." These hitherto completely unexplained phenomena of rotational vertigo follow as necessary consequences of our assumption and provide the proof of the theory — The original sensation of rotation of the body is complicated by the reflexive movements that occur to maintain balance and by the phenomena of ocular vertigo. These require more detailed consideration.

5. We already find sufficient teleological significance for this organ since it enables precise and powerful reflex actions as those that occur during falling, e.g., for maintaining balance. Other considerations, however, point toward yet a different relationship.

Head rotations lead to compensatory rotations of the eyeball. Gräfe (Arch. Ophthalmol. I) showed that rabbits, when rotated in a sagittal direction [plane] (about their transverse axis), produce compensatory <u>torsional movements</u>[8.5] of the eyeball. In humans <u>Nagel</u>[8.6] demonstrated the existence of such compensatory torsional movements for displacements of the head in the frontal plane. These were smaller and their existence had been denied for a long time. N. [Nagel] points out that it is only the position of the head that matters, independently of which muscle action or which joints were involved in producing the movement, or if it was passively produced.

We therefore use compensatory movements of the eyeballs to correct for the displacement of the visual field caused by head rotation. Let us recall that in Flourens' experiments, as well as for electrical and rotational vertigo, intensive nystagmus in the corresponding planes always accompanies the vertigo (op.cit. by means of the canals, according to our understanding). Therefore it is reasonable to

Phänomene des electrischen Schwindels. Es werden eben die Endapparate des Gleichgewichtssinnes durch den Strom afficirt. Schon dass diese desto intensiver werden, je näher die Electrode dem Proc. mastoid. steht, macht es wahrscheinlich, dass es sich dabei um die Reizung eines im Felsenbein befindlichen Organs handelt. Ich habe feinen Draht als Electrode in die Bogengänge eingeführt, und bei Durchleitung galvanischer Ströme intensive Bewegungen des Kopfes gesehen.

4. Unsere Annahme hat eine nothwendige Consequenz. Bei fortgesetzter Drehung des Kopfes (natürlich mit dem Körper) muss die anfängliche rückläufige Bewegung der Endolymphe durch die Reibung an den Wänden aufgezehrt werden. Steht dann plötzlich der Kopf still, so muss die Endolymphe ihrem Beharrungsvermögen zufolge in dem Sinne der Kopfdrehung weiter strömen; es muss also die Vorstellung entstehn, Kopf und Körper würden in einer der früheren Drehung entgegengesetzten Richtung zurückgedreht; und zwar in der Ebene jener Bogengänge, welche am meisten in der Drehungsebene lagen. Rotire ich auf meinen Füssen stehend, um mich selbst, so liegen bei gewöhnlicher Kopfhaltung die horizontalen, bei Neigung des Kopfes auf die eine oder andere Schulter die sagittalen, war das Gesicht nach oben oder abwärts gerichtet, die frontalen Bogengänge vorwiegend in der Drehebene. Es wird also die scheinbare Nachbewegung des Körpers das eine Mal in horizontaler, das andere Mal in sagittaler oder frontaler Richtung geschehen. Das ist nun das von Purkyně aus seinen und des ältern Darwin Versuchen abstrahirte Gesetz: »Dass der Durchschnitt des Kopfes (als eine Kugel), um dessen Achse die erste Bewegung geschah, die Schwindelbewegung bei jeder nachmaligen Lage des Kopfes unveränderlich bestimmt.« Diese bisher so gut wie ganz unerklärten Phänomene des Drehschwindels ergeben sich so als die nothwendigen Consequenzen unserer Annahme und bilden die Probe auf die Rechnung. – Die Complication der einfachen Empfindung von Drehung des Körpers durch die eintretenden Reflexbewegungen zur Erhaltung des Gleichgewichtes und die Phänomene des Augenschwindels erfordern eine ausführlichere Betrachtung.

5. Die teleologische Bedeutung des Apparates für uns würden wir schon darin befriedigend finden, dass er so präcise und energische Reflexactionen möglich macht, wie die sind, die beim Ausgleiten z. B. die Balance erhalten. Andere Betrachtungen weisen aber noch auf eine hievon verschiedene Beziehung hin.

Kopfdrehungen werden durch compensirende Bulbusdrehungen beantwortet. Gräfe (Arch. f. Ophthalm. I.) hat gezeigt, dass bei Kaninchen, wenn sie in sagittaler Richtung (um ihre Querachse) gedreht werden, compensirende Raddrehungen der Bulbi auftreten. Dasselbe hat für frontale Verschiebungen des Kopfes, allerdings in geringerem Ausmaasse, Nagel beim Menschen nachgewiesen, nachdem es lange in Abrede gestellt war. N. macht aufmerksam, dass es dabei nur auf die Stellung des Kopfes ankomme, gleichgiltig, durch welche Muskelwirkung, in welchen Gelenken sie erzeugt werde, oder ob sie passiv hergestellt sei.

Wir haben also Correctionen der Verschiebung des Gesichtsfeldes bei Drehung des Kopfes durch compensirende Bewegungen der Bulbi. Bedenken wir, dass bei den Flourens'schen Versuchen, beim electrischen und Drehschwindel immer mit dem Schwindel (l. c. nach unserer Auffassung der Affection der Bogengänge), intensiver Nystagmus in den entsprechenden Ebenen vorhanden ist,

believe that some of the teleological significance of this organ can be found in these corrections of the of the visual field position.

The first note of Brown of January 19, 1874, is not accessible to me. However, as his later publication is also very short, it is reprinted here in its entirety. [The following was in English in the original text.]

From the Journal of Anatomy and Physiology. Vol. VIII.

On the sense of Rotation and the Anatomy and Physiology of the Semi-circular Canals of the Internal Ear. By Prof. A. Crum Brown, M.D. University of Edinburgh.

For some time I have been convinced that we possess a sense of Rotation quite distinct from all our other senses. By means of this sense we are able to determine – 1st, the axis about which rotation of the head takes place; 2nd, the direction of the rotation; and 3rd, its rate.

In ordinary circumstances we do not wholly depend upon this sense for such information. Light, hearing, and the muscular sense assist us in determining the direction and amount of our motions of rotation, as well as of those of translation; but if we pourposely [sic] deprive ourselves of such aids we find that we can still determine with considerable accuracy the axis, the direction, and the rate of rotation. The experiments that I have made with the view of determining this point were conducted as follows: a stool was placed on the centre of a table capable of rotating smoothly about a vertical axis; upon this the experimenter sat, his eyes being closed and bandaged; an assistant then turned the table as smoothly as possible through an angle of the sense and extent of which the experimenter had not been informed. It was found that, with moderate speed, and when not more than one or two complete turns were made at once, the experimenter could form a tolerably accurate judgement of the angle through which he had been turned. By placing the head in various positions it was possible to make the vertical axis coincide with any straight line in the head. It was found that the accuracy of the sense was not the same for each position of the axis in the head, and further, that the minimum perceptible angular rate of rotation varied also with the position of the axis. It was also found that considerable differences of accuracy exist in different individuals.

The sense of rotation is, like other senses, subject to illusions, rotation being perceived where none takes place. Vertigo or giddiness is a phenomenon of this kind.

When, in the experiments just mentioned, rotation at a uniform angular rate is kept up for some time, the rate appears to the experimenter to be gradually diminishing, and to cease altogether after a time, varying with the position of the head, and different with different individuals; if the rotation be then stopped, he experiences the sensation of rotation about the same axis in the opposite direction. If the position of the head be changed after the prolonged rotation has been made, the position of the axis of the apparent rotation is changed, retaining always the same position relatively to the head as was occupied by the axis of the real rotation. The readiness with which this complementary apparent rotation is produced is not the same for each axis. In such experiments, as long as the eyes are shut, and the axis of rotation kept vertical, a sensation of giddiness is not experienced. That sensation appears to be caused

so liegt es nahe zu glauben, dass in diesen Correctionen der Stellung des Gesichtsfeldes ein Theil der teleologischen Bedeutung unseres Apparates liege.

Die erste Note von B r o w n vom 19. Jänner 1874 ist mir nicht zugänglich, da aber auch seine spätere Publication sehr kurz ist. so lasse ich sie hier vollständig folgen.

[Original English text of the following section appears on the facing pages.]

by the discordance between the testimony of the sense of rotation, and that of some other sense. Thus if I experience a sensation of rotation, it makes no difference to my mind whether that sensation corresponds to a real rotation or not, as long as I have no means of ascertaining independently the existence or non-existence of the real rotation. I am in that position as long as my head is fixed and my eyes shut. But if, while the complementary apparent rotation is felt, I open my eyes, I still feel that I am being turned round, but at the same time I see that external bodies retain their position relatively to me – if I am turning round so are they – and this produces at once a feeling of insecurity or giddiness. Similarly this giddy or insecure feeling is produced, if, while the complementary apparent rotation is felt, the head be moved so that the axis of this rotation is no longer vertical.

The sense of rotation, being a special sense, must necessarily have a special peripheral organ physically constituted so as to be affected by rotation, a sensory nerve, and a central organ. The structure of the semicircular canals of the internal ear is such as to fit them to act as such a peripheral organ, and the experiments of Flourens and of Goltz support this view. The bony canals are filled with liquid, in which float loose connective tissue and the membranous canals with the contained endolymph. Rotation of the head about an axis at right angles to the plane of a canal will then produce, on account of the inertia of the liquid, ect. [sic], motion of the contents relatively to the walls of the canal, and this may be expected to irritate the terminations of the nerves in the ampulla. If the rotation be continued at a uniform rate, fluid friction, friction of the endolymph against the membranous canal, and of the perilymph against the membranous canal and the periosteum, will gradually diminish this relative motion, which will at last cease. We should therefore expect, as we have seen to be the case, that continued uniform rotation should be perceived less and less strongly, and that the sensation should at last die away altogether. The time required for this equalisation of the motion of the canal and its contents will depend upon the rate of rotation and upon the dimensions of the canal and the amount of attachment of the membranous canal to the periosteum. These latter conditions are not the same in the three canals, and therefore we ought to find, as we do, that the rate at which the sense of rotation dies away is not the same for different positions of the head. Again, if the uniform rotation is stopped, the contents of the canal will continue to move on, thus causing an apparent rotation in a direction the reverse of that of the original rotation, and this also will die away owing to friction.

As the three canals are in planes nearly at right angles so [sic] one another, rotation about any axis can be resolved into rotations, each of which will produce the effect described above upon one of the canals, and thus any rotation will have its appropriate sensation.

So far then this view of the function of the semicircular canals seems to explain the phenomena of the sense of rotation, and I find that an explanation almost identical with this was given by Professor Mach, of Prague, and by Dr. Breuer, of Vienna, shortly before I communicated the substance of this paper, as a preliminary note, to the Royal Society of Edinburgh (Proceedings, 19th January. 1874). But this explanation is not sufficient. As far as we know, a nerve current can vary only in intensity and not in kind, so that, if irritated at

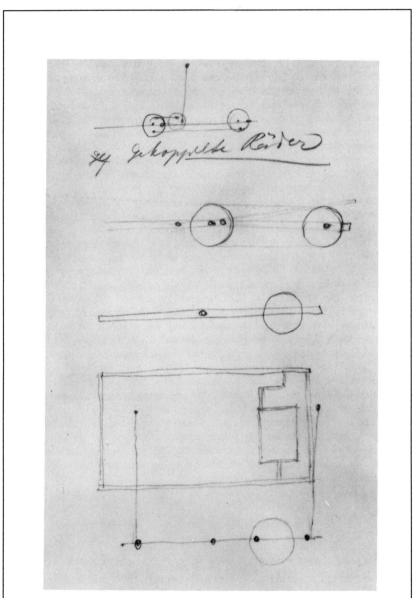

Figure A-6: Rotating chair including "coupled wheels". Sketch of the multi-axis rotating frame with an off-axis chair, showing the use of pulleys to couple chair rotation to frame rotation (from Mach's notebook, 1874; photograph courtesy of E. Mach Archive, Deutsches Museum, Munich).

all, whether by right-handed so by left-handed rotation, the nerve would convey the same message to the central organ. The solution of this difficulty which I proposed is as follows: – Each canal has an ampulla at one end only, and there is thus a physical difference between rotation with the ampulla first, and rotation with the ampulla last, and we can easily suppose the action to be such that only one of these rotations (any that with the ampulla first, in which case, of course, there is a flow from the ampulla into the canal) will affect the nerve terminations at all. [In the preliminary note above referred to, I described a way in which this might be supposed to take place.] One canal can therefore, on the supposition, be affected by, and transmit the sensation of rotation about one axis in one direction only, and for complete perception of rotation in any direction about any axis six semicircular canals are required, in three pairs, each pair having its two canals parallel (or in the same plane) and with their ampullae turned opposite ways. Each pair would thus be sensitive to any rotation about a line at right angles to its plane or planes, the one canal being influenced by rotation in the one direction, the other by rotation in the opposite direction.

Now we have six semicircular canals, three in the one ear and three in the other, and I find in all the animals that I have examined, that the exterior canal of one ear is very nearly in the same plane as that of the other; while the superior canal of one ear is nearly parallel to the posterior canal of the other.

The three axes are therefore – 1st, a vertical*) axis at right angles to the plane of exterior canals; 2nd, an axis which may be roughly defined as passing (in the human subject) through the left eye and the right mastoid process at right angles to the planes of the right superior and the left posterior canal, and 3rd, an axis passing through the right eye and the left mastoid process at right angles to the right posterior and left superior canal.

In different animals there are great differences in relative size and position of the canals, but the relation just mentioned appears to exist in all cases. This relation may be most simply statet [stated] thus: In each ear there is one canal (the exterior) in a plane at right angles to the medial plane, and two other canals (the superior and the posterior) in planes equally inclined to the medial plane. In no other way is it possible to harmonize the bilateral symmetry of the two ears with the condition, that each of the three axes shall have two oppositely turned canals in planes at right angles to it.

2.

After we had found that the skin, muscle, blood, and brain could not absolutely be excluded as sources for movement sensation, but are not sufficient to explain the details, one is obliged to assume the existence of a special end organ for movement sensation. Because these sensations are not already consumed for other purposes, they can unambiguously characterize locomotion. We will now further develop the previously introduced

---

*) In the human subject this axis is not quite vertical when the head is held in its usual position; it becomes so when the face is inclined slightly downwards.

2.

Nachdem wir gefunden, dass Haut, Muskel, Blut, Hirn sich als Quellen der Bewegungsempfindungen zwar nicht absolut ausschliessen lassen, aber zur Erklärung der Einzelheiten nicht ausreichen und dass man zur Annahme eines besondern End organs der Bewegungsempfindungen gedrängt wird, dessen Sensationen, weil sie nicht schon für andere Zwecke verbraucht sind, unzweideutige Merkmale der

hypothesis concerning this organ.

Let us imagine a cavity in a body B, covered with nerve endings and containing some fluid or solid matter A. First, because of its weight it will exert more pressure against one place on the wall than against all others and thereby indicate the orientation of B relative to the vertical. Second, with every acceleration that B undergoes, A generates a relative opposite acceleration (a counterpressure) that combines with gravity, thereby varying the pressure intensity as well as the point of pressure on the wall of the cavity. Third and finally, A will attempt to make a counterrotation for every angular acceleration taken by B. B can determine its orientation and linear acceleration[8.7] through 1 and 2, and its angular acceleration through 3. The labyrinth, including the semicircular canals, is probably such an organ, whose functional scheme we have developed here. It is also probable that specifically separate nerves are used for 1 and 2 than for 3. With respect to the semicircular canals, they seem to us particularly appropriate for the implementation of the principle of the conservation of areas. For each angular acceleration about an axis normal to the plane of a semicircular canal, its contents must exert a rotational moment in the opposite sense, which the nerves of the ampullae convey as an indication of this angular acceleration.

J. Müller pointed out that every sensory nerve is tied to a specific energy, such that the most different stimuli only trigger a single characteristic sensory quality in a nerve, whereas the same stimulus applied to different sensory nerves produces the respective different sensation qualities. According to Müller's proposed principles I – VIII[8.8] (Handbook of Physiology, vol 2, pp. 251 – 258 [actually pp. 250 – 262]), we may now assume that the ampullar nerves have the specific energy to respond to every stimulus with a sensation of rotation. Immediately many phenomena become clear and various observations can be explained by one principle. Flourens' phenomena of rotation become self-explanatory under this assumption, especially if we presume with Löwenberg that cutting the semicircular canals results in stimulation rather than sensory paralysis.[8.9] Goltz' view is basically preserved,

Locomotion liefern können, wollen wir nun die schon vorläufig angedeutete auf dies Organ bezügliche Hypothese näher entwickeln.

Denken wir uns eine mit Nervenenden ausgekleidete Höhle in einem Körper *B*, welche irgend einen flüssigen oder festen Inhalt *A* hat. Derselbe wird erstens vermöge der Schwere auf eine Stelle der Wand stärker drücken, als auf die übrigen und dadurch die Lage von *B* gegen die Verticale bezeichnen. Bei jeder Beschleunigung, die *B* annimmt, wird zweitens *A* eine Gegenbeschleunigung, (einen Gegendruck) ausüben, die sich mit der Schwerebeschleunigung zusammensetzt, wodurch sowohl die Druckintensität als auch die Druckstelle an der Höhlenwind sich ändert. Endlich wird, drittens bei jeder Winkelbeschleunigung, die *B* annimmt, *A* eine Gegendrehung auszuführen suchen. Durch 1 und 2 kann *B* zur Kenntniss seiner Lage und progressiven Beschleunigung, durch 3 zur Kenntniss seiner Winkelbeschleunigung gelangen. Der Vorhof mit den Bogengängen dürfte ein solches Organ sein, dessen Schema wir hier entwickelt haben. Es ist auch wahrscheinlich, dass für 1 und 2 specifisch andere Nerven vorhanden sind wie für 3. Betrachten wir die Bogengänge, so scheinen uns diese besonders geeignet, das Flächenprincip zur Geltung zu bringen. Für jede Winkelbeschleunigung um die zur Bogenebene senkrechte Axe muss der Inhalt ein Drehungsmoment von entgegengesetztem Sinn ausüben, welches die Nerven der Ampullen als Zeichen dieser Winkelbeschleunigung weiter leiten.

J. Müller hat darauf hingewiesen, dass jedem Sinnesnerven seine specifische Energie zukommt, so zwar, dass die verschiedensten Reize nur die ihm eigenthümliche Empfindungsqualität auslösen, während derselbe Reiz in verschiedenen Sinnesnerven eben die verschiedenen ihnen zukommenden Empfindungsqualitäten erregt. Nehmen wir nun nach den von Müller, Handbuch der Physiologie Bd. 2, S. 251 – 258 aufgestellten Grundsätzen I – VIII an, dass die Ampullennerven, die specifische Energie haben, auf jeden Reiz mit einer Drehempfindung zu antworten, so werden uns sofort viele Erscheinungen klar und mancherlei zerstreute Beobachtungen sammeln sich unter ein Princip. Die Flourens'schen Dreherscheinungen verstehn sich unter dieser Annahme von selbst um so mehr, wenn wir mit Löwenberg es für wahr-

although it is subject to certain modifications. The familiar phenomena of rotational vertigo can be explained by the same principle. The rotational vertigo experiment is nothing more than Flourens' experiment. Instead of cutting the semicircular canals, we stimulate the ampullar nerves by subjecting the contents of the semicircular canals to a rotational moment by angular acceleration, thereby permitting the contents to pull on the ampullar nerve. As one sees, this theory easily explains all subjective phenomena occurring during passive movements of the body in such a simple and self-explanatory manner that a detailed discussion seems unnecessary. Furthermore, those phenomena that occur when the head is moved during rotation are easily explained. Consider, for example, for a rotation about a vertical axis that a semicircular canal is aligned such that its plane contains the rotation axis. No component of the rotational moment will be exerted on the contents of this canal. As soon as the semicircular canal plane is tilted away from the rotation axis, however, a component of the rotational moment immediately appears in the semicircular canal and gives rise to the corresponding sensation of rotation. One also sees that during small changes of position the axis of the illusory movement is orthogonal to both the rotation axis and the axis of head tilt. The explanation does not significantly change and follows directly from Poinsot's[8.10] theory of the combination of rotations[8.11], if one considers all of the semicircular canals.

3.

If one follows the principle of specific energies rigorously, one is led to the assumption that different rotational sensations must correspond to different nerves.[8.12] In this way not only rotations about each different axis, but also rotations in opposite directions would be conveyed by separate nerves. Now, it is not very easy to show that the labyrinths fulfill this requirement. One can think of the sensation of rotation about arbitrary axes

---

*) Poinsot, new theory about rotation of bodies. Paris 1859.

scheinlich halten, dass die Durchschneidung der Bogengänge nicht als Lähmung, sondern als Reizung wirkt. Die Goltz'sche Ansicht bleibt im Wesentlichen bestehn, wenngleich sie gewisse Modificationen erfährt. Die gewöhnlichen Erscheinungen des Drehschwindels erklären sich nach demselben Princip. Der Drehschwindelversuch ist eben nichts weiter als der Flourens'sche Versuch. Statt die Durchschneidung der Bogengänge anzuwenden, reizen wir die Ampullennerven, indem wir dem Inhalt der Bogengänge durch Winkelbeschleunigung ein Drehungsmoment ertheilen, vermöge dessen er an den Ampullennerven zerrt. Wie man sieht, erklärt diese Theorie ausserordentlich leicht alle bei passiven Bewegungen des Körpers beobachteten subjectiven Erscheinungen in so einfacher und selbstverständlicher Weise, dass eine besondere Ausführung darüber unnöthig erscheint. Auch jene Phänomene, welche bei Bewegungen des Kopfes während der Rotation auftreten, bieten der Erklärung keine Schwierigkeit. Denken wir uns z. B. bei einer Rotation um eine verticale Axe einen Bogengang, dessen Ebene die Rotationsaxe enthält. Auf den Inhalt dieses.Bogenganges wird keine Componente des Drehungsmomentes entfallen. Sobald aber die Bogengangebene gegen die Rotationsaxe geneigt wird, tritt in dem Bogengang sofort eine Componente des Drehungsmomentes auf und gibt Anlass zur Entstehung der betreffenden Drehempfindung. Man sieht auch, dass bei geringen Lagenänderungen die Axe der scheinbaren Drehung senkrecht steht auf der Rotationsaxe und der Axe der Kopfneigung. Die Erklärung ändert sich nicht wesentlich und folgt in gleicher Weise aus der Poinsot'schen *) Theorie der Zusammensetzung der Drehungen, wenn man auf alle Bogengänge Rücksicht nimmt.

3.

Führt man das Princip der specifischen Energieen consequent durch, so gelangt man zu der Annahme, dass verschiedenen Drehempfindungen auch verschiedene Nerven entsprechen müssen, so zwar, dass nicht nur eine Drehung um eine andere Axe, sondern auch eine Drehung von entgegengesetztem Sinn, von einem andern Nerv geliefert wird. Es ist nun nicht ganz leicht, die Erfüllung dieser For-

---

*) Poinsot, nouvelle théorie de la rotation des corps. Paris 1859.

as combinations of the sensations of rotations about semicircular canal axes, in the same way that the movements themselves are combined. But how does one manage to sense rotations of opposite senses?

At the time of my first communication I could not convince myself that the labyrinths on both sides need to be considered such that only both organs together could convey a complete representation of space. I therefore concluded there are two nerves in each of the three ampullae of each labyrinth, one of which reacts to rotation in one direction, and the other to the other direction. This assumption, which I considered and rejected several times, has now been made by Brown. Now that I have been exposed to this theory objectively, and I am able to judge it more impartially, I have no hesitation in considering it as the much superior one, and I concur with it. I believe this even more strongly on the basis of details of Flourens' experiments that are fully consistent with this notion. I had not known of these results earlier, nor does Brown mention them.

To clarify the combined action of both labyrinths, as Brown thinks of it, we will illustrate the position of the canal planes in both labyrinths. To that end we use the same principle that we already used in an earlier publication for the same purpose*). Using two <u>coordinate planes</u>[8.13] that are easily identified relative to the head, we draw outlines of planes which lie parallel to the semicircular canal planes and which pass through two points R and L (right and left ear). One coordinate plane H (which we will call the horizontal one) is perpendicular to the median plane of the head. It passes through the anterior and lower extensions of the maxilla and through a point located in the median plane of the basilar bone about 7^{mm} anterior to the occipital border in the spongiform part of the bone, and about 15^{mm} above the lower and outer surface of that bone. The second plane V (the vertical) passes through the latter point and is normal to the last plane and to the

---

*) E. Mach and J. Kessel, Contributions to the topography and mechanics of the middle ear. Proc. Vienna Acad. vol. 69.

derung am Labyrinth aufzuzeigen. Die Empfindung der Drehung um beliebig liegende Axen kann man sich zusammengesetzt denken aus den Empfindungen der Drehung um die Bogengangaxen in derselben Weise wie sich die Bewegungen selbst zusammensetzen. Allein wie gewinnt man die Drehempfindungen von entgegengesetztem Sinn?

Zur Zeit meiner ersten Mittheilung konnte ich mich nicht entschliessen, das beiderseitige Labyrinth in der Weise als zusammengehörig zu betrachten, dass nur beide Organe zusammen eine vollständige Raumvorstellung geben. Ich nahm desshalb an, dass in jeder der drei Ampullen eines Labyrinthes zwei Nerven seien, von welchen der eine auf den einen, der andere auf den andern Drehungssinn reagirt. Die Annahme, welche sich mir mehrmals dargeboten hat, und die ich immer zurückgewiesen habe, ist nun von Brown gemacht worden. Nachdem sie mir nun objectiv gegenübersteht und ich dieselbe unbefangener beurtheilen kann, stehe ich nicht an sie für die weitaus bessere zu halten und mich ihr anzuschliessen. Ich muss dies umsomehr thun, als Details der Flourens'schen Versuche, die mir früher unbekannt waren und die auch Brown nicht anführt, sich dieser Annahme in ausgezeichneter Weise anschliessen.

Um die Zusammenwirkung der beiden Labyrinthe, wie Brown sich dieselbe denkt, deutlich zu übersehn, wollen wir die Stellung der Bogengangebenen in beiden Labyrinthen darstellen. Hiezu verwenden wir das Princip, welches schon in einer frühern Arbeit zu demselben Zweck gebraucht wurde*). Wir stellen die Tracen von Ebenen, welche den Bogengangebenen parallel sind und die wir durch zwei Punkte R und L (rechtes und linkes Ohr) legen, auf zwei im Kopf leicht auffindbare Coordinatenebenen dar. Die eine dieser Coordinatenebenen H (wir wollen sie die horizontale nennen) steht senkrecht zur Medianebene des Kopfes und geht durch den vordern und untern Rand des Oberkiefers und durch einen Punkt, der in der Medianebene im Os basilare etwa 7^{mm} vor dem Occipitalrand in der spongiosen Substanz und etwa 15^{mm} über der untern

---

*) E. Mach u. J. Kessel, Beiträge zur Topographie und Mechanik des Mittelohres. Sitzgsber. d. Wiener Akad Bd. 69.

median plane. Because only the orientation and not the position of the semicircular canal planes are of importance, we can choose any planes parallel to these coordinate planes and trace the planes parallel to the semicircular canal planes, following the principle of descriptive geometry.

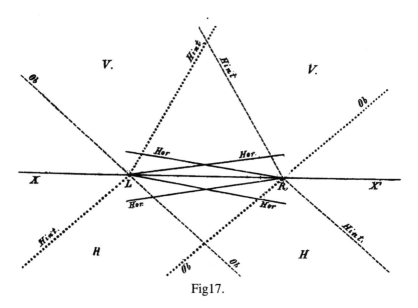

Fig17.

For clarity, we put the points R and L (Fig. 17)[8.14] at the intersection of XX' with the planes H and V. In the figure, the space above XX' designates the plane V, below XX' the plane H. If one folds the sheet to a right angle along XX', the traces present a clear picture of the orientation of the semicircular canal planes. In the figure, Hor stands for horizontal, Ob for upper [anterior], and $Hint$ for the posterior semicircular canal. The drawing should be viewed as seen from behind the head whose hearing organs are shown. As one sees, the horizontal semicircular canals are not tilted very much relative to each other. Also, the posterior semicircular canal of one side and the anterior semicircular canal of the other side are nearly parallel[8.15] to each other. The semicircular canals which are considered as parallel on both sides are designated by the same line styles. Brown correctly places considerable emphasis on this parallel orientation of the two sides. If we mark the rotation direction from the ampulla into the semicircular canal (without passing through the vestibule) with an arrow, and if we assume, as Brown does, that

Aussenfläche dieses Knochens liegt. Die zweite Ebene *V* (die verticale) geht durch letzteren Punkt und steht zur vorigen Ebene und zur Medianebene senkrecht. Da es nur auf die Stellung und nicht auf die Lage der Bogengangebenen ankommt, so können wir irgend

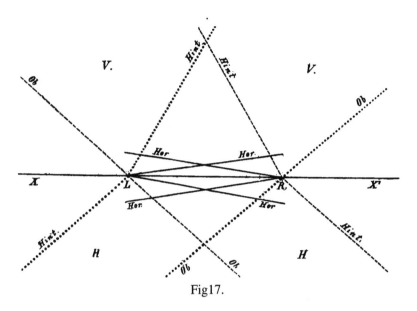

Fig 17.

welche diesen Coordinatenebenen parallele wählen und auf denselben nach dem Princip der descriptiven Geometrie die Tracen der den Bogengangebenen parallelen Ebenen zeichnen. Die Punkte *R* und *L* (Fig. 17) legen wir der Uebersichtlichkeit wegen in den Durchschnitt *XX'* der Ebenen *H* und *V.* In der Figur bedeutet aber das Blatt oberhalb *XX'* die Ebene *V*, unterhalb *XX'* die Ebene *H.* Knickt man das Blatt rechtwinklig um *XX'*, so geben die Tracen unmittelbar ein Bild von der Stellung der Bogengangebenen. In der Figur bedeutet nun *Hor* den horizontalen, *Ob* den oberen und *Hint* den hinteren Bogengang. Man hat sich beim Anblicke der Zeichnung vorzustellen, dass man hinter dem Kopfe steht, dessen Gehörorgane dargestellt sind. Wie man sieht, sind die horizontalen Bogengänge nicht sehr geneigt gegeneinander. Ausserdem sind der hintere Bogengang der einen und der obere Bogengang der andern Seite einander ziemlich parallel. Die als parallel angesehenen Bogengänge beider

a semicircular canal only senses rotation in one direction, we see at once that the right anterior and the left posterior semicircular canals seem to be appropriate to signal rotations in opposite directions about an axis normal to their plane. In the same way the left anterior and right posterior semicircular canals belong together and signal rotations in opposite directions about an axis normal to their plane. In Figure 18 the surfaces H and V on the cube represent the two coordinate planes. We mark the points R and L at the corners of the cube and represent the canal planes with paper. R is depressed and L stands out towards us. The semicircular canals are indicated with the same letters as in the last figure, and the arrows on the planes indicate the rotation sense[8.16] from the ampulla into the semicircular canal.

The two horizontal canals share a common axis passing through the head, parallel to the median plane, and pointing posterior and upward.

The right anterior and left posterior canals indicate rotation in opposite

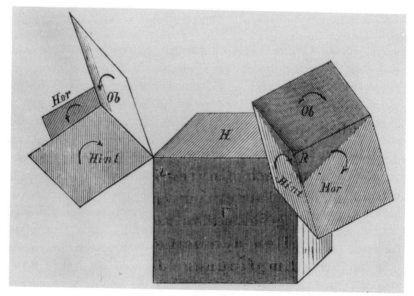

Fig. 18.

Seiten sind durch gleiche Ausführung der Tracen kenntlich. Auf diesen Parallelismus beider Seiten legt nun Brown, wie ich glaube mit Recht, grossen Werth. Bezeichnen wir den Drehungssinn aus der Ampulle in den Bogengang (ohne Passierung des Vorhofs) durch einen Pfeil, und nehmen wir mit Brown an, dass nur bei dem einen Drehungssinn der Bogengang mit Empfindung reagirt, so erkennen wir leicht, dass der obere rechte und der hintere linke Bogengang geeignet scheinen, entgegengesetzte Drehungen um eine zu ihrer Ebene senkrechte Axe zu signalisiren. Ebenso gehören der obere linke und der hintere rechte Bogengang zusammen und geben entgegengesetzte Drehungen um eine andere zu ihrer Ebene senkrechte Axe an. In der Figur 18 bedeuten die Flächen *H* und *V* des Klotzes die beiden Coordinatenebenen. An die Ecken des Klotzes legen wir die Punkte *R* und *L* und stellen an denselben durch Papier die Bogengangebenen vor. Bei *R* sehen wir in eine hohle, bei *L* auf eine convexe Ecke. Die Bogengänge sind mit denselben Zeichen, wie in der vorigen Figur bezeichnet, und die Pfeile auf den Ebenen geben den Drehungssinn von der Ampulle in den Bogegang an.

Es gehören also die beiden horizontalen Bogengänge zusammen zu einer Axe, welche durch den Kopf geht, der Medianebene parallel steht und nach hinten und oben gerichtet ist.

Der rechte obere und der linke hintere Bogengang geben den

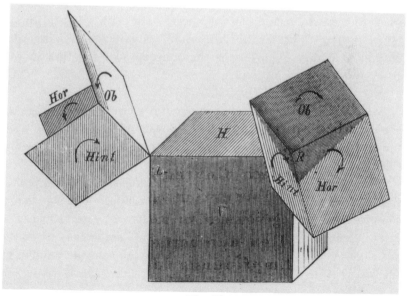

Fig. 18.

directions about an axis that passes through the head pointing anterior upward and to the left; according to B r o w n passing through the left eye and the right mastoid process.

The right posterior and left anterior semicircular canals correspond similarly. Their axis passes through the head in a direction anterior upward and to the right. B r o w n says that it passes through the right eye and the left mastoid process.

Because these three axes are almost orthogonal[8.17] to each other, they are quite appropriately arranged to adequately represent the components for every rotation about an arbitrary axis.

We will now recall those details of F l o u r e n s' experiment that strongly support this interpretation.

F l o u r e n s states that if both of the "vertical lower" [posterior] canals are cut the animals fall onto their back, whereas with an operation of both "vertical upper" [anterior] canals they fall in a forward direction. Indeed, if B r o w n is correct, this result follows from the orientation of the canals.

After cutting the left semicircular canal in a rabbit, F l o u r e n s finds a twist of the head to the left. Combining this result with the experiments in which animals were rotated, one can conclude that with stimulation of the ampullae[8.18] of the horizontal left semicircular canal the animal probably experiences the sensation of rotation to the right about a vertical axis.

When both posterior vertical semicircular canals in the rabbit are cut one observes backward somersaulting. In case of interference with the posterior canals the animal thus feels as though it were falling forward.

According to F l o u r e n s it is very difficult to cut the anterior vertical semicircular canals in rabbits without causing a brain lesion. Nevertheless, F l o u r e n s believes he observed a forward somersault in a few cases. Accordingly, stimulation of the ampullae of both anterior vertical semicircular canals causes a backward falling sensation.

Now we can draw further conclusions easily. With stimulation of the

entgegengesetzten Drehungssinn um eine Axe an, welche durch den Kopf geht nach vorn oben links; nach Brown durch das linke Auge und den rechten proc. mastoid.

Der rechte hintere und der linke obere Bogengang sind wieder zusammengehörig. Ihre Axe geht durch den Kopf nach oben rechts vorn. Brown sagt, sie geht durch das rechte Auge und den linken proc. mastoïd [mastoid].

Da diese drei Axen zu einander nahe rechtwinklig stehn, so sind sie sehr zweckmässig disponirt, um für jede Drehung um eine beliebige Axe immer ausgiebige Componenten zu erhalten.

Wir wollen uns nun noch jene Details der Flourens'schen Versuche in Erinnerung bringen, welche dieser Auffassung sehr günstig sind.

Flourens gibt an, dass bei Durchschneidung der beiden »verticalen untern« Canäle die Thiere auf den Rücken stürzen, während sie bei Operation der beiden »verticalen obern« Canäle nach vorn überfallen. In der That muss, wenn Brown Recht hat, bei der Stellung der Canäle gegeneinander dies das Resultat sein.

Nach Durchschneidung des linken Bogenganges am Kaninchen findet Flourens eine Verdrehung des Kopfes nach links. Halten wir dies Resultat mit den Versuchen an gedrehten Thieren zusammen, so geht daraus hervor, dass das Thier bei Reizung der Ampulle des horizontalen linken Bogenganges wahrscheinlich die Empfindung einer Rechtsdrehung um eine verticale Axe hat.

Bei Durchschneidung der beiden hintern Verticalcanäle am Kaninchen zeigt sich Ueberschlagen nach hinten. Das Thier meint also bei Affection der Ampullen der hintern Bogengänge nach vorne überzustürzen.

Die Durchschneidung der vordern Verticalcanäle an Kaninchen ist ohne Hirnverletzung nach Flourens' Angabe sehr schwer ausführbar. Doch glaubt Flourens in einzelnen Fällen ein Ueberstürzen nach vorn wahrgenommen zu haben. Demnach würde Reizung der Ampullen der beiden vordern verticalen Bogengänge die Empfindung des Umstürzens nach hinten bedingen.

Leicht können wir nun noch weitere Schlüsse ziehn. Bei Rei-

ampulla of the left horizontal semicircular canal we experience rotation to the right about a vertical axis; a clockwise rotation as seen from above. Conversely, if we perform a rotation to the right, the ampulla of the left horizontal semicircular canal will be stimulated. The inertial reaction forces push the contents of the semicircular canals from the ampulla into the semicircular canal (without going through the vestibule). Based on details of Flourens' experiments we may conclude that only a reaction force of the ampulla in the direction against the semicircular canal will be sensed as a rotation in the opposite direction, not, however, a push of the ampulla away from the semicircular canal.[8.19] Brown arrived at a similar conclusion, although apparently for different reasons.

With the help of these results we can explain both of the other phenomena observed by Flourens, and this can be considered as strong evidence for the proposed theory. According to this principle somersaulting backward stimulates both ampullae of the upper (anterior) vertical semicircular canals. On the other hand, if one stimulates the ampullae of the upper semicircular canals a sensation of falling backwards ensues. In fact this is just what we learn from Flourens' experiments, because the animals fall forward in their urge to compensate for the perceived motion.

The third case can be explained just as easily. Falling forward produces a rotational component in both posterior semicircular canals, whose ampullae are pushed in a direction against the semicircular canals. The ampullae of both posterior semicircular canals together signal a forward fall, causing the animals to fall backward in their attempt to compensate, as also observed by Flourens. All of the known facts, down to the last detail, completely support the explanations afforded by our hypothesis.

Flourens has observed rotation of operated animals about the vertical axis and the transverse axis of the head. It is clear how one would have to operate on the animals in order to observe the as yet unnoted rotation about the third, longitudinal axis of the head and trunk. In the rabbit one would have to cut the left posterior and upper semicircular canals, in order that the animal rotates in a clockwise direction about the body's longitudinal axis (as

zung der Ampulle des linken horizontalen Bogenganges empfinden wir Rechtsdrehung um eine verticale Axe, also von oben gesehn Drehung im Sinne des Uhrzeigers. Umgekehrt wird also, wenn wir eine Rechtsdrehung ausführen, die Ampulle des linken horizontalen Bogenganges gereizt werden müssen. Dann geht aber der Trägheitsantrieb des Bogenganginhaltes dem Sinne nach von der Ampulle (ohne Passirung des Vorhofs) in den Bogengang. Wir werden also durch die Details der Flourens'schen Versuche dazu geführt, anzunehmen, dass nur der Bewegungsantrieb der Ampulle gegen den Bogengang als umgekehrte Drehung empfunden wird, nicht aber der Bewegungsantrieb der Ampulle vom Bogen gang weg. Dies hat Brown, wenn auch, wie es scheint, auf andere Gründe gestützt, vermuthet.

Mit Hilfe dieses Resultates können wir aber die beiden andern von Flourens beobachteten Phänomene erklären und dies wird man für eine nicht geringe Stütze der Theorie halten. Ueberstürzen nach hinten reizt nämlich nach dem angegebenen Grundsatz die beiden Ampullen der obern (vordern) verticalen Bogengänge. Umgekehrt wird also Reizung der Ampullen der obern Bogengänge die Empfindung des Umstürzens nach hinten zur Folge haben müssen. Dies lehrt der Flourens'sche Versuch in der That, denn die Thiere stürzen im Bedürfniss die empfundene Bewegung zu compensiren nach vorn.

Gleich gut lässt sich der dritte Fall erklären. Umstürzen nach vorn gibt eine Drehcomponente an beiden hintern Bogengängen, deren Ampullen dadurch gegen die Bogengänge angetrieben werden. Beide Ampullen der hintern Bogengänge zusammen signalisiren also das Vorwärtsstürzen und dies beobachtet auch Flourens, denn die compensationsbedürftigen Thiere fallen nach hinten. Soweit also die Thatsachen bisher bekannt sind, fügen sie sich bis in die Details einer vollständig consequenten Erklärung aus unserer Hypothese.

Flourens hat Drehung der operirten Thiere um die Verticalaxe und um die Queraxe des Kopfes beobachtet. Man sieht, wie man die Thiere operiren müsste, um die dritte bisher vermisste Drehung um die Längsaxe des Kopfes und Rumpfes wahrzunehmen. Man müsste beim Kaninchen den hintern und obern Bogengang der linken Seite anschneiden, damit sich das Thier von hinten gesehn im Sinne

seen from behind). The same operation on the right side would produce a counterclockwise rotation as seen from the same point.

The operations on the left anterior and right posterior, just as on the right anterior and left posterior semicircular canals, should compensate[8.20] for each other. It is truly regrettable that B r e u e r's promising experiments with electrical stimulation[8.21] of the semicircular canals have not yet been concluded. Eventually, it should be possible to stimulate an isolated ampulla by two electrodes on its surface, arranged such that the net current in the brain disappears. The theory proposed here could then be scrutinized free from all the objections raised against vivisection.[8.22]

4.

We have repeatedly noted above that linear as well as angular accelerations are perceived. Accordingly, we can expect to find an organ for linear motion [sensation] in the labyrinth of the ear, similar to the semicircular canals for rotation. This organ must be responsive to acceleration from every side with nearly equal sensitivity. I reached the conclusion, which I had earlier communicated as an assumption, that the sacculus[8.23] of the vestibule constitutes just this organ for the detection of linear acceleration.

Similar considerations led B r e u e r to state the following:

…."In fact, however, rigid bodies with a specific weight different from that of the fluid in which they are somehow suspended will remain behind with every linear movement because of their inertia, i.e., relative to the fluid movement, [the suspended bodies] will move in the opposite direction, and upon stopping in the same direction. The magnitude of these movements depends on the difference of specific weights, the resistance [drag] of the fluid and the manner of suspension. Do we find this in the labyrinth? – Yes, according to the description of the acoustic macula given by H a s s e (Journal of Scientific Zoology, 1867, p. 637)"….

des Uhrzeigers um die Längsaxe des Körpers dreht. Dieselbe Operation auf der rechten Seite würde für denselben Standpunkt die Drehung dem Uhrzeiger entgegen hervorbringen.

Die Operationen am linken vordern und rechten hintern, sowie jene am rechten vordern und linken hintern Bogengang müssten sich compensiren. Es ist lebhaft zu bedauern, dass die Erfolg versprechenden Versuche von B r e u e r mit der electrischen Reizung der Bogengänge bisher noch nicht zum Abschluss gekommen sind. Es müsste endlich doch möglich sein, eine einzelne Ampulle durch die b e i d e n auf dieselbe gesetzten Electroden, wobei die Stromschleifen ins Hirn doch verschwinden, isolirt zu reizen. Die hier vorgebrachte Theorie würde damit einer Prüfung unterzogen werden konnen, welche von den gegen die Sectionen erhobenen Einwürfen vollständig frei wäre.

4.

Wir haben im Vorhergehenden wiederholt die Bemerkumg gemacht, dass Progressivbeschleunigungen ebenso wie Winkelbeschleunigungen empfunden werden. Dem entsprechend werden wir erwarten, ähnliche Organe, wie die Bogengänge es für die Drehung sind, auch für die Progressivbewegung im Ohrlabyrinth vorzufinden. Es muss dies ein Organ sein, welches der Einwirkung der Beschleunigung von jeder Seite her ziemlich gleich zugänglich ist. Ich habe mir die Vorstellung gebildet und dieselbe bei Gelegenheit einer frühern Mittheilung als Vermuthung ausgesprochen, dass der Sacculus des Vorhofs dieses Organ für die Empfindung der Progressivbeschleunigung sei.

B r e u e r von ähnlichen Erwägungen geleitet, spricht sich folgendermaassen aus:

....»Wohl aber werden feste Körper, welche ein von dem der Flüssigkeit verschiedenes specifisches Gewicht haben und irgendwie in derselben suspendirt sind, bei jeder auch geradlinigen Bewegung ihrem Beharrungsvermögen zufolge zurückbleiben, d. h. relativ zur Flüssigkeit die entgegengesetzte, beim Stillstehn des Ganzen dieselbe Bewegung machen. Wie gross diese Bewegungen ausfallen, wird von dem Unterschiede des specifischen Gewichtes, vom Widerstande der Flüssigkeit und von der Art der Suspension ahhängen. Finden wir im Labyrinth dergleichen ? – Ja, wenn wir der Be-

Breuer believes that the inertia of the otolith mass not only conveys the sensation of acceleration but also that of head orientation at rest. He argues for the existence of the latter function based upon the observation by Skrebitzky and Nagel of <u>rotation of the eyeball</u>[8.24], which persists for certain head tilts.

In my first communication I already stressed the need to assume the existence of organs for the perception of linear acceleration that are different from those for angular acceleration. It was also mentioned there that by strictly applying the theory of specific energies, one would require even a third organ, namely for the sensation of [spatial] orientation. This is because when the motion becomes constant the sensation of it disappears. However, the <u>perception of orientation</u>[8.25] – and the consequent eye position [torsion] – remains as long as the tilt.

One difficulty still remains to be overcome. A single kind of nerve would suffice for the perception of orientation and of linear acceleration, if one could assume that simply the pressure on the inner surface of a cavity is sensed. The orientation sensation changes according to the location of the largest pressure, and the sensation of acceleration changes depending on the value of the pressure. However, we know that a continuous linear acceleration is no longer sensed after some time. The acceleration organs therefore fatigue. For the organs of orientation we may not assume such an adaptation. It is difficult to imagine such a non-fatiguing organ. The acoustic nerve seems to behave similarly in many respects. Since we always see that the organs of acceleration perception and of orientation perception demand such distinct properties, we cannot assume them to be present together in a single organ.

If one were transported to a <u>planet of smaller mass</u>[8.26], one would have the initial feeling of continuous sinking. On the basis of the experiments described in this publication this feeling would undoubtedly slowly disappear and one would feel at rest, since a continuous acceleration eventually is no longer sensed. One can hardly propose that only the g

schreibung der Macula acustica von Hasse folgen (Zeitschr. f. wissensch. Zoologie. 1867. p. 637)«....

Breuer meint, dass die Trägheit der Otolithenmasse sowohl die Wahrnehmung der Beschleunigungen als jene der ruhigen Kopfstellungen vermittle. Er führt für die Existenz der letztern Function die von Skrebitzky und Nagel beobachteten Drehungen der Bulbi an, welche bei bestimmter Kopfneigung persistiren.

Ich habe schon in meiner ersten Mittheilung betont, dass man für die Empfindung der Progressivbeschleunigung andere Organe annehmen müsse, wie für jene der Winkelbeschleunigung. Auch wurde dort bemerkt, dass bei strengem Festhalten an den specifischen Energieen sogar noch eine dritte Art Organe, namlich für die Empfindung der Lage, nothwendig sei. Denn wenn die Bewegung gleichförmig geworden, verschwindet die Empfindung. Die Empfindung der Lage – und auch ihre Folge, die Augenstellung – bleibt aber so lange wie die Lage selbst.

Immerhin bleibt hier eine Schwierigkeit aufzulösen. Es würde für die Empfindung der Lage und der Progressivbeschleunigung eine Art von Nerven ausreichen, wenn man annehmen könnte, dass einfach der Druck auf die Innenfläche einer Höhle empfunden wird. Je nach der Stelle des grössten Druckes ändert sich die Empfindung der Lage, je nach der Grösse des Druckes die Empfindung der Beschleunigung. Nun wissen wir aber, dass eine bleibende Progressivbeschleunigung nach einiger Zeit nicht mehr empfunden wird. Die Organe der Beschleunigung erschöpfen sich also. Für die Organe der Lage dürfen wir aber eine solche Erschöpfung nicht annehmen. Es ist schwer, sich ein solches nicht ermüdendes Organ vorzustellen. Der Hörnerv scheint freilich in mancher Hinsicht sich ähnlich zu verhalten. Immer sehn wir aber, dass für die Organe der Beschleunigungsempfindung und der Lagenempfindung so verschiedene Eigenschaften gefordert werden, dass wir sie zugleich in einem Organ nicht annehmen können.

Würde man auf einen Planeten von kleinerer Masse versetzt, so müsste man zunächst das Gefühl des fortwährenden Versinkens haben. Nach den in dieser Schrift mitgetheilten Versuchen kann es aber nicht zweifelhaft sein, dass dies Gefühl allmälig verschwinden und dem der Ruhe Platz machen würde. Denn auch eine

which we normally feel will not be sensed as motion, but as rest, while every larger gravitational acceleration must be sensed as rising and every smaller one as sinking. One must consider that every change of *g* will affect the organs. In order to avoid that situation they tend toward a new state of equilibrium in which they are no longer stimulated. This state of equilibrium is reached earlier with small changes of acceleration, later with larger ones. The organ adapts. Should we furthermore think of an additional organ that <u>refreshes itself so rapidly</u>[8.27] that it always remains functional?

5.

The simple movement sensations, which we have found, are therefore:

1. the six sensations of angular acceleration, corresponding to the six ampullae, arranged in opposing pairs;

2. the sensations of linear acceleration, which also must number at least six and similarly be arranged in opposing pairs, in order to explain the phenomena.

To these the following, in all probability, can be added

3. Sensations of orientation or of equilibrium, whose number and nature we cannot specify at present, thus we do not discuss them further here.

We cannot distinguish these basic elements of sensation during our movements. We sense rotation about this or that axis and rotation sense, and at various accelerations. We also sense linear motion of various direction, sign and acceleration. However, we do not consciously detect the <u>elementary components</u>[8.28] with their respective direction and sign. It is similar to the subject of colors. There are infinitely many wavelengths of light and their corresponding shades of color, however, according to <u>Young</u>'s[8.29] theory,

bleibende Beschleunigung wird zuletzt nicht mehr empfunden. Man kann sich ja auch nicht vorstellen, dass gerade das bei uns übliche *g* allein nicht als Bewegung, sondern als Ruhe, jede grössere Schwerebeschleuniguug aber als Steigen, jede kleinere als Versinken empfunden werden muss. Man wird sich eben zu denken haben, dass jede Veränderung des *g* die Organe afficirt, welche, indem sie dieser Affection zu entgehen suchen, nach einem neuen Gleichgewichtszustand streben, in welchem sie nicht mehr afficirt werden. Dieser Gleichgewichtszustand. wird bei geringeren Aenderungen in der Beschleunigung früher, bei grösseren später erreicht. Das Organ adaptirt sich. Daneben sollen wir uns nun ein anderes Organ denken, welches sich so rasch ernährt, dass es immer in Thätigkeit bleibt?

5.

Die einfachen Bewegungsempfindungen, die wir gefunden haben, sind also:

1. die Empfindungen der Winkelbeschleunigung, und zwar sechs an der Zahl, den sechs Ampullen entsprechend, wovon je zwei einander entgegengesetzt sind;

2. die Empfindungen der Progressivbeschleunigung, welche mindestens auch sechs an der Zahl sein müssen, ebenfalls paarweise entgegengesetzt, wenn die Erscheinungen erklärbar sein sollen.

An diese schliessen sich aller Wahrscheinlichkeit nach noch

3. Empfindungen der Lage oder Gleichgewichtsempfindungen, über deren Zahl und Art wir vorläufig nichts Näheres feststellen können, wesshalb auch hier nicht weiter davon die Rede sein soll.

Diese einfachen Empfindungselemente fühlen wir bei unsern Bewegungen nicht heraus. Wir fühlen eine Drehung um diese oder jene Axe, in diesem oder jenem Sinn, mit verschiedener Beschleunigung, eine Progressivbewegung von verschiedener Richtung, verschiedenem Sinn und verschiedener Beschleunigung, aber die einfachen Componenten mit ihrer Richtung und ihrem Sinn kommen uns nicht zum Bewusstsein. Es ist hier ähnlich wie im Gebiete

there are only three fundamental color sensations. Elsewhere I have pointed out the modifications to Young's theory that I believe are required*). Similarly, there are infinitely many axes in space about which rotation can occur. Subjectively there are only three basic axes – as designated above – corresponding to six fundamental sensations. Whereas Young's fundamental colors could only be assumed and not proven thus far, here the anatomy provides evidence of the organs for these basic sensations, and one will undoubtedly reach the point of stimulating each organ <u>individually</u>.[8.30]

Primitive man was flooded with sensory stimulation from all sides. He had neither time nor opportunity to distinguish what flowed through one or the other nerve bundle. Thus the first concepts he acquires are of things that pertain to certain relatively fixed sets of <u>complexes of stimuli</u>[8.31] that are composed of different sensory perceptions. The first words are names for these things.

Richer experience then teaches him that part of these abstract complexes, such as the sensations that arrive through the eyes, may also occur within different complexes, and are then separated from the initial contexts in which they have been observed alone. In this manner, one abstracts the visible from the object, and by a similar process the form, color, etc. from the visible. The respected language expert <u>Geiger</u>[8.32] tried to prove that all current names for colors originally were names for objects, supposing that primitive man did not see colors, but things.

Further <u>analysis of sensations</u>[8.33], however, primarily falls within the realm of science. It breaks up color into primary colors, or sound into pitch sensations. On a different occasion I tried to show that the track pursued thus far has hardly brought the analysis to completion**). From the overall perceptions one can presumably abstract separate sensation characteristics like form, rhythm, duration, distance, intensity, and pitch. To abstract means

---

*) On intermittent light stimulation. Archiv von Reichert u. Dubois. 1865.
**) Mach, Remarks on the theory of spatial vision. Fichte's Journal of Philosophy, 1864.

der Farben. Es gibt unendlich viele Wellenlängen des Lichtes und zugehörige Farbennüancen, nach der Young'schen Theorie aber nur drei Grundfarbenempfindungen. Auf die Modificationen, welche nach meiner Meinung an der Young'schen Theorie angebracht werden müssen, habe ich anderwärts hingewiesen*). Ebenso gibt es nun unendlich viele Axen im Raume, um welche die Drehung stattfinden kann. Subjektiv aber gibt es bloss drei Grundaxen – die oben bezeichneten – mit ihren zugehörigen sechs Grunddrehempfindungen. Während aber die Young'schen Grundfarben bisher nur vermuthet und nicht aufgezeigt worden sind, weist hier die Anatomie die Organe der Grundempfindung auf und man wird ohne Zweifel dazu gelangen, jedes einzelne Organ für sich zu reizen.

Auf den Naturmenschen stürmen die Erscheinungen von allen Seiten ein; er hat nicht Zeit noch Gelegenheit, zu sondern was auf der einen, was auf der andern Nervenbahn ihm zufliesst. So sind die ersten Begriffe, die er sich erwirbt, Dinge, gewisse mit relativer Constanz auftretende Complexe von Erscheinungen, die sich aus den verschiedensten Sinnesempfindungen zusammensetzen. Die ersten Worte sind Namen für diese Dinge.

Erst eine reichere Erfahrung lehrt, dass Theile dieser Complexe, etwa die durch das Auge zufliessenden Empfindungen in verschiedenen andern Complexen auch vorkommen und sich von jenen, in welchen sie früher allein beobachtet wurden, loslösen lassen. So abstrahirt man von den Dingen das Sichtbare, durch einen gleichen Process von dem Sichtbaren die Gestalt, die Farbe u. s. w. So hat der geachtete Sprachforscher Geiger nachzuweisen gesucht, dass alle jetzigen Farbennamen ursprünglich Namen für Dinge waren, indem der Urmensch eben nicht Farben gesehen hat, sondern Dinge.

Die weitere Analyse der Empfindungen fällt aber in der Regel erst der Wissenschaft zu. Sie löst die Farben in Grundfarben, den Schall in Tonempfindungen auf. Bei anderer Gelegenheit habe ich zu zeigen versucht, dass mit dem bis jetzt Geleisteten die Analyse kaum abgeschlossen ist**). Muthmasslich entspricht Allem, was wir aus unsern Empfindungen abstrahiren können, wie der Gestalt, dem

---

*) Ueber intermittirende Lichtreize. Archiv von Reichert u. Dubois. 1865.

**) Mach, Bemerkungen zur Lehre vom räumlichen Sehen. Fichte's Zeitschrift für Philosophie. 1864.

to sense the common aspects of disparate things. These induced characteristics must exist in all that we compare and must be founded on sensation.

Once we identify every last element of sensation, then perhaps we might also be able to produce them in isolation. From that point there would be no difficulty in analyzing them in every perception, just as we already now can effortlessly sense color in what we see, or tone in sound or noise. Then the age of quantitative psychology[8.34] will not be far off. We will understand the same basic processes which make up all psychological phenomena and which then can be enumerated relative to each other just like elementary physical processes.

6.

We have already noticed a peculiarity associated with movement sensation, which requires further discussion. Pairs can be related to each other as positive and negative. Their organs must be arranged relative to each other such that the function of one cancels the other. If we stick to our hypothesis, then the function of the ampulla of the left horizontal semicircular canal can neutralize the activity of the ampulla of the right horizontal semicircular canal. The same relation applies to the ampullae of the left anterior and right posterior, and to the right anterior and left posterior semicircular canals.

This follows from [the following] experiment. If one stops after a long rotation, the sensation of turning in the opposite direction appears and continues after all accelerations must have already disappeared. The sensation therefore lasts much longer than the stimulus. The continuing sensation, however, can immediately be eradicated by reimposing the initial rotation. With an appropriate choice of velocity one feels no rotation at all. Since one ampulla is stimulated and remains active at the end of rotation while the second ampulla becomes activated with the reintroduction of the rotation and no net effect on sensation occurs, the two ampullae must be related as described above.[8.35] Obviously this notion stands even if our

Rhythmus, der Dauer, der Distanz, der Intensität, der Tonhöhe eine Empfindungsqualität eigener Art. Abstrahiren heisst in Disparatem gemeinsame Bestandtheile herausfühlen. Dieses Herausgefühlte muss nun auch in allem Verglichenen existiren und muss einen reellen Boden haben in der Empfindung. Wenn wir diese letzten Empfindungselemente kennen werden, vielleicht auch im Stande sein werden, sie isolirt wach zu rufen, wird es keine Schwierigkeit haben, sie aus jeder Wahrnehmung herauszufühlen, so wie wir jetzt schon die Farbe aus dem Gesehenen, den Ton aus dem Klang und Geräusch ohne Mühe herausfühlen. Dann wird die Zeit der exacten Psychologie nicht fern sein. Wir werden die gleichartigen Theilvorgänge kennen, aus welchen sich alles Psychologische zusammensetzt und welche sich dann wie physikalische elementare Vorgänge gegeneinander abzählen lassen.

6.

Wir haben an den Bewegungsempfindungen schon eine Eigenthümlichkeit bemerkt, auf die wir hier nochmals zurückkommen müssen. Zwei und zwei stehen zu einander im Verhältniss des Positiven und Negativen. Ihre Organe müssen also in einer derartigen Wechselwirkung sein, dass die Function des einen jene des andern aufhebt. Halten wir unsere Hypothese fest, so kann die Function der Ampulle des linken horizontalen Bogenganges die Thätigkeit der Ampulle des rechten horizontalen Bogenganges aufheben. In der gleichen Beziehung stehn die Ampullen des linken obern und rechten hintern, so wie jene des rechten obern und linken hintern Bogenganges.

Dies geht aus dem Versuch hervor. Hat man eine längere Rotation unterbrochen, so tritt die Empfindung der Gegenrotation ein und währt noch fort, wenn längst alle Beschleunigungen verschwunden sein müssen. Diese Empfindung dauert also viel länger als der Reiz. Die nachdauernde Empfindung kann aber sofort durch die Wiedereinleitung der ursprünglichen Rotation zum Verschwinden gebracht werden. Bei passender Wahl der Geschwindigkeit meint man dann, sich gar nicht zu drehn. Da nun beim Unterbrechen der Rotation die eine Ampulle gereizt wird und fortwirkt, während bei der Wiedereinleitung der Rotation die zweite Ampulle in Func-

hypothesis proves false, and then applies to whichever organs are really the basis of movement sensation.

The long aftereffect of movement sensation and the relation between the ampullae also explains why vertigo does not occur with simple short head movements. The sensation that is induced at the start of movement and continues during the ongoing movement is extinguished by the sensation at the end of the movement.

If we rotate at constant velocity over longer time the effect of the initial rotational moment decays. We [continue to] turn without any further sensation of rotation. If we now interrupt the rotation a moment arises in the opposite direction, which becomes fully effective by means of an outlasting sensation. This comprises rotational vertigo.

7.

I now have to discuss a difference between my interpretation of how the semicircular canals function, and that of <u>Breuer and Brown</u>.[8.36] Since the semicircular canal contents are fluid, one first and naturally assumes, as I also did originally, that the content of the semicircular canal carries out a relative counter-flow during each body turn. This idea occurred naturally to me on the basis of earlier, completely separate experiments. I frequently observed that fluid in a semicircular canal that has been broken open and resealed with glass flows easily when the bony labyrinth is rotated. With such a procedure in a petrous bone, however, one can never be completely sure that the labyrinth is not injured in some place. When I then performed the experiment with a closed glass pipette of the size and form of a semicircular canal, or with a corresponding groove in a brass plate covered with glass, I never observed any relative flow on a centrifuge, even for rapid rotation. On the other hand, a flat water drop of about 1^{cm} diameter with a dusted surface on the disk of the centrifuge readily demonstrates the conservation of areas principle. I thus gave up the original hypothesis of actual flow and assumed that the pure turning moment (the pressure) of the semicircular canal contents, without noticeable rotation, activates the nerve

tion tritt und sich nun gar kein Empfindungseffect einstellt, so müssen die beiden Ampullen in der angegebenen Beziehung stehn. Selbstverständlich bleibt dies bestehn, wenn auch unsere Hypothese fällt, und gilt dann von jenen Organen, welche der Bewegungsempfindung wirklich zu Grunde liegen.

Aus der grossen Nachdauer der Bewegungsempfindung und dieser Beziehung der Ampullen erklärt es sich auch, warum man bei einfachen kurzen Kopfwendungen keinen Schwindel bekommt. Die Empfindung, welche beim Bewegungsbeginn erregt wird und während der ganzen Bewegung nachdauert, wird ausgelöscht durch die Empfindung beim Ende der Bewegung.

Wenn wir uns längere Zeit gleichförmig drehen, erlischt die Wirkung des beim Beginn erregten Drehungemomentes. Wir drehen uns ohne weitere Drehempfindung. Unterbrechen wir nun die Drehung, so entsteht ein entgegengesetztes Drehungsmoment, welches mit seiner nachdauernden Empfindung zur vollen Wirksamkeit gelangt. So entsteht der Drehschwindel.

7.

Ich muss nun eine Differenz zwischen meiner Auffassung der Functionsweise der Bogengange und jener von Breuer und Brown besprechen. Da der Bogenganginhalt flüssig ist, so verfällt man zunächst und natürlich auf den Gedanken, auf den auch ich ursprünglich kam, dass der Inhalt des Bogenganges bei jeder Körperdrehung eine relative Gegenströmung ausführe. Dieser Gedanke lag mir besonders nahe, da ich bei früheren ganz heterogenen Versuchen oft gesehn hatte, dass die Flüssigkeit in einem aufgebrochenen und wieder mit Glas verschlossenen Bogengange bei Drehung und Wendung des Felsenbeins leicht herumfliesst. An einem Felsenbein kann man aber nie ganz sicher sein, ob das Labyrinth nicht irgendwo verletzt ist. Machte ich nun den Versuch mit einem geschlossenen Glasröhrchen von der Grösse und Form eines Bogenganges, oder mit einer entsprechenden Rinne in einer Messingplatte, die mit Glas gedeckt war, so erhielt ich auf der Centrifugalmaschine selbst bei raschen Drehungen niemals eine beobachtbare relative Drehung. Ein flacher Wassertropfen von etwa 1^{cm} Durchmesser auf der Scheibe der Centrifugalmaschine, zeigt hingegen bei bestäubter Oberfläche das

similar to the way pressure stimulates the tactile nerves of skin. Breuer and Brown, on the contrary, assume a substantial flow of endolymph.

One will easily notice that the outlasting duration of apparent motion following the interruption of a rotation cannot be explained by a continuing flow. How unbelievably small the coefficient of friction in question would have to be to permit such a possibility! In any case one must accept the fact that the sensation lasts much longer than the stimulus. Continuous flow is unnecessary as the basis for explanation. Even further! The fluid flow can only act as a stimulus to the extent that the movement required by the principle of the conservation of areas is impeded.

We will now see how the relation between movement and its perception is covered by the different hypotheses. Let us first assume that there is no substantial movement of endolymph. If we are suddenly turned at a constant velocity, acceleration is present only at the beginning. Nevertheless, we feel the continuation of the movement. One might say that acceleration activates the continuous sensation of velocity. This holds true for linear as well as for angular accelerations. This is easy to understand teleologically, since it is most important to perceive the velocity at which we are moved. This is especially critical, for example, when animals are required to use their space organ for measuring the size of a jump, for which the resulting movement parameters, such as the velocity achieved with constant mass, are crucial.

The stimulated movement perception continues with declining intensity. Let us first consider such a short movement that during the course of it we need not pay any attention to a reduction of the stimulated sensation. The duration of the movement vanishes relative to the duration of the outlasting sensation. For that case we can construct a mathematical model of the relation between movement and sensation although, of course, its general validity cannot yet be proven. We assume that the sensations corresponding

Flächenprincip sofort. Ich habe daher die ursprüngliche Annahme einer wirklichen Strömung aufgegeben und angenommen, dass das blosse Drehungsmoment (der Druck) des Bogenganginhaltes ohne merkliche Drehung auf den Nerven wirke, so wie etwa Druck die Tastnerven der Haut erregt. B r e u e r und B r o w n hingegen nehmen eine ausgiebige Strömung der Endolymphe an.

Man wird leicht bemerken, dass die N a c h d a u e r der Scheinbewegung nach Unterbrechung einer Rotation nicht aus einer fortdauernden Strömung zu erklären ist. Wie unerhört klein müsste der in Frage kommende Reibungscoefficient sein, der dies allein ermöglichen würde. Man muss also jedenfalls annehmen, dass die Empfinduug viel länger dauert als die Erregung. Dann wird aber die a n h a l t e n d e Strömung als Erklärungsgrund entbehrlich. Noch mehr. Die Flüssigkeit kann nur in so fern als Reiz wirken, als sie an der durch das Flächenprincip geforderten Bewegung gehindert ist.

Wir wollen nun sehen, wie sich das Verhältniss von Bewegung und deren Empfindung nach den verschiedenen Auffassungen gestaltet. Nehmen wir zunächst an, es finde keine ausgiebige Bewegung der Endolymphe statt. Kommen wir plötzlich in eine gleichförmige Bewegung, so wirkt die Beschleunigung nur am Anfang. Dennoch fühlen wir die Fortdauer der Bewegung. Man kann sagen die B e s c h l e u n i g u n g erregt die fortdauernde Empfindung einer G e s c h w i n d i g k e i t. Dies gilt sowohl für die Progressiv- als auch für die Winkelbeschleunigung. T e l e o l o g i s c h ist dies leicht zu begreifen, denn es handelt sich hauptsächlich darum eine Vorstellung von der G e s c h w i n d i g k e i t zu erlangen, in welche wir bei der Bewegung gerathen sind. Namentlich ist dies wichtig, wenn die Thiere ihr Raumorgan etwa zur Abmessung eines Sprunges sollen verwenden können, bei welchem es auf die erlangte Bewegungsgrösse, also bei gleichbleibender Masse, auf die erlangte Geschwindigkeit ankommt.

Die erregte Bewegungsempfindung wirkt in abnehmender Intensität nach. Betrachten wir zuerst eine Bewegung von so kurzer Dauer, dass wir während derselben auf die Abnahme der erregten Empfindung keine Rücksicht zu nehmen brauchen. Die Dauer der Bewegung soll also gegen die Dauer der Nachempfindung verschwinden. Für diesen Fall können wir uns ein mathematisches Bild des Zusammenhanges von Bewegung und Empfinduug ent

to a series of successive stimuli can be added algebraically. If v represents the variable angular velocity of a head rotation or the velocity of a linear movement, and t the time, then we [can] state the induced sensation of velocity caused by an element of motion as $d\rho = a\,\dfrac{dv}{dt}\,dt$.

If we assume that the duration of the movement vanishes in comparison to the duration of the aftereffect, at the end of the movement we have the sensation

$$\rho = a \int_{t_0}^{t_1}\frac{dv}{dt}dt$$

which itself $= 0$ if v has the same value at t_1 and t_0, that is, the movement begins and ends with the same velocity; since we always have a complete differential[8.37] under the integral sign.

In every other case the value of ρ depends only on the initial and final values of v. Because $\dfrac{dv}{dt}$ is proportional to the pressure p, which the organ in the labyrinth experiences, we can also write

$$\rho = k \int_{t_0}^{t_1} p\,dt$$

where k is a new constant.

One therefore sees that the sensation of movement is related to the varying acceleration in much the same way that the final velocity of a mass depends on the variable pressure, or the swing velocity of a magnetic needle depends on the variable intensity of the induction current. This seems to be teleologically important.[8.38]

Let us assume that the movement of the endolymph vanishes and postulate that the elements of sensation can be algebraically summed as discussed above. Then every increase in velocity corresponds to an increase in sensation of same sign, so that for a short movement the final motion

werfen, dessen allseitiges Zutreffen freilich für jetzt noch nicht nachgewiesen werden kann. Die aufeinanderfolgenden Reizwirkungen
und die entsprechenden Empfindungen betrachten wir als algebraisch summirbar. Ist v die variable Winkelgeschwindigkeit einer
Kopfwendung oder die Geschwindigkeit einer progressiven Bewegung, t die Zeit, so setzen wir die durch ein Element der Bewegung

erregte Geschwindigkeitsempfindung $d\rho = a \dfrac{dv}{dt} dt.$

Nehmen wir die Dauer der Bewegung verschwindend gegen die
Dauer der Nachwirkung, so haben wir am Ende der Bewegung die
Empfindung

$$\rho = a \int_{t_0}^{t_1} \frac{dv}{dt} dt$$

welche selbst $= 0$ ist, wenn für t_1 und t_0 auch v denselben Werth
hat, wenn die Bewegung mit derselben Geschwindigkeit anfängt
und endigt; denn wir haben unter dem Integralzeichen immer ein
vollständiges Differential.

In jedem andern Fall hängt der Werth des ρ nur von dem

Anfangs- und Endwerthe von v ab. Da $\dfrac{dv}{dt}$ proportional dem Druck

p, welchen das Organ im Labyrinth erfährt, so können wir auch
schreiben

$$\rho = k \int_{t_0}^{t_1} p\, dt$$

wobei k eine neue Constante.

Man sieht also, dass die Bewegungsempfindung in derselben
Weise von der variablen Beschleunigung abhängt, wie die Endgeschwindigkeit einer Masse von dem variablen Druck oder die Aus-
schlagsgeschwindigkeit einer Magnetnadel von der variablen Intensität des Inductionsstromes, was eben teleologisch wichtig zu sein
scheint.

Nehmen wir also die Bewegung der Endolymphe als
verschwindend an und setzen die Empfindungselemente
in der besprochenen Weise als algebraisch summirbar
voraus, so entspricht jedem Geschwindigkeitszuwachs
ein Empfindungszuwachs von gleichem Zeichen, der-

sensation (velocity sensation) parallels the total change in velocity.

Let us now assume a substantial endolymph flow. Call the angular velocity of the bony labyrinth $v = \varphi(t)$, and that of the contents u [both relative to inertial space]. If the friction is taken as proportional to the relative velocity one obtains

$$\frac{du}{dt} = k\big[\varphi(t) - u\big]$$

where k is a constant. From this equation it follows that

$$u = e^{-kt}\left[k' + \int e^{kt}\varphi(t)dt\right]$$

in which k' is the constant of integration.

Let us regard the instantaneous sensation of rotation as proportional to the relative velocity, then this proportionality can be expressed as

$$\varphi(t) - e^{-kt}\left[k' + \int e^{kt}\varphi(t)dt\right]$$

The simple relation between the actual final velocity and the movement sensation then gets lost.

We will now assume that the endolymph barely flows, and concentrate on the gradual attenuation of the induced sensation. If the intensity of sensation at the moment of stimulation is 1, it may be expressed over the course of time τ by $\psi(\tau)$. If the velocity of the movement, beginning at $t = 0$, is $\varphi(t)$, then the movement sensation at time t is expressed as

$$\int_{\tau=0}^{\tau=t} \varphi'(t-\tau)\psi(\tau)d\tau$$

One sees that each of the last two assumptions yields results that are already quite complicated. It would be even less pleasant if the two had to be combined. Finally one might also take fatigue of the organ during constant stimulation into account. However, one will certainly note that the first and simplest concept probably meets all requirements.

Note that the movement of endolymph cannot be used to explain the sensation of linear acceleration using the same principle applied for

art, dass für eine kurze Bewegung die endliche Bewegungsempfindung (Geschwindigkeitsempfindung) der totalen Aenderung der Geschwindigkeit parallel geht.

Nehmen wir nun eine ausgiebige Strömung der Endolymphe an. Nennt man die Winkelgeschwindigkeit des knöchernen Labyrinthes $v = \varphi\,(t)$, jene des Inhaltes u und, setzt man die Reibung proportional der Relativgeschwindigkeit, so ist

$$\frac{du}{dt} = k\big[\varphi\,(t) - u\big]$$

wobei k eine Constante. Hieraus folgt

$$u = e^{-kt}\left[k' + \int e^{kt}\varphi(t)dt\right]$$

in welchem Ausdruck k' die Integrationsconstante vorstellt.

Betrachtet man die momentane Drehempfindung als proportional der Relativgeschwindigkeit, so ist dieselbe proportional dem Ausdrucke

$$\varphi(t) - e^{-kt}\left[k' + \int e^{kt}\varphi(t)dt\right]$$

Die einfache Beziehung zwischen der erlangten Endgeschwindigkeit und der Bewegungsempfindung geht hiemit verloren.

Wir wollen nun noch annehmen, dass die Endolymphe nicht merklich strömt, dafür aber Rücksicht nehmen auf die allmälige Abnahme der einmal erregten Empfindung. Wenn die Intensität der Empfindung im Momente der Erregung 1 ist, soll die nach Verlauf der Zeit τ durch $\Psi(\tau)$ vorgestellt werden. Die Geschwindigkeit der Bewegung, welche mit $t = 0$ beginnen soll, sei $\varphi(t)$, dann ist die Bewegungsempfindung zur Zeit t ausgedrückt durch

$$\int_{\tau=0}^{\tau=t} \varphi\,'(t - \tau)\psi\,(\tau)\,d\tau$$

Man sieht, dass die beiden letzten Annahmen, jede für sich, schon sehr complicirte Resultate geben. Noch weniger erfreulich wäre es, wenn beide vereinigt werden müssten. Es könnte endlich auch gefordert werden, auf die Erschöpfung des Organs bei constantem Reize Rücksicht zu nehmen. Doch wird man wohl bemerken, dass wahrscheinlich die erste und einfachste Betrachtungsweise für alle Zwecke ausreicht.

Was die Bewegungen der Endolymphe betrifft, so ist noch zu bemerken, dass man dieselben nicht verwenden kann, um die Em-

rotational movements. The labyrinth has no room to allow continuing linear displacement for the duration of the aftersensations. The canals, in contrast, have the advantage that the assumption of a circulating movement in a closed system poses no such difficulties. Incidentally, as I have heard, certain kinds of animals have labyrinths whose contents are gelatinous, which would make this concept completely impossible. The truth may possibly lie closer to Breuer-Brown's concept for some animals, and closer to mine for others.

8.

It was Hitzig *) who doubted that the inertial accelerations of the content of the labyrinth are large enough to serve as perceptible stimuli. I considered this question myself and looked for an answer, of course, with only very few clues available. First, one must be aware of the high sensitivity of the nerve endings in general. It has already been mentioned that the stimulus strength can be related to the induced sensation energy by analogy to the relationship between the energy of a spark and the energy of a batch of gunpowder that it explodes. We will go further into details below.

The manner in which pressure on a sensitive nerve end organ induces nervous energy is probably not known. We therefore cannot begin with the pressure of the labyrinth content during a given acceleration. However, for some cases one knows the minimal stimulus energy that must be applied to a sensory organ over one second to induce any sensation.

Töpler[8.39] and Boltzmann**)[8.40] used optical methods to determine the excursions of the air in a pipe and thereby could calculate that at the threshold of hearing an energy transfer to the ear of about[8.41]

$\dfrac{1}{3\,\text{Trillion}}$ kilogram-meter of work occurs each second. I measured the air

excursion with the help of flames, and thereby found even a slightly smaller

---

*) Investigations on the brain. p. 255.
**) Pogg, Ann. 1870. vol. 141. p. 351.

pfindung der Progressivbewegungen nach demselben Princip zu erklären, wie jene der Drehbewegungen. Es ist im Labyrinth kein Raum, um progressive Nachbewegungen von der Dauer der Nachempfindung zu gestatten. Bei den Canälen hat man allerdings den Vortheil, dass die Annahme einer in sich zurücklaufenden Bewegung an sich nicht dieselben Schwierigkeiten bietet. Es sollen übrigens, wie ich gehört habe, bei manchen Thierarten Labyrinthe mit gallertartigem Inhalt vorkommen, bei welchen dann diese Auffassung ganz unmöglich ist. Möglich, dass bei manchen Thieren sich mehr die Breuer-Brown'sche, bei andern meine Auffassung der Wahrheit nähert.

8.

Es ist von Hitzig*) in Zweifel gezogen worden, dass die Massenbeschleunigungen des Labyrinthinhaltes ausreichen, um als wahrnehmbare Reize zu wirken. Diese Frage habe ich mir selbst vorgelegt und auch zu beantworten versucht, wofür natürlich nur wenige Anhaltspunkte zu finden sind. Zunächst muss auf die grosse Empfindlichkeit der Nervenendapparate überhaupt hingewiesen werden. Es ist schon daran erinnert worden, dass die Reizbarkeit zu der ausgelösten Empfindungsarbeit beiläufig in dem Verhältniss steht, wie die Arbeit eines Funkens zur Arbeit einer entzündeten Pulverladung. Wir wollen aber weiter ins Detail eingehn.

In welcher Weise der Druck auf einen empfindlichen Nervenendapparat Nervenarbeit auslöst, ist wohl nicht bekannt. Von dem Drucke des Labyrinthinhaltes bei gegebener Beschleunigung können wir also nicht ausgehn. Wohl aber ist für einige Fälle das Minimum der Reizarbeit bekannt, welches in der Secunde einem Sinnesorgan zufliessen muss, damit noch Empfindung erregt werde.

Töpler und Boltzmann**) haben auf optischem Wege die Excursionen der Luft einer Pfeife bestimmt und daraus ableiten können, dass an der Grenze der Hörbarkeit an das Ohr in der Secunde etwa $\dfrac{1}{3\,\text{Billionen}}$ Kilogrammeter Arbeit abgegeben werden. Ich selbst habe mit Hilfe von Flammen die Luftexcursionen gemes-

---

*) Untersuchungen über das Gehirn. S. 255.
**) Pogg. Ann. 1870. Bd. 141. S. 351.

energy for the <u>hearing threshold</u>*).[8.42]

Experiments by <u>Thomsen</u>**)[8.43] allowed him to determine the mechanical light equivalent of a candle which burns 8.2 grams of whale oil per hour. It emits $^1/_{35}$ kilogram-meter of energy per second as radiation. At a distance of 115 m this corresponds to $\dfrac{1}{5740000 \; \textit{Millionen}}$ kilogram-meter per 1 mm². Nevertheless, this provides sufficient energy to see the candle.

We can now attempt a similar calculation for the ear labyrinth. If a mass m having velocity v_0 is to take on the velocity v_1, then the corresponding change of <u>living [kinetic] energy</u>[8.44], and thus the required work is given by $\dfrac{m}{2}\left(v_1^2 - v_0^2\right)$. If the bony labyrinth undergoes a change in velocity then, whether slowly or quickly, its contents must finally undergo the same change in velocity. It is probably the rigid parts, and not the sensitive ones, that must deliver the work required to change the velocity of the contents. Some part of this work, however, is delivered by the nerve end organs themselves, thereby giving rise to the sensation.

If a weight p, given in kilograms, is subjected to a change in its velocity v, given in meters, and if the <u>acceleration</u>[8.45] of gravity is also expressed in meters, then the energy required is $\dfrac{p\,v^2}{2\,g}$.

Let us assume that 0.1 grams of fluid inside the labyrinth experiences a velocity change of 1^{cm} per second by means of the resistance of the end organs. Then the stimulus energy transmitted to the nerve is $\dfrac{1}{1960\,\text{Million}}$ kilogram-meter. Although the numbers were purposely chosen small, the calculated energy is much higher than in the other two cases. If one claims that the hypothetical organs of movement sensation in the labyrinth are no less sensitive than other sensory organs, then the available stimulus values are more than sufficient

---

*) Optical-acoustic experiments. The spectral and stroboscopic investigation of sound-emitting bodies. Prague 1872.

**) Pogg. Ann. Vol. 125. P. 389.

sen und daraus für die Grenze der Hörbarkeit eine noch etwas kleinere Arbeit gefunden*).

Nach den Versuchen von Thomsen**) zur Bestimmung des mechanischen Lichtaequivalentes sendet eine Kerze, welche in der Stunde 8.2 Gramme Walrath verzehrt, in der Secunde $^1/_{35}$ Kilogrammeter Arbeit als Strahlen aus. Davon kommen auf 1 ❏ ^{mm} in 115 Metern Entfernung $\dfrac{1}{5\,740\,000\,Millionen}$ Kilogrammeter. Gleichwohl reicht diese Arbeit hin, um die Kerze zu sehn.

Eine ähnliche Berechnung können wir nun für das Ohrlabyrinth versuchen. Wenn einer Masse m, welche die Geschwindigkeit v_0 hat, die Geschwindigkeit v_1 ertheilt wird, so ist die betreffende Aenderung der lebendigen Kraft, also die aufgewandte Arbeit gegeben durch $\dfrac{m}{2}\left(v_1{}^2 - v_0{}^2\right)$. Erfährt nun das knöcherne Labyrinth eine Geschwindigkeitsänderung, so muss schliesslich, ob allmälig, ob schnell ist einerlei, der Inhalt desselben die gleiche Geschwindigkeitsänderung erfahren. Diese Aenderung der Geschwindigkeit des Inhaltes wird wohl zum Theil durch feste und nicht empfindliche Theile herbeigeführt, welche diese Arbeit aufbringen müssen. Ein Theil dieser Arbeit wird aber von den Nervenendapparaten selbst geliefert und wird umgekehrt Anlass zur Empfindung.

Soll einem in Kilogrammen ausgedrückten Gewicht p eine in Metern angegebene Geschwindiggeitsänderung v beigebracht werden, so ist, wenn die Beschleunigung der Schwere ebenfalls in Metern ausgedrückt wird, die hiezu erforderliche Arbeit $\dfrac{p\,v^2}{2\,g}$.

Nehmen wir nun an, es würde 0.1 Gramm Flüssigkeit des Labyrinthinhaltes durch den Widerstand der Endapparate in einer Secunde eine Geschwindigkeitsänderung von 1^{cm} ertheilt, so ist die den Nerven zufliessende Reizarbeit hiebei $\dfrac{1}{1960\,Millionen}$ Kilogrammeter. Obgleich also die Zahlen absichtlich niedrig gegriffen sind, so fällt doch die berechnete Arbeit sehr viel höher aus, wie in den beiden andern Fällen. Muthet man den hypothetischen Organen der Bewegungsempfindung im Labyrinth

---

*) Optisch-akustische Versuche. Die spectrale und stroboscopische Untersuchung tönender Körper. Prag 1872.
**) Pogg. Ann. Bd. 125. S. 389.

for their stimulation.

As mentioned, as to whether the counter-current may or may not be sensed, it is irrelevant, for a given velocity change of the labyrinth content, whether the total energy delivered to the nerve comes about quickly or slowly. It is only the total change in kinetic energy that is important. This is also apparent if we examine the situation in detail. Let us call the velocity of the [labyrinth] content relative to the wall v and the mass of this content m. We assume friction to be proportional to the relative velocity and denote the coefficient of friction as k. Finally $\underline{V}^{8.46}$ is the initial relative velocity, s the path that the content of the labyrinth takes relative to the wall, and t the time. The following $\underline{\text{equations}}^{8.47}$ are then readily obtained:

$$m\frac{dv}{dt} = -kv$$

$$v = Ve^{-\frac{k}{m}t}$$

$$s = \frac{mV}{k}\left(1 - e^{-\frac{k}{m}t}\right)$$

$$v = V - \frac{ks}{m}$$

It can be seen that, strictly speaking, the contents of the labyrinth never come to rest with our assumptions. However, the entire path, when the content comes to rest, is finite and given by

$$S = \frac{mV}{k}$$

The total energy that must be applied to bring the content to relative rest is therefore

$$k\int_0^S vds = k\int_0^S \left(V - \frac{ks}{m}\right)ds = \frac{mV^2}{2}$$

according to our assumptions. The coefficient of friction k drops out completely. The total energy is always the same [depending only on V]. It is provided by small forces acting over a large distance, and consequently over a longer time if the friction is small. With larger friction it is delivered over a shorter distance and for a briefer time.

keine geringere Empfindlichkeit zu wie den übrig-
en Sinnesorganen, so sind die vorhandenen Reizgrössen
zu ihrer Erregung reichlich genügend.

Wie erwähnt, ist es für die Gesammtarbeit, welche den Nerven
zufliesst, gleichgültig, ob bei einer gegebenen Geschwindigkeitsänderung der Labyrinthinhalt diese rasch oder langsan annimmt,
ob dessen Gegenströmung unmerklich oder merklich ist. Es kommt
eben nur auf die totale Aenderung der lebendigen Kraft an. Dies
zeigt sich auch, wenn wir das Schema des Vorgangs im Detail ver-
folgen. Nennen wir die Relativgeschwindigkeit des Inhalts gegen
die Wand v, m die Masse des Inhalts und k den Coefficienten der
Reibung, welche wir der Relativgeschwindigkeit proportional setzen,
ferner V die anfängliche Relativgeschwindigheit, s den Weg, welchen der Labyrinthinhalt relativ gegen die Wand beschreibt, t die
Zeit, so ergeben sich leicht der Reihe nach die Gleichungen:

$$m\frac{dv}{dt} = -kv$$

$$v = Ve^{-\frac{k}{m}t}$$

$$s = \frac{mV}{k}\left(1 - e^{-\frac{k}{m}t}\right)$$

$$v = V - \frac{ks}{m}$$

Streng genommen kommt, wie man sieht, bei unsern Annahmen
der Labyrinthinhalt nie zur Ruhe. Der ganze Weg aber, nach welchem die Ruhe eintreten würde, ist endlich und gegeben durch

$$S = \frac{mV}{k}$$

Die gesammte Arbeit, welche aufgewandt werden muss, um
den Inhalt in relative Ruhe zu bringen, ist demnach

$$k\int_0^S v\,ds = k\int_0^S \left(V - \frac{ks}{m}\right)ds = \frac{mV^2}{2}$$

wie wir dies auch angenommen haben. Es fällt der Reibungscoefficient k ganz aus. Die Gesammtarbeit ist immer dieselbe, sie wird
mit kleinen Kräften auf einer grossen Strecke und natürlich auch
in längerer Zeit geleistet, wenn die Reibung gering ist. Bei grosser
Reibung wird sie mit grossen Kräften auf kurzer Strecke und in
kurzer Zeit geliefert.

9.

Breuer correctly disputed Goltz' assumption that the ampullae are under more or less tension depending on the height of the fluid column above them. Breuer emphasized that the membranous canals are filled with fluid and are immersed in a fluid of nearly the same specific weight. The pressure on the inner and outer surface of each element of the membranous canal wall will always be balanced. There is, however, one more thing to notice. Just as perilymph can flow downward, that is against the round window, endolymph could do the same and a modified version of Goltz's view would still be valid. I am actually not in agreement with Breuer when he states: "because in closed tubes of small diameter, where air cannot enter to fill the empty space when the water sinks to the deepest places, the water does not push downward at all". Even in completely sealed spaces totally filled with fluid, that fluid must exert a downward pressure, and that pressure must increase from top to bottom. This is easily demonstrated by the buoyancy experienced by small bodies immersed in fluids of lower specific weight. Such particles ascend in the fluid. A pressure-sensitive corpuscle on the wall or within such a cavity can at least sense every change of orientation as a change of pressure. I therefore do not deny that the otoliths, because of their different specific weight relative to the surrounding fluid, seem to be perfectly suited to play such a role.

10.

The sensation of rotation or linear acceleration can be described in psychological terms as a series of spatial position sensations. One might now ask how simple activation of a sensory organ can induce a whole series of sensations. An image for this would be the entire chain of memories that is evoked if only one link of the chain is recalled, or the occurrence of complicated reflex patterns. By the way, it is also possible that velocity is perceived by a process of its own.

9.

Breuer hat mit Recht gegen die Goltz'sche Annahme, nach welcher die Ampullen je nach der Höhe der darüber stehenden Flüssigkeitssäule mehr oder weniger angespannt werden, hervorgehoben, dass die häutigen Bogengänge mit Flüssigkeit gefüllt und, in einer Flüssigkeit von nahe gleichem specifischen Gewicht eingetaucht sind. Der Druck auf die Innen- und Aussenfläche jedes Wandelementes des häutigen Bogenganges wird immer gleich sein. Es ist aber noch Eins zu beachten. So gut die Perilymphe nach abwärts zieht, etwa gegen das runde Fenster, kann es auch noch die Endolymphe thun und eine Modification der Goltz'schen Ansicht wäre noch immer zulässig. Ich bin nämlich nicht mit Breuer einverstanden, wenn er sagt: »denn in geschlossenen Röhren kleiner Dimensionen, wo nirgend Luft zutreten kann, um den durch Senkung des Wassers nach den tiefsten Stellen entstehenden leeren Raum anzufüllen, drückt das Wasser gar nicht nach abwärts«. Auch in vollständig geschlossenen, vollständig von Flüssigkeit erfüllten Räumen, muss die Flüssigkeit abwärts drücken und der Druck von oben nach unten zunehmen. Dies beweist ja augenscheinlich der Auftrieb, den kleine specifisch leichtere Körperchen in einer solchen Flüssigkeitsmasse erfahren. Solche Theilchen steigen ja in der Flüssigkeit auf. Ein druckempfindliches Körperchen an der Wand oder im Innern einer solchen Höhle kann immerhin jede Bewegung und jede Aenderung der Stellung als Druckänderung empfinden. Ich stelle darum nicht in Abrede, dass die Otolithen, vermöge des Unterschieds im specifischen Gewicht gegen die umgebende Flüssigkeit, vorzüglich geeignet scheinen, die angeführten Umstände anzuzeigen.

10.

Die Empfindung der Drehung oder der Progressivbewegung lässt sich psychologisch auflösen in eine Folge von Lagenempfindungen. Man kann nun fragen, wieso eine einfache Erregung eines Sinnesorgans eine ganze Reihe von Empfindungen auslösen kann. Ein Bild dafür wären die Reihen, welche im Gedächtniss ablaufen, wenn an einem Glied der Kette gerührt wird, oder die complicirteren Reflexvorgänge. Es ist übrigens auch möglich, dass der Empfindung der Geschwindigkeit ein Vorgang eigener Art zu Grunde liegt.

I will mention another curious phenomenon that might be quite important. I already mentioned it in my second communication and now see that Brown also discussed it. During the experiments described above it became clear that a feeling of nausea[8.48] developed, particularly when it was difficult to reconcile the movement sensations with the optical sensations. It seemed as if parts of the activity coming from the labyrinth were forced to leave the optical pathways, which were then occupied by other activity, and to find completely different routes.

Years ago, while performing optical experiments, I was attracted to a similar although still very incomplete idea. Namely, it seemed to me that because of the incompatibility of the two retinal images, parts of the optical stimulus would have to flow along different pathways. We would then learn to interpret them as a separate feature of the image that we could call depth sensation, in agreement with Hering. I have also repeatedly experienced a feeling of nausea when I tried to fuse stereoscopic images having large disparity.

11.

Study of Hasse's work brings one to the viewpoint that the organ of hearing developed from a tactile organ by adaptively evolving to periodic stimuli. The connection between the hearing organ and movement is therefore not surprising. The semicircular canals indeed appear much earlier in the order of animals[8.49] than the cochlea. It is possible, as Breuer also surmises, that the so-called hearing organs of the lowest animals are nothing more than organs of movement sensation. This relationship of the hearing organ to movement then might have been inherited from the time in which changing location, grasping and gulping were all one. The direct and immediate initiation of movement by noise (when startled) is also relevant here.

The function of the labyrinth as a balance organ is further supported by morphological evidence. The swimbladder of fish[8.50], which is definitely an organ of equilibrium, exerts its effect directly on the contents of the

Noch eine eigenthümliche Erscheinung will ich erwähnen, welche vielleicht nicht unwichtig ist. Ich habe derselben schon in meiner zweiten Mittheilung gedacht und sehe jetzt, dass auch Brown sie bespricht. Es hat sich bei den beschriebenen Versuchen wiederholt gezeigt, dass ein Ekelgefühl sich hauptsächlich dann einstellte, wenn es schwer war, die Bewegungsempfindungen mit den optischen Eindrücken in Einklang zu bringen. Es sah so aus, als ob ein Theil des vom Labyrinth ausgehenden Reizes gezwungen worden wäre, die optischen Bahnen, die ihm durch einen andern Reiz verschlossen waren, zu verlassen und ganz andere Bahnen einzuschlagen.

Vor Jahren schon, bei Gelegenheit optischer Versuche, bin ich zu einer ähnlichen, allerdings noch sehr unvollkommenen Vorstellung gedrängt worden. Es schien mir nämlich, als ob vermöge der Unvereinbarkeit beider Netzhautbilder, ein Theil des optischen Reizes in andere Bahnen abfliessen würde, welchen wir dann als ein eigenes Merkmal des Gesehenen betrachten lernen und den wir mit Hering Tiefenempfindung nennen wollen. Auch beim Versuch, Stereoscopbilder mit starken Differenzen zu combiniren, habe ich wiederholt ein Ekelgefühl beobachtet.

11.

Bei der Lectüre der Studien von Hasse drängt sich die Ansicht auf, dass das Gehörorgan sich aus einem Tastorgan durch Adaptiren an periodische Reize entwickelt hat. Die Beziehung des Gehörorgans zur Bewegung kann dann nicht überraschen. In der That treten die Bogengänge schon viel tiefer in der Thierreihe auf, als die Schnecke. Möglich, dass, wie Breuer auch vermuthet, die sogenannten Gehörorgane der niedersten Thiere nichts wie Organe der Bewegungsempfindung sind. Diese Beziehung des Gehörorgans zur Bewegung ist dann wohl ein Erbstück aus jener Zeit, in welcher Ortsbewegung, Greifen und Schlingen noch Eins war. Die leichte und unmittelbare Bewegungsauslösung durch Geräusch (bei Schreck) gehört auch hieher.

Die Function des Labyrinthes als Gleichgewichtsorgan wird noch gestützt durch die morphologischen Verhältnisse. Die Schwimmblase der Fische, welche doch entschieden ein Gleichgewichtsorgan ist,

labyrinths through the so-called hearing ossicles. Of course, such analogies can also be misleading if inappropriately chosen. For example, years ago I considered the tensor tympani as a muscle for accommodation, and only later convinced myself by means of experiments that it does not have this function*). Going from the ear of higher vertebrates to the ear of the fish one would be inclined to consider the swimbladder as a cavity in the eardrum, and ascribe an acoustic function to it. The reverse route could have led to the idea that the <u>muscles of the tympanic cavity</u>[8.51] serve no acoustic function, but are probably somehow responsible for the ventilation of the middle ear, which seems more correct in my opinion. Even though these analogies do not constitute proof, their consideration might nevertheless be useful.

Conclusions

Here we summarize the most important results of the investigations by the following statements:

1. There are special movement sensations of linear and angular body motions and probably also special sensations of body orientation.

2. Linear and angular acceleration serve as the stimuli for these sensations.

3. These sensations, however, can be perceived as the attained linear or angular velocity.

4. The sensations continue with decreasing intensity when the stimulus dies out.

5. They also decline during long-lasting stimuli.

6. No negative afterimage appears.

7. The sensations stand in opposing pairs, positive and

---

*) Mach, On the theory of the hearing organ. Proc. Vienna Acad. 1863. — new edition Prague, Calve 1872. – E. Mach and J. Kessel, On accommodation of the ear. Proc. Vienna Acad. vol. 66.

wirkt direct durch die sogenannten Gehörknöchelchen auf den Labyrinthinhalt. Selbstverständlich können solche Analogieen, unglück-
lich angewandt, auch täuschen. Ich habe z. B. vor Jahren den
tensor tympani als einen Accommodationsmuskel aufgefasst und mich
erst später durch Versuche überzeugt, dass er diese Function nicht
hat*). Geht man von dem Ohr der höhern Wirbelthiere aus und
kommt man auf das Ohr der Fische, so wird man geneigt sein, die
Schwimmblase fur eine Trommelhöhle zu halten und ihr eine akustische Function zuzuschreiben. Der umgekehrte Gang hätte dazu
führen können, zu erkennen, dass die Muskel der Trommelhöhle keine
akustische Function haben, sondern dass sie etwa die Lüftung der
Trommelhöhle zu besorgen haben, was mir jetzt richtiger scheint.
Wenn diese Analogieen auch nicht beweisend sind, so mag ihre
Beachtung doch immer nützlich erscheinen.

Schluss

Indem wir nun die wichtigsten Ergebnisse der Untersuchung zu-
sammenfassen, gelangen wir zu folgenden Sätzen:

1. Es gibt besondere Bewegungsempfindungen der
   Progressivbewegung und Drehung des Körpers und
   wahrscheinlich auch besondere Empfindungen der
   Lage.
2. Die Progressiv- und Winkelbeschleunigung wirkt
   als Reiz dieser Empfindungen.
3. Diese Empfindungen lassen sich aber vorstellen
   als Empfindungen einer erlangten Progressiv- oder
   Winkelgeschwindigkeit.
4. Die Empfindungen dauern allmälig an Intensität
   abnehmend fort, wenn der Reiz erlischt.
5. Sie erschöpfen sich auch bei Fortdauer des Reizes.
6. Ein negatives Nachbild tritt nicht auf.
7. Je zwei dieser Empfindungen stehn im Gegensatz

---

*) Mach, zur Theorie d. Gehörorgans. Sitzgsber. d. Wiener Akademie 1863.
— Neu aufgelegt Prag, Calve 1872. – E. Mach u. J. Kessel, über die Accommodation d. Ohres. Sitzgsber. d. Wien. Akademie. Bd. 66.

negative, such that the excitation of one compensates for the presence of the other.

8. The movement sensations cannot be explained by the activity of sensitive elements in the bone and connective tissue, nor through activity of the skin, muscle, blood or brain, although some contribution of these factors to the knowledge of locomotion cannot be completely excluded.

9. The organ of movement sensation demonstrably lies in the head.

10. The assumption that the brain constitutes this organ cannot be completely excluded, but it presents great difficulties in explaining the details of the phenomena.

11. The assumption that a part of the labyrinth is the organ of movement sensation, namely that the six ampullae of the semicircular canals correspond to the six, pairwise opposing, basic sensations of turning, explains all of the experiments described here, as well as Flourens' experiments, down to the finest detail.

Evidently, conclusions 1-10 remain valid even if the hypothesis put forth in 11 should prove to be wrong, because they do not depend upon it at all.

In order to prove hypothesis 11, I had already suggested in my first communication that it would be desirable for the following experiments to be performed faultlessly: mechanical imposition of a rotation moment on a semicircular canal or electrical stimulation of a single ampulla. Breuer had already conducted such experiments before the publication of my paper. He had also recognized their importance. Unfortunately, they are so difficult to perform that Breuer himself does not acknowledge his observations thus far as proof. Also, note that for electrical stimulation both electrodes must be placed on one ampulla to overcome the objection that any effect on the brain plays a significant role. Demonstration of Flourens' phenomena without vivisection is especially important with respect to Böttcher's viewpoint. Furthermore, electromagnetic stimulation of the labyrinth, which

des Positiven und Negativen, so dass die Erregung der einen die andere noch vorhandene aufhebt.

8. Die Bewegungsempfindungen lassen sich nicht erklären durch die Wirkung der sensiblen Elemente der Knochen und des Bindegewebes, nicht durch die Wirkung der Haut, der Muskel, des Blutes oder des Hirns, wiewohl ein Mitwirken dieser Factoren bei Erkenntniss der Locomotion nicht vollsändig ausgeschlossen werden kann.

9. Das Organ der Bewegungsempfindungen hat seinen Sitz nachweisbar im Kopfe.

10. Die Annahme, dass das Hirn dies Organ sei, lässt sich nicht vollständig ausschliessen, bietet aber sehr grosse Schwierigkeiten in der Erkläruug des Details der Erscheinungen.

11. Die Annahme, dass ein Theil des Labyrinthes Organ der Bewegungsempfindung sei, namentlich dass die sechs Ampullen der Bogengänge den sechs paarweise entgegengesetzten Grundempfindungen der Drehung entsprechen, erklärt alle hier beschriebenen und auch die Flourens'schen Versuche bis ins Detail.

Wie man sieht, bleiben die Sätze 1 – 10 bestehn, auch wenn die in 11 ausgesprochene Hypothese fällt, da sie sich keineswegs auf letztere stützen.

Zum Nachweise der Hypothese 11 wäre es wünschenswerth, wenn die schon in meiner ersten Mittheilung vorgeschlagenen Ver-. suche, mechanisch dem Bogengang ein Drehungsmoment zu ertheilen oder eine einzelne Ampulle electrisch zu reizen, tadellos ausgeführt werden könnten. Breuer hat solche Versuche schon vor Publication meiner Arbeit ausgeführt. Er hat sie also ebenfalls als wichtig erkannt. Leider sind dieselben so schwierig auszuführen, dass Breuer selbst die Beweiskraft der bisher von ihm gemachten Beobachtungen nicht anerkennt. Auch ist noch zu bemerken, dass bei electrischen Reizungen beide Electroden auf die eine Ampulle gesetzt werden müssen, wenn nicht der Einwurf, dass Hirnaffectionen im Spiele seien, ein grosses Gewicht haben soll. Herstellung

I suggested in my first communication, does not appear impossible to me.

Finally, an important experiment would be to rotate animals with severed acoustic nerves. They should experience no rotational vertigo. Goltz*) already observed a loss of the sense of equilibrium in such animals. How unfortunate that rotational tests were not also performed. Furthermore, Breuer observed no rotational vertigo in a pigeon with completely removed canals while he rotated it in his hand. I must leave such experiments to others because I do not have the knowledge necessary to perform vivisection, and it would require an unreasonable amount of time to acquire this knowledge.

Note Added in Proof.

While this publication was in press I incidentally sought to determine one's sensitivity to angular acceleration. Imagine an inverted T (L), which is suspended from its vertical handle by two cables, while the observer sits on the horizontal arm of the T and grasps the handle. The horizontal arms are significantly extended and are fitted with hooks to support weights. By the addition of weights one increases the apparatus' moment of inertia, and by bringing the two points of suspension closer together one decreases the torsional moment. By means of these two manipulations the period of oscillation of the apparatus can be arbitrarily increased.

The observer, sitting on the T with eyes closed, performs pendular torsional swings following a simple push. The equation given on p. 31 is valid if in that equation x is the angular excursion away from the equilibrium

---

*) Contributions to the theory of the functions of nerve centers in the frog. Berlin 1869.

der Flourens'schen Phänomene mit Vermeidung der Vivisection ist den Böttcher'schen Ansichten gegenüber von besonderm Werth. Auch die electromagnetische Erregung des Labyrinthes, wie ich sie in meiner ersten Mittheilumg angedeutet habe, scheint mir nicht unmöglich.

Ein wichtiger Versuch endlich würde darin bestehn, Thiere mit durchschnittenem Acusticus in Rotation zu versetzen. Diese sollten keinen Drehschwindel bekommen. Goltz*) hat den Verlust des Gleichgewichtssinnes an solchen Thieren schon beobachtet. Schade, dass nicht auch Drehversuche schon angestellt worden sind. Auch Breuer hat an einer Taube mit vollständig entfernten Bogengängen bei Drehung in der Hand die Schwindelerscheinungen vermisst. Solche Versuche muss ich Andern überlassen, da mir zur Ausführung von Vivisectionen die Kenntnisse fehlen und die Erwerbung derselben mit einem unverhältnissmässigen Aufwande an Zeit verbunden wäre.

Nachtrag.

Während. des Drucks dieser Schrift habe ich eine beiläufige Bestimmung der Empfindlichkeit für die Winkelbeschleunigung versucht. Man denke sich ein umgekehrtes T (L), welches an dem verticalen Stiele bifilar aufgehängt ist, während der Beobachter auf den horizontalen Armen des T sitzend den Stiel umklammert. Die horizontalen Arme sind bedeutend. verlängert und zur Aufnahme von Gewichten mit Haken versehn. Durch Hinzulegen von Gewichten vermehrt man das Trägheitsmoment des Apparates und durch Annäherung der beiden Aufhängepunkte vermindert man das Torsionsmoment. Mit Hilfe beider Operationen kann also die Schwingungsdauer des Apparates beliebig vergrössert werden.

Der mit geschlossenen Augen auf dem T sitzende Beobachter führt nun auf einen blossen Anstoss hin pendelförmige Torsionsschwingungen aus, für welche die S. 31 aufgestellten Formeln gel-

---

*) Beiträge zur Lehre von den Functionen der Nervencentren des Frosches. Berlin 1869.

position, a the maximum of this excursion, and T the period of a complete oscillation. The amplitude of the oscillation gradually decays and the observer indicates when he no longer senses it, while an assistant measures the duration of the oscillation and reads the amplitude of the excursion as indicated by a pointer fastened to an arm of the T above a protractor.

For myself the torsional swings became undetectable when the values $a = 10°$ and $T = 14$ seconds were reached. Dr. Dvořak no longer noticed the swings when the values were $a = 13°$ and $T = 16$ seconds. Accordingly the threshold[9.1] for angular accelerations lies between 2 and 3°.

Furthermore, during the weakest torsion swings, I see that any bright flecks and little clouds in the dark visual field, if perceivable at all, also take part in the movement.

––––––––––––

Printing error
p. 64 line 13 from below should read »however« instead of »therefore«.

––––––––––––

ten, wenn in denselben x die Winkelexcursion aus der Gleichgewichtslage, a das Maximum dieser Excursion und. T die Dauer einer ganzen Schwingung vorstellt. Die Schwingungen nehmen allmälig an Excursion ab und der Beobachter gibt an, wann er dieselben nicht mehr fühlt, während ein Gehilfe die Schwingungsdauer zählt und die Excursionen abliest, welche ein am Arme des T befestigter Zeiger über einem Gradbogen beschreibt.

Für mich wurden die Torsionsschwingungen unmerklich, wenn $a = 10°$ und $T = 14$ Secunden war. Herr Dr. Dvořak fühlte die Schwingungen nicht mehr, wenn $a = 13°$ und $T = 16$ Secunden war. Demnach liegt der Schwellenwerth der Winkelbeschleunigung zwischen 2 und 3°.

Auch bei den leisesten Torsionsschwingungen sehe ich, so lange dieselben überhaupt merklich sind, alle hellen Fleckchen und Wölkchen des dunklen Gesichtsfeldes die Bewegung mitmachen.

Druckfehler.
S. 64 Z. 13 von unten ist zu lesen »aber« statt »also«.

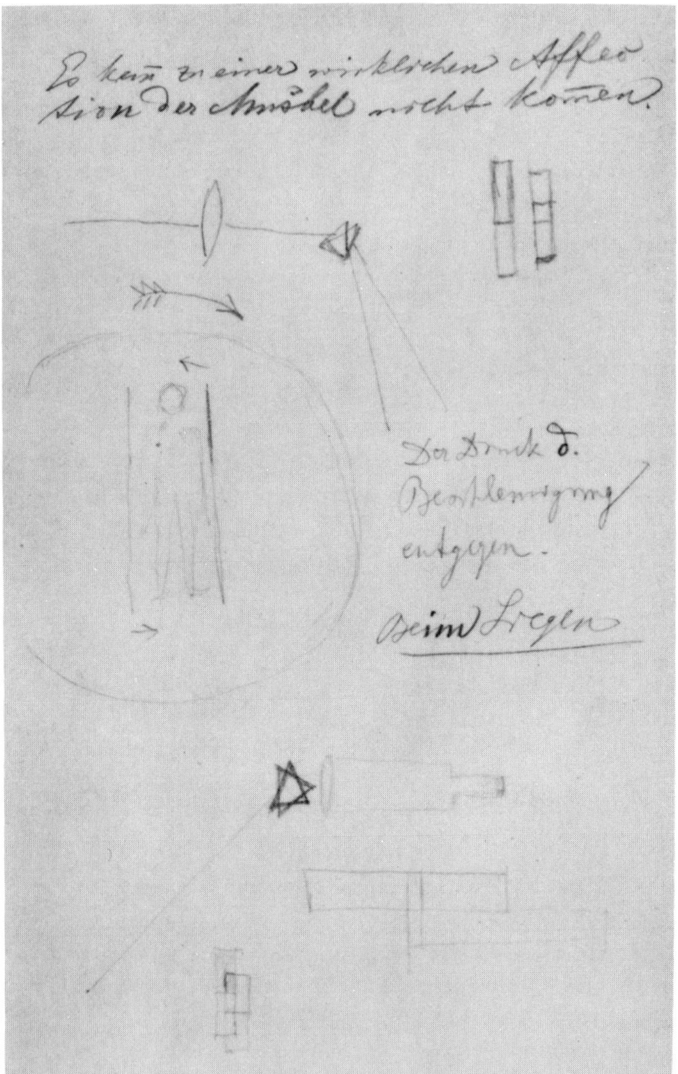

Figure A-7: Sketch of the rotating disk showing a subject restrained by a movable box which could apply pressure to the lateral parts of the body (see pp. 75-76). "It can not cause an actual influence on the muscles. Pressure is opposite to acceleration <u>when lying</u>." This figure also shows sketches related to Mach's continuing interest in optics and visual illusions. (from Mach's notebook, 1874; photograph courtesy of E. Mach Archive, Deutsches Museum, Munich).

Footnotes

Preface

1 **Henn, Volker** (1943–1997): Professor of neurology in Zurich and a leading investigator in oculomotor and vestibular neurophysiology. His strong interest in the history of science and his admiration of Ernst Mach led to this translation (Büttner-Ennever et al. 1998; Cohen 1999; Straumann 1999).

2 **CD-ROM**: This book is accompanied by a CD-ROM that contains a hypertext version (PDF-format) of this book and of the following additional documents (reproduced with kind permission):

Historical:

Breuer, J. (1925). *Autobiographie Josef Breuer.* Wien.

Cohen, B. (1984). Erasmus Darwin's observations on rotation and vertigo. *Hum. Neurobiol.,* 3: 121–8.

Grüsser, O. J. (1984). J.E. Purkyně's contributions to the physiology of the visual, the vestibular and the oculomotor systems. *Hum. Neurobiol.* 3: 129–144.

Henn, V. (1984). E. Mach on the analysis of motion. *Human Neurobiol,* 3:145–148.

Henn, V., & Young, L. R. (1975). Ernst Mach on the vestibular organ 100 years ago. ORL J. *Otorhinolaryngol. Relat. Spec.* 37: 138–148.

Oberndorf, C. P. (1953). Autobiography of Josef Breuer [translated and edited]. Int. J. Psychoanalysis 34: 64-67

Textbook chapter:

Young, L. R. (1983). Perception of the body in space: Mechanisms. In *Handbook of Physiology – The Nervous System III,* (pp. 1023–1066). Bethesda: American Physiological Society.

Obituaries of Volker Henn:

Büttner-Ennever, J., Büttner, U., & Brandt, T. (1998). Volker Henn. *Audiol. Neurootol.,* 3: 61–62.

Cohen, B. (1999). Tribute to Volker Henn. In B. Cohen, & B. J. M. Hess (eds.), *Otolith Function in Spatial Orientation and Movement.* (pp. xiii–xiv). New York: Annals of the New York Academy of Sciences.

Straumann, D. (1999). In memoriam Volker Henn. *J. Vest. Res.* 9: 155.

To view these (*PDF-*) documents, the software *Adobe Acrobat Reader* (release 4.0 or later) needs to be installed on your system. For installation, we recommend downloading the most recent release from

the provider's WEB site: http://www.adobe.com. *Release 5.0* can also be installed from this CD (depending on your operating system, go to the folder *PC*, *MAC*, or *UNIX* inside the folder *Acrobat* and run the appropriate file).

To explore the documents, open the start page *START_ME.pdf* in the root directory of the CD-ROM (on some systems, the start page will open automatically after inserting the CD). The start page contains links (indicated by underlining) to all documents. To select a link, click on the underlined text, which will open the corresponding pages. To return to the previous view, click the *Go To Previous View button* ◆ in the navigation toolbar, or press the *ALT+left arrow* key on the keyboard (*OPTION+left arrow* for Mac systems). For further help, see *Navigation Help* on the start page or *Reader Help* in the Help menu of the Acrobat Reader.

Foreword

0.1 **Mach, Ernst** (1838–1916): In 1873/74, within 3 months, three researchers with vastly different backgrounds published papers in which they outlined a theory of motion perception. The physicist E. Mach, the physician J. Breuer, and the chemist A. Crum Brown each achieved recognition in other fields of scientific research. Mach's contribution is unique in that he also reflects on how he approached the subject and thereby gives an illustration of the "Psychology of Enquiry." This also serves as the subtitle of his book *Knowledge and Error* (1905), which became influential for a whole generation of scientists and philosophers.

In his many and widely read publications Mach contributed to physics, physiology, and philosophy. His name is now chiefly known by the public as the unit for the speed of sound, "Mach 1."

Short Biography

1838 Born in Turas, Moravia (Czech). Father teacher, mother housewife, two sisters

1856 Natural sciences studies at University of Vienna

1860 PhD

1861 Lecturer in physics (Privatdozent)

1864 Professor of mathematics at Graz University (southern Austria)

1867 Professor of physics at Prague University

1895 Professor of philosophy at Vienna University

1896 Stroke with left-sided palsy

1900 Retirement

1916 Died in Vaterstetten (near Munich)

References

Barbour, J., Pfister, H. (eds; 1995) *Mach's Pinciple: from Newton's Bucket to Quantum Gravity*. Boston, Basel, Berlin: Birkhäuser. [Einstein Studies, vol 6]

Blackmore, J. T. (1972) *Ernst Mach – His Work, Life and Influence*. Berkeley: Univ. Calif. Press. [Biography, does not specifically discuss vestibular work]

Bradley, J. (1971) *Mach's Philosophy of Science*. Oxford: Oxford University Press.

Henn, V. (1984) E. Mach on the analysis of motion. Human Neurobiol, 3:145–148. [Short biographical paper specifically in relation to vestibular research]

Holton, G. (1992) Ernst Mach and the fortunes of positivism in America. Isis, 83:27–69. [reprinted in Holton, G. (1993) *Science and Antiscience*. Cambridge: Harvard University Press]

Ratliff, F. (1965) *Mach Bands: Quantitative Studies on Neural Networks in the Retina*. San Francisco: Holden-Day.

Thiele, J. (1963) Ernst Mach-Bibliographie. Centaurus, 8:189-237. [Another bibliography was published by the same author as 'Ernst Mach Bibliographie' in a Collection of Reprints by Ernst Mach. Amsterdam: Bonset, 1996, pp. 16–42]

English Translations: Most journal papers are available in German only. The following gives a list of books which had been translated:

The Science of Mechanics. A Critical and Historical Exposition of its Principles (1893). Translated from the 2nd German ed. by T. J. M. McCormack. Chicago: The Open Court Publishing Co.- 6th ed. (1988). The Open Court Publishing Co.

Popular Scientific Lectures (1895). Translated by T. J. McCormack. Chicago: The Open Court Publishing Co. Reprint (1986). Chicago: The Open Court Publishing Co.

Contributions to the Analysis of the Sensations. Translated by C. M. Williams (1897). Chicago: The Open Court Publishing Co. Reprint (1984). Chicago: The Open Court Publishing Co

History and Root of the Principle of the Conservation of Energy (1911). Translated from the German and annotated by P. E. B. Jourdain. Chicago: The Open Court Publishing Co.

The Principles of Physical Optics: A Historical and Philosophical Treatment (1926). Translated by J. S. Anderson and A. F. A. Young. London: Methuen & Co.

Knowledge and Error: Sketches on the Psychology of Enquiry (19??) Reidel Publishing Co.

Space and Geometry: In the Light of Physiological, Psychological, and Physical Inquiry (1983). Chicago: The Open Court Publishing Co.

0.2 **Purkyně, Johannes Evangelista** (1787–1869): Physiologist who held university positions in Breslau (now Wroclav, Poland) and Prague. Purkinje investigated active versus passive rotational movement with eyes open or closed. He found that passive rotation (on a rotatory chair) with eyes closed lead to motion aftereffects that are not present after rotation with eyes open or if the rotation was active (by stepping movements). If he brought the head to a new position in the presence of motion aftereffects, he found that the plane of rotation moved with the head. He therefore concluded that the sense of equilibrium has to be localized to the head. He assumed that pressure gradients within the head convey this information without postulating a specific sensory organ. On pp. 79–82 and 90–93 of this book Mach extensively cites Purkinjě. Other scientific contributions from Purkinjě primarily relate to vision and microscopic investigations of the brain ('Purkinjě' cells of the cerebellar cortex). (Purkinjě 1820; Grüsser 1984)

0.3 **Flourens, Marie Jean Pierre** (1794–1867): Experimental physiologist, professor of comparative anatomy at the College de France in Paris. He was an accomplished experimentalist pioneering behavioral studies after focal brain lesions. Although he did observe specific deficits, he maintained that higher cerebral functions cannot be further localized within the hemispheres.

Flourens lesioned single semicircular canals in animals and subsequently observed uncontrolled head movements in the planes of the respective canals. He mostly refrained from interpretation and tentatively attributed this to painful hearing, as the then current theory was that the canals were detectors for directional hearing. The dissociation between the carefully done experiments and the lack of theory is remarkable, as, e.g., Mach gave long citations of the work of Flourens in support of his own theory of motion, after Flourens had described the results of the experiments in such a careful way. Mach quotes Flourens extensively on pp. 41–45. (Fourens 1825, 1830, 1842)

0.4 **Goltz, Friedrich Leopold** (1834–1902): Physiologist who held positions in Königsberg, Halle and Strasbourg. Most famous for his experimental work on reflex mechanisms and the labyrinth. He was one of the first to differentiate between the cochlea, serving an acoustic function, and the semicircular canals, which serve equilibrium. He supposed that the ampullae of the canals constitute an organ that detects motion and the direction of gravity. This laid the groundwork for the later work by Breuer, Mach, and Crum Brown who described actual physiological mechanisms. One of his pupils was Richard Ewald[S.1] who introduced the technique of canal plugging that became critical for abolishing a single canal's function. (Goltz 1870)

0.5 **Breuer, Josef** (1842–1925): General practitioner in Vienna who made several seminal contributions to physiology and psychiatry. He started his career by working under the guidance of <u>Ewald Hering</u>[S.2], became a lecturer for medicine (Privatdozent), but soon went into private practice. All experiments concerning labyrinthine function were performed at home, literally on his kitchen table [see CD-ROM for original German autobiography (<u>Breuer 1925</u>) and slightly abridged English translation (<u>Oberndorf 1953</u>)].

Mach had first reported his theory of motion detection at the Vienna Academy of Physicians (November 1873); Breuer heard about it and asked for the publication of a short report and a talk about his own work, which was granted in the same month. While Mach's greatest contribution was the presentation of a coherent theory backed by experiments, Breuer as a physician summarizes the current knowledge, adds his own experiments (stressing eye movements and resultant visual phenomena), and reports about animal experiments that he performed (mostly stimulation and lesions of the canals). Mach extensively cites Breuer (1874, pp. 97–100). He refers to this work as being done simultaneously and independently and discusses his contributions and interpretations throughout the book. His Studies on Hysteria, together with <u>Freud</u>[S.3], (1895) may be considered as the seminal study on which psychoanalysis is founded. (Breuer 1868, 1874; Breuer and Freud 1895).

0.6 **Brown, Alexander Crum** (1838–1922): [in modern literature usually referred to as 'Crum Brown, Alexander']. Professor of chemistry at Edinburgh University. His notation of chemical bonds was readily accepted and basically is still in use today. Atomic linkage was indicated by lines, double bonds by double lines. Such a notation was much superior to the very awkward descriptions and diagrams which were used otherwise.

Crum Brown performed a series of experiments rotating blindfolded subjects on a turntable, and correctly associated the resultant motion sensation with consequent activation of nerves in the canals. He also built an elaborate analog model to illustrate how the canals on both sides, with their approximately orthogonal orientation, could detect angular acceleration in any plane and direction. (Brown 1874, full text quoted by Mach on pp. 100–102 of this book)

0.7 **Boettcher, A.** (1831–1869): Professor of pathology in Dorpat. His major publication concerning the labyrinth, to which Mach refers, is Boettcher (1869).

Introduction

1.1 **Darwin, Erasmus** (1731–1802): The "elder" Darwin refers to the grandfather of the more famous <u>Charles Darwin</u>[8.49] ('The origin of species by means of natural selection, or the preservation of favoured races in the struggle for life' 1859). E. Darwin was one of the leading general practitioners of his time. In addition he gained a reputation for poetry, some explicitly written to educate the public on topics of natural science (*Botanic Garden* 1791; *The Loves of Plant* 1789). His main publication was *Zoonomia* (1794), in which he treats vertigo, and describes rotational experiments on himself (2nd ed. 1796). He observed reflexive eye movements during rotation accompanied by sensation of rotation. Darwin was member of the <u>Lunar Society</u>[S.4]. (Darwin 1794; <u>Cohen</u> 1984)

1.2 **Purkyně, Johannes Evangelista** (1787–1869): see 0.2.

1.3 **Ritter, Johann Wilhelm** (1776–1810): A pharmacist by training with a major interest in electricity. He was the first to separately collect hydrogen and oxygen by electrolysis, did the first experiments leading to electroplating, and found that the breakdown of silver-chloride by light was far more effective in the short-wave color spectrum. He was led to the conclusion that beyond the color violet there are other rays that can trigger certain reactions, a spectrum that we now call ultraviolet.

The Voltaic column or chain is a simple form of a battery, invented by Volta in 1800. Ritter describes, in rather vague terms, how he treated a large number of patients by placing two electrodes on different parts of the body. Diseases that he treated in such a way included loss of hearing or vision, coldness of feet, digestive problems, rheumatism, and others. He postulates that the current triggers chemical reactions. In the above-mentioned paper Ritter does not explicitly refer to motion sensation. Parenthetically, Mach indicates on page 4 that he knew of that research only from citations, but had not seen the full paper (Ritter 1803).

1.4 **Flourens, Marie Jean Pierre** (1794–1867): see 0.3.

1.5 **Goltz, Friedrich Leopold** (1834–1902): see 0.4.

1.6 **Semicircular canals**: Anatomical structure, being part of the inner ear which divides into an acoustic part (cochlea, or organ of Corti) and a labyrinthine part subserving the detection of angular (semicircular canals) and linear acceleration (otoliths). The three semicircular canals are approximately orthogonal to each other. The membraneous canals are filled with a fluid (endolymph) that is not free to flow but is constrained by a barrier, the cupula within each ampulla, which is

richly innervated with hair cells that are very sensitive to mechanical stimulation. During angular acceleration, a pressure difference builds up across the cupula, bending it slightly and exciting the hair cells to give rise to activity changes in the labyrinthine part of the 8th cranial nerve.

The following figures are from Henn, V. (1987) Vestibular system. In: G Adelman (ed.) *Encyclopedia of Neuroscience.* Boston: Birkhäuser. (Reprint with kind permission.)

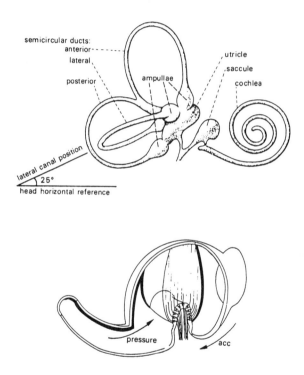

1.7   **Labyrinth:** It consists of the semicircular canals (for detection of angular acceleration) and the otoliths (for the detection of linear acceleration). The figure shows a cast of a left labyrinth as seen from front, positioned in such a way that the lateral canal is horizontal (plastic cast of the semicircular canals of a rhesus monkey, by J. I. Simpson, New York University; with kind permission). The anterior canal lies above the level of the lateral canal.

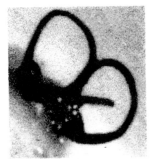

1.8 **Hearing disturbance:** At the time of Flourens, the prevailing and unchallenged assumption was that the semicircular canals have a function in directional hearing. Only after Ménière's seminal clinical studies was the connection between the symptoms of vertigo or dizziness with disease of the inner ear slowly accepted.

1.9 **Chance occurrence:** "I was rounding a sharp railway curve once when I suddenly saw all the trees, houses, and factory chimneys along the track swerve from the vertical and assume a strikingly inclined position. What had hitherto appeared to me perfectly natural, namely, the fact that we distinguish the vertical so perfectly and sharply from every other direction, now struck me as enigmatical. Why is it that the same direction can now appear vertical to me and now cannot? By what is the vertical distinguished for us?

The rails are raised on the convex or outward side of the banked track in order to insure the stability of the carriage against the centrifugal force, the whole being arranged so that the combination of the force of gravity and the centrifugal force of the train produces a force perpendicular to the plane of the rails.

Let us assume, now, that under all circumstances we somehow sense the direction of the total resultant mass-acceleration from wherever it may arise, as the vertical. Then both the ordinary and the extraordinary phenomena will both be explained."

Figure taken from Mach (1897). This lecture was subsequently included in the next edition of *Popular Scientific Lectures* [3rd ed., revised and enlarged. Chicago: The Open Court Publishing Co.].

1.10 **Specific energy**: This term was coined and is explained by Johannes Müller (1826) stating that stimulation of afferent nerves leads to a specific sensation, independent of how they were stimulated. For example, although the physiological stimulus for the eye is light, pressure or electrical stimulation also leads exclusively to visual phenomena. Müller puts forward the following theses:

I   External factors cannot lead to perceptions that we could not also experience, without external stimulation, by sensing the activity of our nerves.

II   The same internal cause induces different sensations in accordance with the special sensitivity of the respective sensory organ.

III   The same external cause induces different sensations in accordance with the special sensitivity of the respective sensory organ.

IV   The specific sensations for each sensory organ can simultaneously originate from several internal and external factors.

V   Sensory perception does not imply that a physical quality or condition of the external body is conducted to be consciously perceived, but it is the conduction of a certain quality in a sensory nerve to reach consciousness, triggered by the external factor, and these qualities are different for the different sensory nerves, the specific sensory energies.

VI   One sensory nerve can only be sensitive for one kind of sensory perception and not for those from other sensory organs, and therefore cannot take over functions of another sensory nerve.

VII   It is not known whether the cause for the different energies of sensory nerves are caused by the nerves themselves, by factors in the brain, or by the medulla through which they are connected. However, there is no doubt that the central sensory areas of our brain are capable of specific sensory perceptions independent of the afferent nerves.

VIII   The sensory nerves in first place only detect their own states, or the sensorium detects the states of sensory nerves. Since the sensory nerves are part of the body, they display qualities that other bodies have: they are

extended in space; they can be stimulated; they can be influenced by chemical agents, heat, and electricity. By these virtues they can signal to the sensorium the condition, quality, and alterations in the external world. This is caused by a change in their states through external factors.

(Müller's *Principles of Sensory Function* 1826, pp. 250–262; not as erroneously given by Mach as pp. 251–258 in the chapter "About the Senses.")

Müller, Johannes (1801–1858): Professor of anatomy and physiology in Berlin. He separated himself from the close connection to philosophical schools common by that time. He relied less on physiological experiments than on careful anatomical observation, which he then interpreted physiologically. His pupils, whose work introduced a major paradigm shift in physiology, included Emil Du-Bois-Reymond, Ernst Brücke, Heinrich Helmholtz[6.7], Rudolf Virchow, and Erich Haeckel. His greatest influence came from several seminal publications: Müller (1826) was most influential, followed by his textbook Müller (1834– 1840).

1.11 **100 Thaler:** At age 26, as a professor of mathematics at Graz University (Austria), Mach earned 700 thaler annually, which he did not consider sufficient to marry and support a family. This salary is about 3 times as high as that of a laborer and a little higher than that of a schoolteacher.

In 1875, at age 37, Mach earned 1370 thaler. At that time he lived in Prague. He was professor and chairman at the Physics Institute, although the name "chairman" might be misleading, since at that time an institute usually consisted of only one professor. The cost of the turntable was therefore about 7% of his annual income. For comparison, the following are some wages and prices from the year 1875. Average annual incomes for laborers were: construction 275 thaler; textiles 150 thaler; printing 415 thaler; average income for school teachers 550 thaler.

1 kg potatoes	0.018 thaler
1 kg rye flower or bread	0.10 thaler
1 kg sugar	0.38 thaler
1 kg salt	0.07 thaler
1 kg veal	0.31 thaler

At Mach's time three different monetary nominations were used. 1 Gulden = 0.57 Thaler = 1.74 Mark. (Florins is the French expression for Gulden.)

1.12 **Relevant literature:** In the original, all references are given in their respective original language. To facilitate understanding, the

references have been translated to English. The original citations can be found in the German text version.

The Mechanical Foundations

2.1 **Mechanical process:** Although today no scientist would doubt this statement, the situation was completely different at the beginning of the 19th century. Organic matter and biological laws were usually considered to be fundamentally different from inorganic matter, and thus not amenable to description in terms of physics. A "vital force" was postulated, which was considered unexplainable by human intellect with the consequence that biological research would always remain futile.

A major shift in the understanding of biology as a science occurred in the mid-19th century. Some of the milestones were the synthesis of urea, an organic molecule, by Wöhler (1828); the formulation of the law of the preservation of energy by Helmholtz (1845); and Darwin's evolutionary theory *On the Origins of Species by Natural Selection* (1859). Clearly, scientific method could succeed in biology.

In the context of movement sensation, compare the careful description of phenomena by Purkyně or Flourens with their notable lack of any explanatory theory. This attitude suddenly changes with the work of Helmholtz, Hering, Mach, Breuer, and Crum Brown, to name only a few physiologists who investigated the vestibular and oculomotor systems.

Mach vehemently opposed the separation of perception from physical reality. This view was later referred to as the Mach principle. It expressed his position that physical reality is not defined by inherent qualities, like mass, but by measurements and relationships to other objects.

Bradley (1971) traces the roots of Mach's philosophy of sense perception from Descartes and Berkely, and relates it to Hume and Kant in the requirement for physical experience to precede feeling. In his *Popular Scientific Lectures* (1896) Mach makes his point clearly: "The world consists of colors, sounds, temperatures, pressures, spaces, times and so forth, which *now* we shall not call sensations, nor phenomena, but because in either term an arbitrary, one sided theory is embodied, but simply *elements*."

2.2 **Galileo, Galilei** (1564–1642): Mach was a great admirer of Galileo and his contributions to understanding dynamic motions and developing the concepts of acceleration and inertia. Ever the admirer

of careful experiments, Mach especially appreciated Galileo's careful and complete exposition of the mere facts (Mach, *The Science of Mechanics*, p.140). Galileo did not conceal anything behind the expression "force" which might be left open to interpretation.

In the context of Mach's work the experiments with pendulums and free-falling bodies are relevant. In his legendary experiments at the leaning tower of Pisa where he dropped objects of different weight and measured the time until they hit the ground, Galileo found that, neglecting air resistance, they all fell at the same rate. This was a decisive new finding in stark contrast to the ideas of Archimedes, still in vogue at that time, stating that the velocity of falling is proportional to a body's weight. Mach noted that Galileo, as well as others before Newton, failed to distinguish between weight and mass. Galileo showed that bodies rolling down one inclined plane rise to very nearly their initial height on a second plane. This led to the theorem that a body will maintain its motion if no external forces are applied, which eventually contributed to the current interpretation of planetary movements. Mach was formally critical of Galileo's "law of equal heights" as being an inductive leap from the non-ideal rolling ball experiment.

2.3 **Newton, Isaac** (1642–1727): Mach stood in awe of Newton's contributions, and especially Newton's concentration on actual facts. Newton's famous utterance "hypothesis non fingo" (I do not frame hypotheses) fit perfectly into Mach's philosophy. In the context of Mach's work Newton's three laws are relevant:

First law: An object at rest remains at rest, and a body in motion remains in motion in a straight line unless acted upon by external forces.

Second law: Changes in motion are proportional to the applied force, and are in the direction of the force. (Force is defined in terms of mass and acceleration. Mach disagreed with Newton's concept of mass as an intrinsic *"quantity of matter"* measured by volume and density.)

Third law: For every action there is an equal and opposite reaction. (Mach restated the third law to include the definition of mass ratios in terms of accelerations on two mutually interacting masses.)

2.4 **Body**: Strictly speaking this statement is true only for the interaction of free bodies, and does not allow for the presence of other external forces, such as those arising from electromagnetic or gravitational fields.

2.5 **Force**: Mach finds it useful to relate force formally to the volitional act of moving something by applying pressure. He defines force as any circumstance that determines motion, with direction the same as the

motion that would occur, and magnitude equivalent to the weight that would produce the same motion. See also: <u>moving force</u>[2.23].

2.6 **Pressure:** Actually Mach is referring to force; however, at times he uses this terms in a rather casual way.

2.7 **Gravity:** The importance of terrestrial gravitation as a special case of universal gravitation was conceived by Newton, building on the astronomical observations and speculations of Copernicus, Brahe, Kepler, Galileo, and Huygens. Mach admired Newton for his statement of the fact of universal gravitation and the importance of action at a distance, without explaining how it could occur. The basic mechanism of gravity remains a major issue in physics. Mach wondered about the velocity of propagation for gravitation, and spoke in favor of ether as its medium. In explaining how inertial reaction forces produced by acceleration resembled gravitational forces, Mach anticipated Einstein's *equivalence principle*, which proved that the two forces are indistinguishable.

2.8 **Conservation of the center of gravity:** The center of gravity is a single point in a distributed body, located such that the body reacts as if all of the gravitational forces on the elements were exerted through that point. As Mach explains by the "law of the center of gravity" that point is not varied by the action of internal forces— only by external forces.

Normally the center of gravity lies at the center of mass—or the average position of all the mass. However, if the gravitational field is not uniform across a body, the center of gravity will be displaced toward the stronger part of the field. Thus for an object orbiting the earth, like the moon, the gravitational field is stronger on the parts closer to the earth, and the center of gravity lies closer to the earth than the center of mass. This explains why the moon, which is not exactly spherical, remains oriented with its same surface, containing a slight bulge, always facing the earth. Designers for "gravity gradient stabilization" of an elongated satellite deliberately use this phenomenon to keep one end pointed toward the earth.

2.9 **Conservation of areas:** With this principle, Mach refers to a special case of the law of conservation of (angular) momentum. It has since fallen into disuse. The principle of the parallelogram of forces introduced by <u>Varignon</u>[2.13] and <u>Newton</u>[2.3] two hundred years earlier, and which we would now cast in vector notation, is used to show how *internal forces* have no influence on "the sum of mass areas"—or the angular momentum.

Referring to Fig 1, if the initial velocities are zero, the area OAB represents the displacement AB times the radius OA, which is the angular momentum if the displacement made in unit time.

In his book *The Science of Mechanics,* Mach (1893) further explains these terms, especially in chapter III "The Laws of the Conservation of Momentum, of the Conservation of the Center of Gravity, and of the Conservation of Areas."

2.10 **D'Arcy, Patrick** (1725–1779): French Marshall and Geometer. Member of the French Academy of Science, Paris. His work contributed to the theory of electricity, gravity, and hydraulics.

2.11 **Euler, Leonard** (1707–1783): One of the most important mathematicians of all time. He systematized analysis, calculus, and trigonometry and showed the importance of the number *e* (2.718...). In oculomotor research, he is commonly known for 'Euler angles' used to describe eye positions and rotations.

2.12 **Fig. 1:** In the original publication this figure has some freckles, so that some of the lettering is slightly affected. The same figure, but without the freckles, had been included in 'The Science of Mechanics' and is reproduced here.

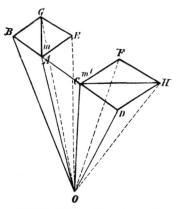

2.13 **Varignon, Pierre** (1654–1722): French professor of mathematics in Paris. Member of the Academy of Science in Paris. He made important contributions in the fields of mechanics and geometry.

2.14 **Lagrange, Joseph** (1736–1813): Mathematician in Turin, Berlin, and Paris recognized for "Lagrange polynomials" and other contributions. Following the French Revolution he was a member of the committee that introduced the metric system of measurement. In this context, Mach notes that Lagrange applied the D'Alembert Principle in 1788.

2.15 **D'Alembert, Jean le Rond** (1717–1783): French mathematician, formulated D'Alembert's Principle that deals with connected masses in static equilibrium and allows one to calculate the forces and accelerations of a connected body. The *virtual work principle* given

here states that the work associated with internal, or *equilibrated*, forces is zero.

In his book *The Science of Mechanics* Mach claims that the convenience of using D'Alembert's Principle comes at the expense of the insight gained by direct application of Newton's law and consideration of moments of inertia.

2.16 **Stevinus, Simon** (1548–1620): Mach attributes the original idea for the principle of virtual displacements to Stevinus in the 16th century. When applied to a system of pulleys it shows why the product of force and displacement remains constant for each counterweight.

2.17 **Torque:** The turning moment caused by twisting or by exerting a force on a lever, as in tightening a nut on a bolt with the help of a wrench. Another term for torque used by Mach is 'moment of rotation'.

2.18 𝕬: Mach uses the old German Frakur letters to designate torque. 𝕬, 𝕭, 𝕮 stand for A, B, C.

2.19 **External disturbance:** This is merely another way of expressing Newton's first law in rotational coordinates—a body which is not rotating will remain non-rotating in the absence of any external torque.

2.20 **Moment of inertia**: Similar to the property of mass, which represents the tendency of a body to resist movement when acted on by a force. The moment of inertia represents the tendency of an extended body to resist rotation when twisted by a torque. It is defined by the relationship between torque and angular acceleration about an axis. Euler showed that the moment of inertia about any axis could be calculated by the sum of the products of masses of a body times the square of the distance of each mass from that axis of rotation.

2.21 **Mass:** Now defined operationally, as the ratio of the force on a body to its acceleration. Newton defined it as an intrinsic quality of a body, the product of volume and density. Mach found Newton's definition to be circular and instead insisted that it can only be defined operationally by the relation between force and acceleration. Galileo and his contemporaries did not distinguish mass from weight, although weight applies only in a gravitational field.

2.22 **Moment of rotation**: Another term for torque.

2.23 **Force:** Mach finds it useful to relate force formally to the volitional act of moving something by applying pressure. He defines force as any circumstance that determines motion, with direction the same as the motion that would occur, and magnitude equivalent to the weight that would produce the same motion. **Moving force:** Although "force" seems intuitive, the distinction between "internal forces" (which sum to zero) and external ones from contact or fields is less evident. Mach defined "moving force" in his rewriting of Newton's second law as "the product of the mass-value of a body into the acceleration induced

in that body" (Bradley, 1971; translated from Mach [1883] *Die Mechanik*, pp. 303-304).

2.24 **Poggendorff's falling machine:** Mach's diagram of Poggendorff's falling machine fails to show the extra string holding the two weights on the left in place initially, which is burned at the beginning of the experiment.

Poggendorff, Johann Christian (1796-1877): German Physicist, founded and edited the *Annalen der Physik und Chemie,* the foremost scientific journal of the 19th century in Europe.

2.25 **Pressure** *pp/(2P+p)* is probably a typographical error as well as confusion between force and pressure. The phrase should read "force *gpp/(2P+p)*."

2.26 **Moon gravity:** The moon has gravity only 1/6 of that on earth. During the flights to the surface of the moon, beginning with Apollo 11 in 1969, the astronauts all reported a feeling of lightness, despite their heavy space suits, and managed to hop and bound across the moon's surface with little difficulty. Contrary to Mach's prediction, they never felt any sinking sensation on the moon, nor do astronauts normally report sinking feelings in the weightlessness of orbital flight even with the eyes closed. The only exceptions are for some cases of extravehicular activity, when the astronaut is outside the spacecraft and staring down at the distant earth, or under special cases of imagination of a free fall. The explanation for the absence of falling sensation seems to reside in the familiarity with the environment and with strong surrounding visual cues that provide a stationary reference frame.

2.27 **Sea sickness:** Mach was quite correct in anticipating the occurrence of symptoms of motion sickness when people are exposed to changed gravitational fields. Roughly two-thirds of all astronauts experience space motion sickness when first exposed to the weightlessness of orbital flight. In the case of the moon landings, the symptoms had all abated during the trip of several days to the moon, and no symptoms were reported on the lunar surface.

2.28 **Hirn, Gustav Adolph** (1815-?): Civil engineer near Colmar, Alsace. He contributed several studies on the theory of heat (see Hirn 1858).

2.29 **Constant velocity rotation**: A constant downward velocity of the wheel requires a constant upward velocity of the man in order to remain stationary in space. One way of considering the problem is to note that the man produces work by pushing down on the descending wheel with a force equal to his body weight. The work is just the integral of this force times the displacement of the wheel.

$$W = \int f(x)dx$$

Another way of looking at this is to imagine a series of short intervals with the man descending on the wheel and then climbing up an equal height, repeatedly. The work required is equal to the body weight times the incremental height for each interval. As the intervals are reduced in time and distance, one approaches this case in the limit, in which the downward wheel velocity is cancelled by the upward climbing velocity.

3.30 **Hovering bird:** Apparently the typographical errors and loose language in this paragraph have confused Mach's example. He presumably meant to state that the mass of the bird was m, and that its weight was mg. When hovering at constant height it required an upward force from the surrounding air of just mg. The air surrounding the bird which was affected by the beating of the bird wings itself has the mass m', This air mass is accelerated downward at acceleration g', so that Mach can consider the incremental velocity of the air each unit time to be g'. Because the force exerted by the bird on the air is just equal and opposite to the force exerted by the air on the bird, $mg = m'g'$. Since the power required to support the bird is inversely proportional to the mass of the surrounding air m', large wings and a large effective mass of air reduces this power. When the bird rests on the ground, the effective mass is that of the entire earth, which essentially doesn't move under the resting bird.

2.31 **Animal free in space:** Although the preservation of the center of gravity assures that an animal cannot translate its center of mass in the absence of any external forces, it is certainly possible to rotate in space. The falling cat is a well-known example, as it manipulates the moment of inertia of its body parts and swings its tail in order to land feet down, without ever violating the conservation of angular momentum. Astronauts have learned to perform these maneuvers in spacecraft.

2.32 **Same relative position:** The demand that the body parts return to the same relative position is a non-necessary restriction. The center of gravity is not influenced by relative movements of body parts.

2.33 **Center of gravity**: see 2.8.

2.34 **Action:** As the pistons and arms move forward, the rest of the engine is pushed back.

2.35 **Page motor:** Consists of two electromagnetic cylinders (A, B) in series, through which a piston (c) with two iron cores moves. If a current flows alternately through each of the two cylinders, the piston moves back and forth. This movement is then transferred to a flywheel (R).

Page, Charles G. (1812–?): Professor of chemistry and pharmacy in Washington, later chief examiner at the United States Patent Office. He

contributed in the fields of electromagnetic and electrogalvanic theory and also built several electrical machines.

2.36 **k:** Note that k does not appear in Fig 4. Mach refers to the two connectors at the very top. Mercury filled ducts were the usual way of providing electrical connection to rotating bodies before reliable slip rings became available.

2.37 **Moon day:** Mach appears to have combined two ideas here, without clarification. The period of the moon's rotation about the earth, one month, depends on the distance of the moon from the earth, such that the centripetal acceleration equals that of earth's gravity at that distance. The same face of the moon is always visible from the earth primarily because of the moon's bulge, which offsets the center of gravity (close to the earth) from the center of mass. The analogy to the flywheel and the friction relates to the tides on the earth, which occur because of the gravitational pull of the moon.

2.38 **Schmidt gyro:** Although the term has dropped out of usage, this is merely the spinning top, a common toy.

2.39 **Pocket watch:** The balance wheel oscillates against a spring restraint, and is loosely coupled to the body of the watch, so that sudden movements of the watch do not initially get transmitted to the balance. In Mach's day the mechanism would have been widely known.

2.40 *ac:* Refers to the distance from the gimbal bearing *a* to the frame (*c* does not show up in Fig. 6).

2.41 **Foucault, Jean Bernard Leon** (1819–1868): French physicist who measured the velocity of light, and demonstrated the earth's rotation by a swinging pendulum.

Foucault's Pendulum provides a dramatic and convincing demonstration of the earth's daily rotation. It was first shown to the public in the Pantheon in Paris in 1851. A 60 meter long pendulum set into oscillation along an earth meridian (without any lateral or circular motion) will maintain its plane of rotation relative to "inertial space," that non-rotating space, non-accelerating relative to the "fixed stars," in which Newton's laws are simply applied. The pendulum will be seen to rotate its plane of oscillation, at the rate of one revolution per day, in a clockwise direction in the Northern hemisphere, and in a counterclockwise direction in the Southern hemisphere, while it will not show apparent motion along the equator. In general, the angular velocity of the Foucault pendulum is equal to the sine of the latitude times the angular velocity of the earth.

The heavy disk in Mach's text similarly assumes an angular velocity relative to inertial space of the sine of the latitude times the angular velocity of the earth. Once inverted, the disk angular velocity

as seen by an observer at that location is twice the Foucault angular velocity.

2.42 **Following version**: In his lecture "On Sensations of Orientation" (Ueber Orientierungsempfindungen, Wien 1897; included in the 3rd edition of his book *Popular Scientific Lectures*, Chicago: The Open Court Publishing Co.) Mach put in a footnote to say that he wants to withdraw this paragraph. He argues (p. 296):

> We have consequently a very simple means for determining whether one is actually the subject or not of uniform and imperceptible rotations. If the earth rotated much more rapidly than it really does, or if our semi-circular canals were much more sensitive, a <u>Nansen</u>[S.5] sleeping at the North Pole would be waked by a sensation of rotation every time he turned over. [Assuming a threshold of 2 deg/s, the earth would have to be rotating at a speed of 1 deg/s, or one revolution every six minutes for Nansen to have detected the rotation.] Foucault's pendulum experiment as a demonstration of the earth's rotation would be superfluous under such circumstances. The only reason we cannot prove the rotation of the earth with the help of our model, lies in the small angular velocity of the earth and in the consequent liability to great experimental errors.

In his lecture, Mach refers to further examples in his book *The Science of Mechanics*. Chicago: The Open Court Publishing Co., 1893, p. 302 [translated from the 2nd German ed.: *Die Mechanik in ihrer Entwicklung*. Leipzig: Brockhaus, 1889, p. 280]:

> The trade-winds, the deviation of the oceanic currents and of rivers, Foucault's pendulum experiment, and the like, may also be treated as examples of the law of areas. Another pretty illustration is afforded by bodies with variable moments of inertia. Let a body with the moment of inertia θ rotate with the angular velocity α and, during the motion, let its moment of inertia be transformed by internal forces, say by springs, into θ', α will then pass into α', where $\alpha\theta = \alpha'\theta'$, that is $\alpha' = \alpha(\theta/\theta')$. On any considerable diminution of the moment of inertia, a great increase of angular velocity ensues. The principle might conceivably be employed, instead of Foucault's method, to demonstrate the rotation of the earth, [in fact, some attempts at this have been made, with no very marked success].

2.43 **Disk**: Just as in the previous example, the disk picks up a fraction of the earth's rotational velocity proportional to the sine of its latitude, if it is not located along the equator. With an ideal, frictionless

suspension, the heavy disk then maintains the orientation of its axis in inertial space, as the earth rotates under it.

2.44 **Origins of motion sensation:** Mach strictly separates the logical problem of motion detection, which is accessible through physical laws, from the biological realization, which can evolve differently for different animal species. In other words, the formal functional description will remain unchanged, even if biological research adds many more details or even detects new principles of sensory mechanisms. This stands in sharp contrast to others in the field at that time who began with the physiological findings.

The Phenomena Observed in Moving Men and Animals

3.1 **Curve in a train:** see 1.9.

3.2 **Centrifugal force:** Einstein later showed in the Equivalence Principle that no physical device can distinguish between gravity and linear acceleration. The vector difference between the two is now referred to as specific force, or force per unit mass. For the case of travel around a curve discussed by Mach, the linear acceleration is centripetal, directed toward the center of the turn. The centrifugal force is added to the weight to produce a net gravito-inertial force vector directed outward from the vertical.

3.3 **Knight, Thomas Andrew** (1759–1838): English vegetable physiologist and horticulturist. He contributed with purely physiological experiments; however, his main object was always the utilitarian aspect of plant physiology. He was an original member of the British Horticultural Society (established 1804).

3.4 **Purkyně, Johannes Evangelista** (1787–1869): see 0.2.

3.5 **Pointer:** Indicated in Fig 7, above the axis α.

3.6 **Tilt from the vertical:** Further research on the subjective vertical found that most subjects slightly overestimate small tilt angles—the Müller (1916) Effect—and that almost all people underestimate their tilt at angles greater than about 30 deg—the Aubert (1861) Effect. Further discussion in Howard (1982).

3.7 **Duration of post-rotatory sensation:** To a first approximation, the strength of the post-rotatory sensation - that of rotating in the opposite direction following sustained constant rotation - follows an exponential decay having a time constant of 15–20 sec, with great individual variability. The duration of the sensation is proportional to the log of the initial speed of rotation $t_{th} = \tau \ln(A_0 / M)$, where t_{th} is the time

when threshold M is reached. τ the time constant and A_0 the initial rotation speed. On occasion, secondary sensation, or rotation in the original direction, is also reported. Further discussion of the relationship to the dynamics of the semicircular canals follows on page 118.

3.8 **Optical**: in 19th century science the word "optisch" (optical) was taken to include visual processes, and extended well beyond the physical characteristics of light and lenses.

3.9 **Entire visible room**: The perception that the external visual world is turning when the semicircular canals are stimulated is now commonly referred to as the oculogyral illusion. It has a threshold considerably lower than that for subjective sensation of rotation or for vestibular nystagmus.

3.10 **Falling to the side:** Simply stated, if the original rotation, seen from above, was clockwise about a vertical axis with the head pitched forward, the post-rotatory sensation will be counterclockwise—from right to forward to left. If the head is then pitched back up to the erect position, the rotation axis remains stationary in the head, and the rotation sensation becomes a roll rate, counterclockwise about the horizontal (nasal-occipital) axis, from right to up to left.

3.11 **Subjective vertical**: In this experiment the net gravito-inertial force during the steady rotation of the apparatus is a combination of downward weight and outward centrifugal force, so that the subject, who is seated facing the center of rotation, feels himself tilted backward. When the apparatus is decelerated to a stop, not only does the centrifugal acceleration disappear and eliminate the backward tilt illusion, but a transient tangential acceleration appears that is directed against the previous motion and produces a component of force toward the previous motion direction. The subject interprets this transient component, along with the steady downward weight, as a lateral tilt of the apparent vertical, and feels that he is rolled to the side, with his head tilted toward the previous motion.

Current research has shown that the linear acceleration must be prolonged for several seconds, as in the case of a catapult takeoff from an aircraft carrier, before the change in direction of the gravito-inertial force is interpreted as a rotation relative to the vertical. The accepted explanation is based on the lack of a confirming angular acceleration cue for very brief linear accelerations (Graybiel and Brown 1951; Young 1983).

3.12 **Moving around the surface of a cone**: Inertia keeps the inner frame and the subject from turning, at first, while the axis a describes a circle. The subject is therefore counter-rotating against the motion of the main

frame R. The outward directed centrifugal force describes a circle as it turns around the subject. Since the instantaneous apparent vertical is the vector sum of the weight and this centrifugal force, it too describes a circle around the subject. From the subject's point of view, this is interpreted as a tilt, whose direction circles around the vertical, as though he were leaning against a cone and moving around it, while always facing the same direction in space.

3.13 **Outlasting sensation**: The result of a mechanical event in the semicircular canal and further neural processing. The cupula has a mechanical time constant of about 8 s, e.g., decays to 1/e of its initial value in 8 s, and consequently the afferent activity in the vestibular nerve outlasts the end of acceleration according to this exponential decay. Effective lengthening of this time constant is performed neurally and can be seen in the lengthened response of neurons in the vestibular nuclei. This further processing has been associated with "velocity storage" (Raphan et al. 1979). As a result, the cupula return to its rest position lags the acceleration, and the subjective rotation further lags the cupula deflection (Goldberg and Fernandez 1984; Henn and Young 1975; Melvill-Jones et al. 1964; Waespe and Henn 1977).

3.14 **Sensation:** would soon cease for that matter if he were lying in any other position.

3.15 **Cross coupling:** Also called "strange sensation" by Mach. The phenomena described here are now termed "cross-coupled accelerations," and occur any time an out-of-plane head movement is made in a rotating environment. The resulting acceleration is about an axis orthogonal to the plane containing the first two rotations, and equal to the cross product of the two angular velocity vectors. In addition, an outlasting turning sensation is developed about the head axis that had previously been normal to the plane of rotation. The disorientation which occurs to pilots when making a pitching head movement during a constant rate turn is well known as a cause of aviation accidents, if pilots do not strictly rely on their instruments.

Cross-coupling should not be confused with the Coriolis-phenomenon.[5.6]

3.16 **Balance:** Mach evidently is comparing his experiments to the similar ones of Purkyně, who used a simple seesaw. A balance allows one to isolate the subject from lateral rotation during the vertical oscillation.

3.17 **Acceleration:** Mach's notation for acceleration, in terms of cm or m, leaves out the usual denominator sec^2. Mach also did it quite extensively for other units throughout the book.

3.18 **Weight acceleration**: Mach uses the term "Schwerebeschleunigung," or "weight acceleration," to refer to vertical acceleration, and not

literally to changes in weight. In later studies, Melvill-Jones and Young (1978) found that the threshold to detection of vertical acceleration was reached when a certain *velocity*, of about 22 cm/sec, was exceeded. Mach's indication of inability to indicate position during vertical oscillation was confirmed also by Malcolm and Melvill-Jones (1974).

3.19 **Extralabyrinthine graviceptors**: Although most research on human perception of the vertical has been based upon the assumption that the sensory signals stem from the otolith organs, recent studies by Mittelstaedt (1996) introduce the distinct possibility of important graviceptors in the lower body, in or near the kidneys.

3.20 **Motion sickness:** Although the "sensory-motor conflict" theory covers most conditions leading to motion sickness, simple overstimulation of the linear acceleration sensors can also produce symptoms. In essence this theory holds that motion sickness results from the appearance of different and inconsistent sensory signals regarding self-motion, either between different senses (e.g., visual vs. vestibular) or within an organ (e.g., semicircular canal vs. otolith). It also can appear consequent to a conflict between any sensory spatial orientation signal and the expectation of motion, based upon an internal model of the situation or knowledge of the muscle actions undertaken by the subject (Oman 1990; Reason and Brand 1969).

3.21 **Atwood, George** (1746–1807): Lecturer in mathematics and physics at Trinity College in Cambridge, later also working in the treasury. At this time he was famous for his teaching and experimental skills. His best-known work, *A Treatise on the Rectilinear Motion...* (1784), is a textbook of Newtonian mechanics. The ingenious apparatus, known as "Atwood's Machine" and to which Mach refers here, was designed to demonstrate the laws of uniformly accelerated motion due to gravity. It was first described in this treatise.

3.22 **Schmidt, Gustav Johann Leopold** (1826–1883): Lecturer and Professor of Mechanical Engineering at the Polytechnicum in Prague.

3.23 **Linear accleration adaptation:** The sensation of velocity lags behind actual velocity during constant acceleration, illustrating a measure of adaptation. First order otolith afferent neurons also show some decline in firing rate during constant tilt (Fernandez and Goldberg 1976).

3.24 **Linear acceleration sled:** Current vestibular research using linear acceleration 'sleds', on the ground and in space, suffers from the same problems of vibration interfering with threshold measurements, and has led to the development of ultra quiet air bearing devices and smooth hydrostatic actuators for aircraft simulators.

3.25 **Acceleration is $a\varphi/t$:** This is the Coriolis[S.6] acceleration of the outward moving observer with velocity a/t on a platform rotating with angular velocity φ.

3.26 **Railway tracks:** In Mach's time the maximum railway speed was too low to make this an issue. Even with current high-speed railways the Coriolis[S.6] forces when traveling North and South are too small to require alterations of the track.

3.27 **Stop:** of the outer frame.

3.28 **Linear:** read centripetal.

3.29 **Flying away:** In this experiment the subject's centripetal acceleration suddenly stops, although his rotation continues just as though he were still on the turning centrifuge. With the withdrawal of the centripetal acceleration, but continuing sense of rotation, the subject then feels himself accelerated oppositely, or outward, along the radial arm. Normally rotation on a centrifuge entails both angular velocity and linear acceleration. With this procedure, the linear motion is removed while maintaining the angular motion, thereby producing an exclusively linear acceleration aftereffect.

3.30 **Extralabyrinthine graviceptors:** see 3.19.

3.31 **Motion sensation:** The constant centripetal acceleration is not ignored during constant motion, but rather it is gradually re-interpreted as tilt relative to gravity—a phenomenon known as "g-tilt," which is often used in moving base flight simulators.

3.32 **Graefe, Friedrich Wilhelm Ernst Albrecht von** (1828–1870): Professor of ophthalmic surgery in Berlin. He introduced the operation of iridectomy for glaucoma, refined several surgical procedures, and had many contributions in the evolving field of neuro-ophthalmology.

3.33 **Czermak, Johannes Nepomuk** (1828–1873): Assistant of Purkyně, taught at the universities of Vienna, Budapest, and Jena. He described a mirror to investigate the larynx.

3.34 **Breuer, Josef** (1842–1925): see 0.5.

3.35 **Pyramid box experiment:** Mach describes a simple method to rotate a small animal, like a rabbit, on the centrifuge inside a four-sided pyramid along several axes of rotations. If side A (in Fig. 8) is bolted to the platform of the centrifuge, the animal is placed in the normal, or prone, position. If side B is bolted to the platform, the animal is in the upright standing position, while if side C is bolted, the animal is in the left-ear-down position.

3.36 **Origins of motion sensation:** see 2.44.

Flourens' Experiment

4.1 **Flourens, Marie Jean Pierre** (1794—1867): see 0.3

4.2 **Middle ear:** "Vorhof" literal translation of the Latin "Atrium." It remains unclear what is meant by "nerves of the middle ear."

4.3 **Cutting of the semicircular canals:** Presumably by drilling through the bony canal and cutting the membranous canal, may also destroy the ampulla. Only later Ewald[S.1] introduced the technique of canal plugging, which spared the ampulla. Because of the absence of post-experiment histological controls at that time it was not easy to determine the exact anatomical nature of the lesion. For instance, Mach assumed that with canal cutting, the ampulla would be irritated and excited. In fact, if it was effectively destroyed, it would have lead to an eradication, rather than an increase, of afferent nervous activity.

It is now known that destruction of any of the nerves leading to the ear, for hearing or for balance, would not produce pain. It remains unclear to what extent the striking movements of an animal were assumed to be caused by pain.

4.4 **Pain:** Violent head movements were induced as a result of the sudden vestibular stimulation followed by loss of afferent nerve activity. This was interpreted as an expression of pain rather than vestibulo-collic and vestibulo-ocular reflexes. It followed from the accepted theory at that time that the semicircular canals serve a function in directional hearing, and that such manipulation leads to painful hearing[S.7] and subsequent violent head movements. In fact the cutting of the canals led to the destruction of their ampulla.

4.5 **Specific energy:** see 1.10.

4.6 **Harless, Emil** (1820–1861): Professor of physiology in Munich.

4.7 **Czermak, Johannes Nepomuk** (1828–1873): see 3.33.

4.8 **Brown-Séquard, Charles Edouard** (1818–1894): Born in Mauritius, he worked in France and the U.S. before becoming the successor of C. Bernard at the Collège de France in Paris. The neurological symptoms, which occur after one half of the spinal cord had been lesioned, is widely known as the "Brown-Séquard Syndrome."

4.9 **Vulpian, E Felix Alfred** (1826–1887): Professor of pathological anatomy in Paris.

4.10 **Schiff, Moritz** (1823–1896): German physician and neurophysiologist, whose many different positions included being Director of Ornithology at the Natural Museum in Frankfurt (Senckenberg Museum), professor of anatomy and pathophysiology in Berne, professor of physiology in Florence, and finally professor of physiology in Geneva. His original experimental work covered the whole field of physiology, with

concentration in neuromuscular research. He wrote a textbook *Muscle and Neurophysiology* in 1858–59.

4.11 **Distribution of fluid pressure:** Although it is certainly true that a fluid-filled flexible tube surrounded by a gas or any less dense fluid would expand toward the bottom, as Goltz suggested; this has little to do with the function of the semicircular canals. The presence of perilymph surrounding the membranous canals (with virtually the same density as the endolymph inside the canals) determines that the transmural pressure is independent of the position of the canal relative to gravity, and thus the bottom of the canal does not bulge out.

4.12 **Hasse, Karl** (1841–?): Anatomist in Würtzburg and professor and director of the Anatomical Institute in Breslau. Mach refers to one of his early works *Anatomische Studien* published in Leipzig 1870–72.

4.13 **Cyon, Élie** (1843–?): A determined critic of Mach and Goltz, Cyon was an assistant for anatomy and physiology in the faculty of physical mathematics in St. Petersburg, and later worked in Paris at the behest of Claude Bernard. In his thesis and subsequent publications he set forth a complex theory of the function of the ear as an organ for the perception of space and time. Although not fully presented until 1878, it was evidently known to Mach in 1873 (Cyon 1873). He moved from the findings of Flourens and Purkyně to a conception of geometric and arithmetic knowledge having origin in the ear. In his 1911 treatise *L'Oreille: Organe d'orientation dans le temps et dans l'espace* (*The Ear: Organ of Orientation in Space and Time*) he was ruthless in his criticism of Mach as an opportunistic physicist with a penchant toward metaphysics.

In particular in his comments on this book, he accuses Mach of rushing to the Vienna Academy of Science with the idea of fusing Cyon's concept of the semicircular canals as sense organs for the three directions of space with Mach's own concept that they measured rotation. He states that this book, "published the following year, surpassed all that one might fear in this respect. Therefore I feel obliged...to furnish the experimental proof of the complete insufficiency of these interpretations" (Cyon 1911, pp. 44–45). He goes on to present a variety of experiments involving the draining of perilymph from the semicircular canals, which fails to produce the Flourens head shaking, and uses these to refute the theories of Crum Brown, Breuer, and Mach concerning the role of endolymph movement on rotation sense. He ridicules Mach for drawing an analogy between the rotation sensations and vertigo discussed by Purkinje and the animal postural reactions of Flourens, and further for the idea that acceleration, rather than velocity, is the adequate stimulus for motion sensation. He even quotes and refutes Mach's famous initial

observation concerning the apparent tilt of trees and houses seen from a train going around a curve. "For a non-metaphysical spirit this observation would be simply interpreted, but Mach was too much under the influence of Goltz' endolymph hypothesis to recognize the true state of things" (Cyon 1911, p. 53).

4.14 **Curschmann, Heinrich** (1846–?): Medical doctor and director of the Municipal Hospital in Hamburg, later professor and director of the Medical University Clinic in Leipzig. Among his numerous publications, Mach refers to his early work (1874) as listed in Mach's own bibliography.

4.15 **Boettcher, A** (1831–1869): see 0.7.

4.16 **Berthold, Emil** (1836–?): A contemporary of Mach, professor of ophthalmology and otolaryngology in Königsberg, who published several similar papers prior to those of Breuer, Crum Brown, and Mach. He declared: "The semicircular canals therefore have the function to mediate the coordination of movements via reflexes. They serve this function together with two other sensory organs, vision and touch. Whether the semicircular canals also have an acoustic function, had not been determined so far."

Phenomena Reminiscent of Flourens' Experiment

5.1 **Flourens, Marie Jean Pierre** (1794–1867): see 0.3.

5.2 **Ménière, Prosper** (1799 - 1862): French physician, appointed as director of the Institut Sourds-Muets in 1838 (Imperial Insitute for Deaf-Mutes). Today his name is connected with a disease of the inner ear, Ménière's disease. Symptoms are sudden attacks of tinnitus and vertigo, and with recurrent attacks, a gradual loss of hearing and vestibular function. Up until his time there were only sporadic reports that connected the symptom of vertigo with inner ear disease. The then current opinion was that vertigo is caused by congestion of blood in the brain. The labyrinth was considered to serve an acoustic function, specifically directional hearing. (Ménière 1861; Atkinson 1961; Ruben 1989)

5.3 **Exner, Sigmund** (1846–1926): Professor of physiology in Vienna, student of Brücke[5.4] and Helmholtz[6.7].

5.4 **Brücke, Ernst** (1819–1892): Professor of Physiology in Königsberg, later in Vienna. Breuer[0.5] particularly speaks of him with high esteem.

5.5 **Türk, Ludwig** (1819–1868): Neurologist in Vienna. He found fiber degeneration in the direction of projection after focal lesions in the brain or spinal cord. Thus he was able to trace descending and

ascending fiber tracts in the spinal cord. He was one of the first to describe the segmental cutaneous distribution of afferent dorsal root fibers.

5.6 **Wundt, Wilhelm** (1832 - 1920): Physician who became interested in psychophysics. After having been appointed Professor of Physiology in Leipzig he established an Institute for Experimental Psychology that was the first one of its kind worldwide. His books cover all fields of psychology. (Wundt 1873–74; Blumenthal 1975).

5.7 **Magendie, François** (1783–1855): Physician and physiologist. His most important discovery was the different functions of the spinal nerves: the anterior roots serving motor action, and the posterior roots conveying sensory information from the periphery. He did this experimental work independent of Charles Bell, who challenged his priority for this discovery.

5.8 **Schiff, Moritz** (1823–1896): see 4.10.

5.9 **Mitchell, Weir** (1829–1914): The name in the text is given as Mitschell, but this is probably a spelling error. Born in Philadelphia, he studied medicine, became an army surgeon, and described his keen observations, especially concerning pathophysiology of pain, in numerous publications. This very colorful figure dedicated the second half of his life to poetry.

5.10 **Henle, Jakob** (1809–1885): Physiologist in Berlin.

5.11 **Ritter, Johann Wilhelm** (1776–1810): see 1.3.

5.12 **Purkyně, Johannes Evangelista** (1787–1869): see 0.2.

5.13 **Galvanic stimulation:** By placing electrodes on the skin over the mastoid process behind the ear, one delivers electrical current, usually of the order of ± 4 mA. Although the mechanism of action is still not well understood, it seems likely that by this procedure one directly stimulates the vestibular nerve, thereby bypassing the sensory apparatus in the labyrinth. This method still has some limited clinical application.

When Breuer used galvanic stimulation, he was evidently attempting to selectively stimulate individual canals by external galvanic stimulation, which does not seem possible. However, placing of electrodes in the vicinity of the ampulla of a single canal can experimentally stimulate the hair cells of that individual canal. (Dieterich et al. 1999; Moore et al. 1990; Pfaltz 1969; Suzuki and Cohen 1966; Suzuki et al. 1969).

5.14 **Smee's battery:** Serial battery usually containing 6-24 elements with zinc cathodes and platinum anodes, submersed in sulfuric acid (Smee 1840).

Smee, Alfred (1818–1877): Surgeon in London, mostly known for his work on electro-metallurgy (Smee 1849). He also pioneered in the new territory of electrical physiology.

5.15 **Hitzig, Eduard** (1838–1907): Professor of psychiatry in Zürich, and later in Halle. His most influential discovery, along with Theodor Fritsch, was the local and specific electrical excitability of the motor cortex. These experiments disproved the holistic theory of brain function that all cortical areas are equivalent. Mach also refers to his experiments concerning galvanic stimulation[5.13].

5.16 **Galvanic stimulation**: see 5.13.

5.17 **Illusory movements:** A continuing debate in oculomotor physiology concerns the manner in which one compensates for eye movements, so that the world normally does not appear to move when the eyes do. The "outflow theory," associated with Helmholtz[6.7] (1866), holds that the intended eye movement is taken into account in canceling the retinal displacement, whereas the 'inflow theory', associated with Sherrington (1918), maintains that measurement of the actual eye movements are used for this purpose. The association of illusory movements of objects in the visual field supports the outflow theory.

5.18 **Escamoteur:** A device used by magicians to make things disappear. In this case it was presumably a source of electric current that stunned the fish.

5.19 **Ruhmkorff apparatus:** Induction apparatus based on the principle that turning current on or off in one wire induces a current in a second wire (usually realized as two coils of different size). The Ruhmkorff apparatus was especially designed to deliver high currents to discharge with a spark. Therefore this apparatus was often used to trigger explosions. A similar device was later used to generate the ignition spark in automobiles (see Shiers 1971).

Rühmkorff, Heinrich Daniel (1803–1877): German inventor (without academic rank), learned and studied in Germany, France, and England before opening a shop in Paris. Particularly known world-wide to physicists for the quality of his electrical instruments.

5.20 **Cobitis barbatula L**: Freshwater fish, zoologically similar to carp.

Comparison of Movement Sensations with Other Sensations

6.1 **Mechanical work:** Potential energy is converted to other forms as the mass sinks, following the law of the conservation of energy. In *The Science of Mechanics* Mach (1893) explores the historical context of J.

R. Mayer's contribution relating heat and energy to the work on conservation of energy to the work of <u>Helmholtz</u>[6.7] and others.

6.2 **Work:** This term as used by Mach means the mechanical stimulus in a general sense, not the technical definition of work as a form of energy.

6.3 **Relay:** Electrical device that uses a small current to switch a larger one. In modern terms, Mach would have probably referred to an amplifier.

6.4 **Plateau, Joseph Antoine Ferdinand** (1801–1893): Leading Belgian physicist in Ghent who is known both for his pioneering work on physiological psychology and optics. He went on to study molecular forces in neutrally buoyant liquids. He became completely blind in 1843 (apparently as a result of an experiment in 1829 when he stared into the sun) but continued conducting research. The term "Plateausches Netz" [Plateau grid] (p. 89) remains obscure.

6.5 **Constant intensity**: One can differentiate several conditions. During prolonged constant auditory stimulation there is obviously some decrease in the perceived loudness of sound. However, with pathological conditions, like tinnitus, the loudness might not change.

6.6 **Oppel, Johann Joseph** (1815–?) Professor of mathematics and physics in Frankfurt. He worked on the perception of auditory and visual sensations and illusions.

6.7 **Helmholtz, Hermann Ludwig Ferdinand von** (1824–1894): One of the most influential scientists of the 19th century. Originally an army surgeon, a career choice that was the only possibility at that time to obtain a free education, Helmholtz quickly turned to physiology and physics. He was professor of physiology in Heidelberg and later director of the Federal Institute of Physics and Technology in Berlin. His major contributions were:

1. Establishment of the law of conservation of energy (1845), performed entirely independent of R Mayer;

2. Invention of the ophthalmoscope (1851);

3. Publication of the standard work on vision: *Treatise on Physiological Optics* (3 vols, 1856–67; translated 1924, and reprinted 1962, New York: Dover). Besides a comprehensive historical overview, Helmholtz described many experiments he performed using mainly psychophysical methods. He proposed the trichromatic theory of color vision, and he provided an extensive review of theoretical and experimental aspects of eye movements.

4. Publication of a comprehensive work on hearing *On the Sensations of Tone* (1863; translated 1875 by A. J. Ellis, London: Longmans, Green, & Co.);

5. Seminal contributions to electromagnetic theory:

6. Finally, Helmholtz made efforts to educate the public about science. His many lectures are edited as *Popular Lectures* (1881; translated by E. Atkinson [1884], *Popular Lectures on Scientific Subjects*. London: Longmans, Green, & Co.).

See Cahan (1993) for an extensive bibliography.

6.8. **Sensation decay:** The simple "torsion pendulum" model of the semicircular canals as transducers of angular acceleration to produce the sensation of angular velocity (Van Egmond et al. 1949) failed to account for this adaptation phenomenon pointed out by Mach. It predicts a constant cupula deflection and constant perceived velocity for a sustained acceleration. However, the actual velocity sensation reaches a peak after about 25 sec and then slowly decays, as shown by Guedry and Lauver (1961). The adaptation effect was discussed by Groen (1961) and modeled independently by Young and Oman (1969) and by Malcolm and Melvill-Jones (1970). Vestibular adaptation as a neural phenomenon is discussed in Young (1983) and appears in peripheral 8th nerve recordings (Fernandez and Goldberg 1983).

6.9 **Misprint:** See remark on page 127. Mach states that "therefore" (as in the original text) should be replaced by "however."

Further Investigations of the Phenomena

7.1 **Hole only under one foot:** In this experiment the spatial distribution of pressure under the soles is altered. When the piston is withdrawn, pressure is reduced under the hole and increased under the remainder of the foot.

This phenomenon is utilized in the "g-seat" of some flight simulators, where pressure distribution on the pilot's seat is altered to enhance motion cues.

7.2 **Body rising:** Scuba diving provides another example. The water pressure increases at a rate of 1 atmosphere per 10 m depth. Subjects with normal vestibular function can easily orient themselves upright or head down under water and usually do not experience any spatial disorientation, despite the pressure gradient. Labyrinthine defective subjects, however, must increasingly rely on visual and pressure cues for orientation, and they become easily disoriented under water (Graybiel et al. 1968).

7.3 **Force X on unit surface:** X indicates pressure, so that $X\,df$ is an element of force over the area df, and $\iint X\,df\,dt$ represents the total

impulse delivered to the body over a certain time that, divided by mass, is equivalent to the change of velocity in that time.

7.4 **Skin**: There can indeed be some influence from skin sensation when it mimics natural movement. The deliberate adjustment of local pressure cues is used in the field of flight simulation in order to produce the onset cues associated with sudden changes in acceleration. (Kron 1975; Berthoz 1997, pp. 36–37).

7.5 **Skinning**: Mach probably refers to patients who lost parts of their skin, e.g., after severe burning. Skin sensation can play a role if, under natural conditions, such information proves to be useful. For example, movement of an animal causes head and eye nystagmus to stabilize gaze. When a pigeon flies, there is normally a flow of air against its breast. The gain of normally generated optokinetic nystagmus can be increased by simultaneously blowing air against its breast.

7.6 **Special innervation to maintain position:** There is, of course, a special innervation to correct for disturbances in muscle position, namely the muscle-spindle and the muscle-tendon spinal reflexes.

7.7 **Perception that the body sinks:** The force that supports our body just opposes our weight (mass times gravity g) minus the inertial force (mass times downward acceleration a). When an elevator accelerates downward, for example, we initially feel lighter, weights in our hands would tend to rise slightly, and we would experience less force against our feet. Our internal accelerometers, particularly the otolith organs, also sense the new gravito-inertial force g-a, and lead to a sense of downward acceleration. In Mach's experiments with reduced weights, only the muscles and local tactile and proprioceptive senses registered and reacted to the "weight loss," but the internal graviceptors continued to indicate that the subject was stationary in a one gravity environment.

7.8 **Head rotation**: This is essentially an example of the well-known reaction to removal of a sustained load, producing an involuntary muscle contraction. The common demonstration takes place with a subject standing in a doorway, pressing his lowered hands outward against the doorjamb. When he walks away the arms involuntarily rise. In the case of this Mach experiment the situation is slightly complicated by the vestibulo-collic reflex, which induces head movements in a direction to stabilize the head in space during involuntary body motions. Although the vestibulo-collic reflex is associated with body motions, it is not necessary for the perception of such motion.

7.9 **1865 publication:** It is unclear to what publication Mach refers.

7.10 **Fessel, Friedrich** (1821–?): Physics teacher and mechanic in Cologne working for Professor Plücker in Bonn. He is best known for his skills

in constructing scientific apparatus. Fessel's flyweel apparatus was a device supporting a rotating disk within a gimballed structure. He used it for demonstrating the precision of gyroscopic motion and stability. The machine illustrated below is discussed in Plücker (1853a,b).

Ann. d. Phys. u. Chem. Bd. 90. St. 2

7.11 **The principle of conservation of areas:** see 2.9.

7.12 **Therapeutic application**: <u>Erasmus Darwin</u>[1.1] considered such an application earlier. He placed the subject in a supine position on a turntable with the head directed away from the rotation axis. His speculation was that during spinning, blood would be shifted toward the brain by centrifugal force. Although this might be true, it creates the even more severe problem that the same force would now prevent venous blood from returning to the heart.

Until the early 19th century, rotating chairs were commonly used to tranquilize patients with psychiatric disorders (See Fig. 1 of <u>Cohen</u> 1984, and Fig. 7 of <u>Grüsser</u> 1984). Artificial gravity, which has been considered a viable countermeasure to the deconditioning effects of long duration space flight, would involve centrifugation of astronauts oriented with the head toward the axis of rotation. Recently, a short-arm (2 m radius) centrifuge has been investigated with respect to therapeutic alteration of lower limb circulation in patients with spinal cord injuries (Cardus 1997).

7.13 **Popper, Josef** (1838–1821): A great friend of Mach's, Popper studied at the Technical Hochschule in Prague and made his mark in many fields, both scientific and governmental. He wrote on physics, technology, anti-Semitism, and human rights, and is recognized for his early "societal engineering," including the idea of a guaranteed annual wage (Blackmore 1972).

7.14 **Blood loses weight**: Mach recognized the potential for blood to serve as the seismic mass to detect acceleration. However, this effect would be very small, since it has nearly the same density as the surrounding fluid. Therefore, blood movement would be a poor indicator of body motion. However, Mach is more concerned with the possibility of redistribution of the blood volume during acceleration as an indicator of body movement.

7.15 **Fatigue eye muscles:** Purkinje's argument here is clearly inadequate. The eye muscles are unique in that they are the only voluntary muscles of the body that do not fatigue under normal conditions (smooth muscles and the heart muscle also do not fatigue). This might be due to their very large size-to-load ratio. The reactions Purkinje notes are partly due to the fall off in the VOR and partly to alterations in slow phase nystagmus, even in the light, which lapses during inattention—a kind of psychological fatigue.

7.16 **Visual vertigo:** Mach uses this term to refer to vertigo resulting from motion sensation caused by seeing apparently moving objects. The outside world remains stationary, however, as evidenced by repeatedly appearing objects in the visual field and by other senses. These conflicting sensations could lead to vertigo. In modern terms, this phenomenon is called "oscillopsia."

7.17 **Finger against the eye:** As the reader can readily establish, the movement of objects depends upon where one presses on the eye. Mach's example, of objects moving in the opposite direction to the push, is best seen when the eye is already deviated fully to one side, so that the finger push is at a point far from the visual axis.

7.18 **Rapid jerks:** These fast eye movements that Mach mentions are the fast phases of nystagmus—a term applied to any rhythmical rapid eye movement. It comes from the Greek for slumping and rapid backward jerking of the head when sleepy. Vestibular nystagmus described by Purkyně consists of alternations between the slow compensatory phase and the fast resetting phase. Nystagmus can also be induced by moving visual scenes (optokinetic nystagmus), and occurs in a variety of pathological conditions.

The function of nystagmus is to stabilize gaze during active or passive head movements. For perfect compensation of rotational head movements, the slow phase eye movement velocity is exactly equal and opposite to the head movement. This is often achieved during rotation in the light, when vestibular and visual stimuli combine. In darkness, with only vestibular stimulation, nystagmus gain is usually below unity. Because the dynamic characteristics of nystagmus in different directions (horizontal, vertical, and torsional) are not the

same, the eye movement direction is also not always exactly compensatory.

Eye movements also compensate for linear translations of the head, but this nystagmus tends to be variable and depend upon perceived viewing distance (Paige 1989; Paige and Tomko 1991; Yue et al. 1995).

7.19 **Suddenly moves forward:** Melvill-Jones showed that for active rapid head movement the first eye movement is usually anti-compensatory, a fast phase, in the direction of head turn.

7.20 **Muscle action cannot develop:** It was not known at that time that the vestibulo-ocular reflex has a very short latency (30 ms), and that its gain increases with stimulus frequency. Therefore, with very rapid head movements, "muscle action" does develop and is the decisive factor in inducing eye movements that nearly compensate for the head movement. Any residual retinal slip is perceived as object motion.

7.21 **No visual impression:** Breuer$^{0.5}$ refers to "saccadic suppression" whereby stationary objects are not seen during saccades or during fast phases of nystagmus unless they are at least an order of magnitude brighter than the background. A simple demonstration shows that the suppression is not complete, and that it depends only on the visual cancellation of a retinal smear. An optokinetic drum or any rapidly moving visual field will appear as a blur unless a rapid eye movement is made *in the direction* of the field motion, in which case the drum appears sharply for a moment.

7.22 **Nystagmic movements:** see 7.18.

7.23 **Afterimage:** Since the afterimage of a bright source remains fixed to the retina for some time, it has long been used as an indicator of eye position that moves with the eye. During rotation the after-image appears to move relative to the subject, even in the presence of a visual field, as noted by many. Yasui and Young (1975) found that the presence of the afterimage actually increased the speed of the compensatory movements during vestibular nystagmus, as the subjects apparently attempted to pursue their own eye-fixed image.

7.24 **Turning in the direction opposite to the drum:** This visually induced sensation of self-motion is now referred to as circular vection. It depends on wide-field stimulation, especially of the peripheral visual field. The striped drum is an optokinetic drum. An optokinetic drum often used for neurological bedside clinical examination is smaller, rotated in front of the subject, and does not induce circular vection.

7.25 **Moving railway:** Commonly referred to as *linear vection.* Like the circularvection phenomena discussed previously, linear vection develops slowly in the presence of a large uniformly moving visual field. It can be enhanced by the presence of a confirming true motion

cue or delayed by an opposite motion cue. Linear vection has become a powerful tool for inducing a sense of motion, whether in a flight simulator or in an amusement ride. Like circular vection, its strength decreases with prolonged exposure.

7.26 **Appears to be stationary**: To observe the described phenomenon fully, the period of constant velocity rotation should be longer than about 30 s (Henn et al. 1980).

7.27 **Powerful nausea**: Visual input alone can lead to nausea. Examples are high accelerations of a visual stimulus that covers the whole visual field. As such a stimulus is apt to induce circular vection; the missing labyrinthine confirmation of an angular acceleration poses a sensory conflict. Sensory-motor conflict is thought to be involved in all forms of motion sickness. The same stimulation with low visual acceleration (less than 5 deg/s^2) minimizes the conflict, and therefore does not lead to nausea.

Another example, called the pseudo-Coriolis effect, occurs after the subject is exposed to whole-field visual stimulation and develops circular vection. If he then turns his head about another axis, some disorientation and sensation of nausea might develop. It is reminiscent of the original (misnamed) Coriolis effect, which is actually attributable to cross-coupled angular acceleration, rather than Coriolis forces.

7.28 **Hitzig, Eduard** (1838–1907): see 5.14.

7.29 **Small oscillations in the corner of the eye:** It had been aptly described as "shimmering nystagmus" by M. B. Bender (1905–1983), a neurologist at Mt. Sinai School of Medicine.

7.30 **Movement perception:** Although the slow phase of compensatory eye movements and the perceived rotational velocity frequently follow similar time courses, whether the stimulus is vestibular, optokinetic, or electrical, neither one can now be said to be caused by the other. Frequent examples of dissociation exist, including the observation that movement perception and slow phase eye movements can be in opposite senses if the subject looks at a moving field, the center of which is seen as moving in an opposite direction by use of a small mirror (Koenig et al. 1987)

7.31 **Rollmann, Wilhelm** (1821-?): Professor of Mathematics in Strahlsund. Mach probably refers here to Rollmann (1868).

7.32 **Cube:** Commonly known as a "Necker cube." Necker was a crystallographer and observed apparent switching of depth when he examined rhomboid crystals through a microscope.

7.33 **Plateau, Joseph Antoine Ferdinand** (1801–1893): see 6.4.

7.34 **Eye movements play no part:** During angular or linear movements of the head, the eyes move to stabilize gaze in space. However, over

certain ranges of accelerations and especially at lower frequencies, eye movements can be visually suppressed by fixation. In this demonstration, with fixation suppression of any compensatory eye movement, only the head motion is taken into account in the interpretation of the moving retinal images.

7.35 **Perceptual space**: Mach touches on the issue of perceived object distance and "size constancy" here. The perceived object distance is used to relate any retinal shift, or any head movement, to the perceived object displacement in space. (Gillingham et al. 1996; Sedgwick 1986).

7.36 **Movement sensations**: It is now understood that the same vestibular signals that lead to movement sensations also may lead to compensatory eye movements, and that visual vertigo may occur with or without such eye movements.

7.37 **Mechanical strains**: The brain itself is not sensitive to mechanical deformation or pain. Therefore this mechanism is ruled out, even as a hypothesis. However, the meninges, the layers that surround the brain and spinal cord, are sensitive and can induce pain, if inflamed with meningitis, for example. Also, excessive dilatation of blood vessels might lead to pain, as current hypotheses for the mechanisms of migraine suggest.

7.38 **Wundt, Wilhelm** (1832–1920): see 5.6.

7.39 **Pitching head movements**: In this critical experiment Mach eliminates neck muscle activation as well as neck joint angle sensation as the source of vertigo that occurs with pitching head movements during or following yaw rotation about a vertical axis. Mach first determines the location of the normal axis of head rotation for pitching movements. He then fixes the head in a frame and passively oscillates the body about this same axis, both during and following rotation of the whole apparatus about the vertical, and fails to produce the vertigo phenomena. The cross coupling[3.15] and vertigo described elsewhere require a change of the planes of the semicircular canals, and thus a head movement, leading to Mach's conclusion that movement sensation is sensed by an organ in the head.

7.40 **When the eyeball is moved by an outside force**: There had been some debate how the visual world remains stable[5.17] despite eye movements. A heroic means of separating the intended eye movement (or any other movement for that matter) from the actual movement was achieved by temporary paralysis with curare. (Stevens et al. 1976).

7.41 **Fixed relative to the head**: Mach supposes that the eye rotation axis, like the subjective sensation of rotation, remains generally fixed relative to the head after repositioning the head, which is true as a first approximation. Recent studies have shown that the axis of vestibular and optokinetic nystagmus does indeed change somewhat with head

position, and tends to align with the vertical. (Merfeld 1966, Gizzi et al. 1994)

Theory of the Phenomenon

8.1 **Lateral line organs**: Small organs in the skin of fish, encompassing sensitive hair cells. Flow of water over the organ stimulates the hair cells and helps to control swimming behavior of the fish. (Recent review: Popper and Platt 1993, pp. 99–136.)

8.2 **Hitzig, Eduard** (1838–1907): see 5.14.

8.3 **Sagittal canals:** Strictly speaking, there are neither sagittal nor frontal pairs of semicircular canals. The anatomical orientation of the vertical canal pairs describes an X-shape in the "left anterior–right posterior" (LARP) and the "right anterior–left posterior" (RALP) directions. Furthermore, the horizontal canals lie not precisely in an earth-horizontal plane for the normal upright head position, but are pitched upwards by about 15 deg (Blanks et al. 1975).

8.4 **Darwin, Erasmus** (1731–1802): see 1.1.

8.5 **Torsional movements:** The relationship between head movements and eye movements is species-dependent, and is a function of the visual field of the animal. Pitching head movements produce ocular torsion of the lateralized eye of the rabbit, and vertical (doll's eye) movement in the human (Graf and Simpson 1981).

8.6 **Nagel, Albrecht** (1833–1895): Student of Graefe[3.32], then professor of ophthalmology in Tubingen.

8.7 **Linear acceleration:** Mach mentions the sacculus as a receptor for linear acceleration (p. 110), but does not mention the utriculus in this context. At his time, the common name for the utriculus still was "macula acustica" (see also p. 111). In the current view, both the sacculus and utriculus are peripheral sensory organs for linear acceleration. Hair cells in both structures are oriented in many directions, so that omni-directional acceleration detection is possible. Since hair cells can also be deflected by pressure waves, the sacculus seems to act as a detector of low frequency sound in some animals (Curthoys et al. 1999).

8.8 **Müller, Johannes** (1801–1858): see 1.10.

8.9 **Sensory paralysis:** Neurological term indicating a loss of sensation, by analogy to motor paralysis that describes loss of motor function. Specifically, in this context it means the loss of afferent activity in the vestibular nerve.

8.10 **Poinsot, Louis** (1777–1859): French mathematician, successor of J. L. Lagrange at the École Polytechnique in Paris. Mach refers here to his

Théorie nouvelle de la rotation des corps (1834), where he treats the motion of a rigid body geometrically. These combined rotations are now routinely treated using vector cross products.

8.11 **Combination of rotations:** The phenomenon known as "cross-coupled angular acceleration" results from simultaneous rotation about two perpendicular axes. The result is angular acceleration about a third, orthogonal axis, equal to the product of the two original angular velocities. This phenomenon is distinct from the Coriolis phenomenon, which refers to the unexpected linear acceleration associated with translation within a rotating environment. It is also unrelated to the change of position after stopping of rotation (Purkyně effect), where the directions of aftereffects remain invariant in head coordinates.

8.12 **Inhibition and excitation of nerves:** Mach obviously assumed that nerves could only be excited. Only much later was it appreciated that vestibular afferent fibers maintain a resting discharge rate, which can be either inhibited or excited, and are thus capable of signaling acceleration in two opposing directions. In rhesus monkeys, for example, the resting discharge of single vestibular afferent fiber is on average 100 pulses per sec (pps).

The discharge rate may increase substantially (up to 500 pps) during excitation by large acceleration in one direction, but it can only be inhibited to zero by acceleration in the opposite direction. This basic asymmetry is normally resolved by use of signals from both labyrinths. Since each nerve carries information about acceleration in both directions, the loss of one labyrinth can usually be compensated, if accelerations are restricted to a modest level.

It can be concluded that Mach did not consider spontaneous activity in neuronal structures, which would allow a bi-directional modulation. However, his deliberations are independent of the details of biological realization. This strict separation between functional relations and biological realization is one of the strengths of Mach's approach.

8.13 **Coordinate planes:** Mach introduces a stereotaxic coordinate system here long before its earliest application in stereotaxic surgery by Horsley and Clarke (1908).

8.14 **Labyrinth model:** Further instructions about how to make a cardboard model of the labyrinth are given in: Lithgraw (1920).

8.15 **Nearly parallel:** The relative orientation of the posterior and anterior canals, as well as the actual plane of the horizontal canals in humans was determined with some accuracy by Blanks et al. (1975). They differ from orthogonality and binaural parallel orientation by about 20 deg.

8.16 **Rotation sense:** Mach designates the anatomical directions of flow from the canals to the ampullae in Fig. 18. Only later it was learned

that the physiological excitatory directions are determined by the hair cell orientation, with excitation in the direction of the kinocilia.

The hair cells of the ampulla of the lateral (horizontal) semicircular canal are excited by rotation of the head towards the ipsilateral side, causing endolymph pressure to build up in a direction consistent with flow from the semicircular canal toward the ampulla (ampullopetal). Directions are reversed for the vertical canals where a pressure developed consistent with flow away from the ampulla (ampullofugal) leads to excitation, and ampullopetal flow leads to inhibition of afferent nerve activity (Hudspeth and Corey 1977). Therefore, anterior canal hair cells are excited with a somersaulting movement in a forward downward direction, while posterior canal cells are excited with a backward nose-up movement (see Goldberg and Fernandez, p. 986).

We now know that almost no endolymph flow actually occurs, and that the canal signals are transduced by minute bending of hair cells lying in the nearly stationary sub-cupular space.

8.17 **Almost orthogonal**: see 8.16.

8.18 **Stimulation of the ampulla:** Although Mach defines the direction for excitatory stimulation of the ampullae in terms of flow of endolymph from the ampulla toward the canal, this in fact is the direction for inhibition of the afferent nerve. Therefore, his statement that stimulation of the left horizontal canal leads to a sensation of turning to the right, although consistent with Mach's definition, is opposite to the current conventional definition of the excitatory direction.

8.19 **Away from the semicircular canal**: see 8.16.

8.20 **Compensating sensation:** This is in fact not true. Cutting a pair of canals that lie in the same plane, e.g., the left anterior and right posterior (LARP) canal, results in an insensitivity for head rotations that take place predominantly along rotation axes normal to the affected plane. After such an operation, one can observe excessive head movements with increased angular velocities in the direction of the lost canals, an effect that attenuates after some days.

8.21 **Electrical stimulation**: see 5.13.

8.22 **Vivisection:** The objections to vivisection refer to the potential danger of injuring the brain during the operation, resulting in an "unclean experiment," as mentioned earlier when discussing the Flourens' experiments. From the way Mach describes the procedure of stimulating the ampulla, it is clear that this still would involve surgery, but the electrical stimulation would permit one to evoke the vestibular sensation under a much more controlled condition.

8.23 **Sacculus**: see 8.7.

8.24 **Rotation of the eyeball:** Commonly referred to as ocular counterrolling (OCR), or eye torsion. It increases with head roll, on average up to 6–8 deg in humans, and is an important clinical indicator of otolith function. Eye rotation is actively controlled by the brain in accordance with Listing's law (Fetter et al. 1997; Helmholtz 1867; Hepp et al. 1997).

8.25 **Perception of tilt sensation:** There are indications that both counterrolling and the perception of tilt can adapt over time; however, this process has a long time constant. Physically, accelerating translatory movements and gravity are indistinguishable, and lead to the same deflection of hair cells. However, because of the differences in duration of the stimuli, long-duration, non-fatiguing responses are associated with head tilt and shorter duration signals are thought to be associated with linear acceleration. A debate remains as to whether the distinction between acceleration and tilt is made centrally, on the basis of multi-sensory integration, or peripherally within the otolith system, requiring separate receptors with high and low frequency characteristics. Although Mach is wrong in his assumption of separate organs for sensing tilt and translatory acceleration, his view is close to that of specialization of function within the otolith sensory system (Hess and Angelaki 1999; Mayne 1974; Merfeld et al. 1991).

8.26 **Planet of smaller mass:** Mach's reasonable speculation could not be tested until the Apollo astronauts walked on the moon in 1969 (see 2.26). They adapted to the lunar gravity by changing their locomotion to loping rather than running. They reported any loss of orientation only when deep in the shadows of a crater with the reduction of visual references. Astronauts inside a spacecraft have rarely been able to induce a sense of falling—but the sensation is more common during outside (extravehicular) activity. Without any close visual reference (except the distant earth) some extravehicular astronauts have reported a feeling of falling. The explanation for the absence of a sense of sinking seems to lie in the adoption of alternate sensory orientation cues, initially visual and tactile, to substitute for the missing otolith signals. After several days in space most astronauts begin to rely on their internal orientation, that the direction of their feet represents a kind of local vertical.

8.27 **Refreshes itself so rapidly:** Mach imagines a non-fatiguing and therefore non-adapting additional organ, which would be able to detect the constant gravitational acceleration.

8.28 **Elementary components:** The vector components of rotation lie normal to the planes of the semicircular canals. The corresponding components of translation are determined arbitrarily in the primary planes of the saccule and utricle.

8.29 **Young, Thomas** (1773–1829): Prolific physician and physicist. He showed that the full range of color can be seen by combining just three basic colors, a trichromatic system. This idea was later adopted by Helmholtz and is generally referred to as the Young-Helmholtz color theory (Helmholtz 1867).

In the next sentence Mach refers to a short paper of his own, quoted in a footnote. He performed a series of experiments whereby he looked at a rotating cylinder that was covered with squares of either black and white, or different colors. By modifying the size of the squares and the angular velocity of the cylinder, he was able to generate short duration visual stimuli of various kinds. In the discussion of the paper he mentions that we seem to perceive four colors, namely red, yellow, green, and blue instead of only three (Young-Helmholtz).

Actually, the two views can be combined, since we have photoreceptors for three different wavelengths in the retina, but the neuronal code processes color preferentially as opposing colors, namely red/green and yellow/blue.

Experiments by Land suggesting that two colors are sufficient raised more questions than it answered (Land 1959, 1964). The nature of color vision is still an active field of research.

8.30 **Independent stimulation:** Canals had been stimulated separately by electrodes implanted in the labyrinth (Money and Scott 1962). Another useful method is canal plugging, introduced by Ewald[S.1], which leads to the selective loss of afferent modulation in single canals without affecting resting discharge in the nerve. Since any combination of canals can be surgically plugged, one is able to investigate responses of only single canals or canal pairs.

8.31 **Complexes of stimuli:** Mach's contributions to philosophy of science are enormously important, but are not discussed in this little book except in passing. Here, for example, his comments concerning the way man recognizes a complex of stimuli rather than its individual components foreshadows his seminal contribution on logical positivism (Bradley 1971; Holton 1993).

8.32 **Geiger, Lazarus** (1829–1870): Philosopher and philologist in Frankfurt-on-Main known for his aim to prove that the evolution of human reason is closely related to the evolution of language.(Geiger 1868).

8.33 **Analysis of sensation:** "Contributions to Analysis of Sensation" is the title of a later book by Ernst Mach (1886), which was very influential.

8.34 **Quantitative psychology:** Mach's influence on the developing field of psychology foreshadowed the modern approach to psychophysics, in which sensation could be measured and quantified. Mach's ideas were shared closely by William James, who visited him several times. The

emphasis on sensation and perception passed from James to Jacques Loeb, to B. F. Skinner and then to S. S. Stevens, the father of modern psychophysics (Holton 1993; Skinner 1979; Stevens 1975).

8.35 **Two ampullae related as described above:** see: 8.16.

8.36 **Breuer and Brown:** Mach is often associated with Breuer and Brown in the mistaken notion that continuous endolymph flow stimulates the labyrinth nerve. Here he clearly corrects this error and presents an observation consistent with the current understanding of the cupula and hair cells as stiff pressure transducers, requiring only infinitesimal deflections.

8.37 **Complete differential:** In modern terms, such a differential would be called *exact*.

8.38 **Teleologically important:** This term seems to refer to the widespread occurrence of the process of integration with respect to time. Mach has discovered that the semicircular canals follow this principle in generating a sensation of velocity when stimulated by acceleration. His two physical examples both relate to the natural time integration occurring when a force or torque produces acceleration, and thereby velocity. The magnetic needle example entails an electromagnetic torque produced by current in a coil surrounding the needle, and is valid only if the magnet is very lightly damped.

8.39 **Töpler August Joseph Ignaz** (1836–1912): Professor and director of the Physikalisches Institut [Department of Physics] of the Technische Hochschule Dresden, who is best known for his work on electromagnetic phenomena. Prior to that, when Professor in Graz, Boltzmann[8.40] worked in his laboratory on optical methods for determination of dielectric constants, which led to their study of acoustic pressure in the ear (Töpler and Boltzmann, 1870).

8.40 **Boltzmann, Ludwig E.** (1844-1906): Austrian pioneer of modern physics, Bolzmann extended kinetic theory, and is associated with the Maxwell-Boltzmann distribution of the fraction of particles having a given energy. Mach evidently admired his work on statistical mechanics. He was attacked by the logical positivists, probably including Mach's followers, who opposed atomistic theories, and, suffering form depression, committed suicide while on vacation in 1906. Mach refers to results to his work with Töpler[8.39] on optical measurements of air currents (Töpler and Boltzmann 1870).

8.41 **1 Trillion:** The unit 1 billion means 10^{12} in German as in most of the world. In the U.S., however, 10^{12} is 1 trillion, whereas 1 billion equals 10^9.

8.42 **Hearing threshold:** The threshold for subjective sensation of motion has been related to a pressure difference across the cupula of only 1.25 x 10^{-4} dyne/cm^2. This may be compared with estimates of hearing

threshold pressures in the cochlea of 2 x 10^{-4} dyne/cm^2 (Oman and Young, 1972).

8.43 **Thomson, Hans Peter Jörgen Julius** (1826–1909): Danish chemist known for his work in thermochemistry. Thomson performed extensive calorimetric measurements relating heat and energy in chemical reactions.

8.44 **Living energy**: In *The Science of Mechanics*, Mach explains that "living force" (*vis viva*) was a term introduced in 1695 for mv^2 by Leibniz and that *living power* used by Belanger for $\frac{1}{2} mv^2$ is now called *kinetic energy*.

8.45 **Acceleration**: see 3.17.

8.46 **V**: Note that *v* is newly defined, and not the same as *v* on page 118. Here *v* is the velocity of the endolymph relative to the wall, whereas on page 118 *v* was the angular velocity of the bony labyrinth.

8.47 **Equations**: in the second formula there is a stroke-like bar after the constant *k* in the original text, which we believe is an error in the print.

8.48 **Nausea**: see motion sickness, 3.20.

8.49 **Order of animals**: Although Charles Darwin's *The Origin of Species* appeared only 16 years earlier than Mach's current book, in 1859, the notion of the order of animals in evolution was apparently firmly established. Darwin mentions nothing about organs for motion sense directly, but alludes to the possible superiority of "savages" in eyesight and in other senses.

In Darwin's only passing reference to the labyrinth, he quotes Renegger, who observes: "the cavities in the skull for the reception of the several sense-organs are larger in the American aborigines than in Europeans; and this probably indicates a corresponding difference in the dimensions of the organs themselves."

Darwin, Charles (1809–1882): British naturalist and most influential biologist. Developed a general theory of evolution and of the natural selection of species that revolutionized biology. Grandson of Erasmus Darwin[1.1].

8.50 **Swimbladder of fish**: The view of swimbladder function has changed considerably since Mach in 1875. According to J. Rosowski (Harvard Medial School), the swim bladder has become associated with the sense of hearing and sound production as well as buoyancy (Alexander 1966; Henson 1974), and is not associated with vestibular function. In general the sacule of fishes is believed to serve an auditory function. The "Weberian" ossicles (specializations of the cranial vertebrae) are believed to couple sound induced displacements of the air-filled bladder to the lymphs of the inner ear much like the terrestrial ossicles couple motions of the tympanic membrane to the cochlea. This

function is well reviewed in Henson (1974) and in Popper and Fay (1973).

8.51 **Muscles of the tympanic cavity:** Considering the functions of the muscles of the tympanic cavity (M. tensor tympani), Mach was quite mistaken. According to current knowledge, they regulate the sound translation of the middle-ear ossicles from the eardrum to the oval window of the inner ear. Ventilation of the middle ear, in contrast, is accomplished by the Eustachian tube connecting the middle ear and the pharynx, and is regulated by pharynx muscles.

Conclusions

9.1 **Threshold for oscillation:** Given in deg, instead of deg/s^2. Current values obtained with more sophisticated techniques yield lower values: 0.2 deg/s^2 for yaw and 0.5 deg/s^2 for pitch and roll. The acceleration thresholds for apparent motion of a lighted point (oculogyral illusion) are as low as 0.1 deg/s^2 (see Guedry 1974).

Secondary Footnotes

S.1 **Ewald, Richard** (1855–1921): Physiologist at the University of Strasbourg. After studying medicine, he became an assistant in the physiological laboratory of Goltz[0.4], and was promoted to professor in 1886. Ewald invented the method of surgically plugging canals: if during angular acceleration the fluid in a semicircular canal is blocked from flowing or building up a pressure gradient across the cupula, then an acceleration in the plane of that canal will not modulate afferent activity. As the resting discharge of afferent activity is not altered, the animal does not display spontaneous nystagmus at rest. This is a decisive difference from a lesion of the labyrinth, where one observes strong spontaneous nystagmus as a result of the elimination of the resting discharge of the nerve on the lesioned side. (Ewald 1892).

S.2 **Hering, Ewald** (1834-1918): Professor of physiology in Vienna (worked with Joseph Breuer), Prague (successor of Purkinjě and contemporary of Mach), and Leipzig. Next to Helmholtz[6.7], he was one of the most distinguished physiologists of the 19th century who did seminal work on the physiology of breathing, liver function, and memory, as well as vision. He often took a position opposite to that of Helmholtz. Hering particularly stressed that many functions of the

nervous system are inborn, along with some shaping during development. (Hering 1886).

S.3 **Freud, Sigmund** (1856–1939): Studied medicine in Vienna, where he spent most of his professional life until he was forced to emigrate because of his religion as well as his profession. Freud left a voluminous work ranging from anatomical research conducted as a student, and several neurological investigations including a book on aphasia, to the seminal study on hysteria published with J Breuer[0.5]. All his subsequent publications relate to psychoanalysis.

S.4 **Lunar Society:** Members of the Society met for dinner on the Monday nearest to the full moon to give them light on their way home. Members were:

>   Matthew Boulton (steam engine)
>   Erasmus Darwin (physician)
>   Thomas Day (political thought)
>   Richard Lovell Edgeworth (educator and inventor)
>   Samuel Galton   (naturalist)
>   Robert Augustus Johnson (educator)
>   James Keir (chemist)
>   Jospeh Priestly (chemist)
>   William Small (chemist)
>   Jonathan Stokes (naturalist)
>   James Watt (steam engine)
>   Josiah Wedgewood (ceramics industry)
>   Johne Whitehurst (industrialist)
>   William Withering (botanist)

This assembly of illustrious names, which had no president and kept no records, reflects the time of transition from a rural to an industrial society. One can only speculate as to what extent novel ideas and concepts were developed and cross-fertilized among the members of this society. (Schofield 1963).

S.5 **Nansen, Fridtjof** (1861-1930): Norwegian explorer and naturalist, widely known from his popular travel and expedition books. Nansen undertook several expeditions to the Arctic Ocean and (in 1895) got as far as 86°14' latitude, further towards the North Pole than anyone before him. In 1896 Nansen became professor of zoology, later of oceonagraphy, and was appointed ambassador in London 1906-1908.

After the First World War, Nansen engaged himself politically, especially for the return of prisoners of war. His actions were highly regarded and he received the Nobel Prize for Peace in 1923. He was appointed high commissioner of the League of Nations, 1921-1923. He was responsible for creating a special passport for stateless persons,

which subsequently was accepted as travel documents by most countries, and became popularly known as the 'Nansen Passport'.

S.6 **Coriolis, Gustave Gaspard** (1792-1843): Professor at the École polytechnique in Paris. Today mostly known from the term Coriolis force, a mechanical (inertial) force that acts on bodies moving in a reference frame that is rotating with respect to an inertial system.

S.7 **Painful hearing:** At the time of Flourens[0.3], the prevailing and unchallenged assumption was that the semicircular canals have a function in directional hearing. Only after Ménière's[5.2] seminal clinical studies was the connection between the symptoms of vertigo or dizziness and disease of the inner ear slowly accepted.

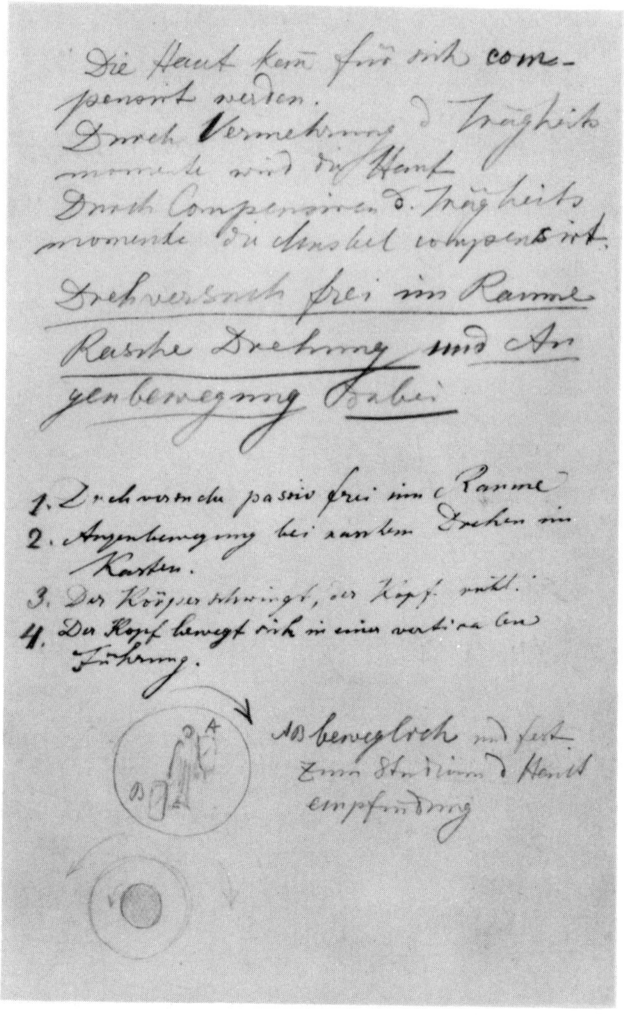

Die Haut kann für sich compensirt werden.
Durch Vermehrung d. Trägheitsmomente wird die Haut
Durch Compensiren d. Trägheitsmomente die Muskel compensirt.
Drehversuch frei im Raume
Rasche Drehung und Augenbewegung dabei
1. Drehversuche passiv frei im Raume
2. Augenbewegung bei raschem Drehen im Kasten.
3. Der Körper schwingt, der Kopf nicht.
4. Der Kopf bewegt sich in einer verticalen Führung.
AB beweglich und fest
Zum Studium d. Hautempfindung

Figure A-8: Notes on experiments regarding skin and muscle senses. "The skin can be compensated separately. By increasing the moment of inertia the skin will be compensated, and by compensating the moments of inertia the muscles will be compensated. <u>Turning experiment free in space. Rapid turning accompanied by eye movements</u>. 1. Turning experiment passively free in space 2. Eye movements in the case of rapid turning in the box. 3. The body swings, not the head. 4. The head moves in a vertical guide. *AB* moveable and fixed for the study of skin sensation." (from Mach's notebook, 1874; photograph courtesy of E. Mach Archive, Deutsches Museum, Munich).

References

Alexander, R. M. (1966). Physical aspects of swimbladder function. *Bio. Rev.* 41: 141–176.

Atkinson, M. (1961). Ménière's Original Papers, reprinted with an English translation together with commentaries and biographical sketch. *Acta Oto-Laryngologica*, Suppl 162: 1–78.

Atwood, G. (1784). *A Treatise on the Rectilinear Motion of Rotation of Bodies With a Description of Original Experiments Relative to the Subject.* Cambridge.

Aubert, H. (1861). Eine scheinbar bedeutende Drehung von Objekten bei Neigung des Kopfes nach rechts oder links [An apparent considerable rotation of objects with tilting the head to the right or to the left]. *Virchows Arch.* 20: 381–393.

Berthold, E. (1875). Ueber die Function der Bogengänge des Ohrlabyrinths. *Arch Ohrenheilkunde*, 9 (new series, vol 3): 77–95.

Berthoz, A. (1997). *Les sens du movement.* Paris: Editions Odile Jacob.

Berthoz, A., Pavard, B., & Young, L. R. (1975). Perception of linear horizontal self-motion induced by peripheral vision (linearvection): Basic characteristics and visual-vestibular interactions. *Exp. Brain Res.* 23: 471–489.

Blackmore, J. T. (1972). *Ernst Mach; His Work, Life, And Influence.* Berkeley: University of California Press.

Blackmore, J. T. (1992). *Ernst Mach, a Deeper Look: Documents and New Perspectives.* Dordrecht, The Netherlands: Kluwer Academic Publishers.

Blanks, R. H. I., Curthoys, I. S., & Markham, C. H. (1975). Planar relationships of the semiciruclar canals in man. *Acta Otolaryngol* 80: 185–196.

Blumenthal, A. L. (1975). A re-appraisal of Wilhelm Wundt. *American Psychologist* 1081–1088.

Boettcher, A. (1869). *Ueber Entwicklung und Bau des Gehörlabyrinths nach Untersuchungen an Säugethieren. 1. Theil [On development and anatomy of the hearing labyrinth in mammals, part 1.]* Dresden: E Blochmann & Sohn.

Bradley, J. (1971). *Mach's Philosophy of Science.* Oxford: Oxford University Press.

Breuer, J. (1868). Die Selbststeuerung der Athmung durch den Nervus vagus [The autonomous control of breathing by the vagal nerve, i.e. the Hering-Breuer reflex.] *Sitzungsb Akad Wiss, math-naturwiss Klass, Abt II*, 58: 909–937.

Breuer, J. (1874). Ueber die Funktion der Bogengänge des Ohrlabyrinths [About the functions of the semicircular canals of the ear labyrinth.] *Med Jahrbücher 2nd series*, 4: 72–124.

Breuer, J., & Freud, S. (1895). *Studien über Hysterie [Studies on Hysteria.]* Leipzig-Wien: Franz Deuticke.

Breuer, J. (1925). *Autobiographie Josef Breuer.* Wien.

Brown, A. C. (1874). On the sense of rotation and the anatomy and physiology of the semicircular canals of the internal ear. *J. Anat. Physiol.*, 8: 327–331.

Büttner-Ennever, J., Büttner, U., & Brandt, T. (1998). Volker Henn. *Audiol. Neurootol.*, 3: 61–62.

Cahan, D. (1993). Hermann von Helmholtz and the Foundations of Nineteenth-Century Science. Berkeley: University of California Press.

Cardus, D., & McTaggert, W. G. (1997). Cardiovascular effects of a sustained -Gz force in the horizontal position. *Aviat. Space Environ. Med.*, 12: 1099–103.

Cohen, B. (1984). Erasmus Darwin's observations on rotation and vertigo. *Hum. Neurobiol.*, 3: 121–8.

Cohen, B. (1999). Tribute to Volker Henn. In B. Cohen, & B. J. M. Hess (eds.), *Otolith Function in Spatial Orientation and Movement.* (pp. xiii–xiv). New York: Annals of the New York Academy of Sciences.

Curthoys, I. S., Betts, G. A., Burgess, A. M., MacDougall, H. G., Cartwright, A. D., & Halmagyi, G. M. (1999). The planes of the utricular and saccular macculae of the guinea pig. In B. Cohen, & B. J. M. Hess (Eds.), *Otolith Function in Spatial Orientation and Movement* (pp. 27–34). New York: Annals of the New York Academy of Sciences.

Cyon, E. D. (1873). Die Funktionen der halbzirkelförmingen Bogengänge. *Archiv f. Physiologie und Anaotomie von du Bois-Reymond Engelmann 8.*

Cyon, E. D. (1878). Recherches experimentales sur les fonctions des canaux semi-circulaire et sur leur rôle dans la formation de la notion de l'espace. *Bibl. De l'École des Hautes-Etudes Paris. Section des Sciences Naturelles* 18.

Cyon, E. d. (1911). *L'Oreille: Organe d'oreintation dans le temps et dans l'espace.* Paris: Librarie Félix Alcan.

Darwin, C. (1859). *The Origin of Species,* Modern Library Edition. New York: Random House.

Darwin, E. (1794). *Zoonomia.* London: J Johnson.

Dichgans, J. T., & Brandt, J. J. (1973). Optokinetic motion sickness and pseudo-Coriolis effects induced by visual stimuli. *Acta Otolaryngol.* 76: 3339–3348.

Dieterich, M., Zink, R., Weiss, A., & Brandt, T. (1999). Galvanic stimulation in bilateral vestibular failure: 3-D ocular motor effects. *Neuroreport* 10: 3283–3287.

Einstein, A. (1916). Ernst Mach. *Phys. Zs.* 17: 101–104.

Enriques, F. (1906). *Problemi della scienza* [Engl. transl.: Royce, K. (1914): Problems of Science, London-Chicago, Open Court]. Bologna: Zanichelli.

Ewald, J. R. (1892). *Physiologische Untersuchungen ueber das Endorgan des Nervus octavus* [Physiological investigation of the endorgan of the eighth nerve]. Wiesbaden: J.F. Bergmann.

Fernandez, C., & Goldberg, J. M. (1971). Physiology of the peripheral neurons innervating the semicircular canals of the squirrel monkey. The response to sinusoidal stimulation and dynamics of the peripheral vestibular system. *J. Neurophysiol.* 34: 661–675.

Fernandez, C., & Goldberg, J. M. (1976). Physiology of peripheral neurons innervating otolith organs of the squirrel monkey. I. Response to the static tilts and to long-duration centrifugal force. *J. Neurophysiol.* 39: 970–984.

Fetter, M., Haslwanter, T., Misslisch, H., & Tweed, D. (1997). Three-Dimensional Kinematics Of Eye, Head, and Limb Movements. Amsterdam: Harvard Academic Publishers.

Flourens, M. J. P. (1825). *Expériences sur le système nerveux.* Paris: Crevot.

Flourens, P. (1830). Expériences sur les canaux semi-circulaires le l'oreille, dans les oiseaux. *Mem. Acad. Roy. Sci.* 9:455–497.

Flourens, P. (1842). *Recherches expérimentales sur les propriétés et les fonctions du systeme nerveux dans les animaux vertébrés.* Paris: Crevot. 2nd ed. (1842): Balliere. [Only the 2nd edition fully described the relevant experiments in which single semicircular canals were lesioned; chapter XXVII.]

Fritsch, T., & Hitzig, E. (1870). Ueber die elektrische Erregbarkeit des Grosshirns [About the electrical excitability of the cerebral cortex]. *Arch. Anat. Physiol. Leipzig* 300–332.

Fukuda, T. (1961). Studies on human dynamic postures from the viewpoint of postural reflexes. *Acta Oto-Laryng. Suppl.* 161,: 1–52.

Geiger, L. (1868). *Ursprung und Entwicklung der meschlichen Sprache und Venunft (vol. i).* Stuttgart.

Gillingham, K. K., & Previc, F. H. (1996). Spatial disorientation in flight. In R. L. DeHart (ed.), *Fundamentals in Aerospace Medicine 2nd ed.* (pp. 309–397). Baltimore: Williams and Wilkins.

Gizzi, M., Raphan, T., Rudolph, S., & Cohen, B. (1994). Orientation of human optokinetic nystagmus to gravity: A model-based approach. *Exp. Brain Res.* 99: 347–360.

Goldberg, J. M., & Fernandez, C. (1984). The vestibular system. In *Handbook of Physiology* (pp. 977–1022). Bethesda, Maryland: American Physiological Society.

Goltz, F. (1870). Ueber die physiologische Bedeutung der Bogengänge des Ohrlabyrinths. [On the physiological significance of the semicircular canals of the ear labyrinth]. *Pflügers Arch Physiol.* 3: 172–192.

Graf, W., & Simpson, J. I. (1981). Eye muscle geometry and compensatory eye movements in lateral-eyed and frontal-eyed animals. In B. Cohen (ed.), *Vestibular and Oculomotor Physiology* (pp. 20-30). New York: Annals of the New York Academy of Sciences.

Graybiel, A., & Brown, R. H. (1951). The delay in visual reorientation following a change in direction of the resultant force on a human centrifuge. *J. Gen. Psychol.* 45: 143–150.

Graybiel, A., Miller II, E. F., Newsom, B. D., & Kennedy, R. S. (1968). The effect of water immersion on perception of the oculographic illusion in nomral and labyrinthine defective suubjects. *Acta Oto-Laryngol.* 19: 524–530.

Groen, J. J. (1957). Adaptation. *Pract. Otol-Rhino-Laryngol.* 19: 524–530.

Grüsser, O. J. (1984). J.E. Purkyně's contributions to the physiology of the visual, the vestibular and the oculomotor systems. *Hum. Neurobiol.* 3: 129–144.

Guedry, F. E., & Lauver, L. S. (1961). Vestibular reactions during prolonged constant angular acceleration. *J. Appl. Physiol.* 16: 215–220.

Guedry, F. E. J. (1974). Psychophysics of vestibular sensation. In H. H. Kornhuber (ed.) *Handbook of Sensory Physiology, Vestibular System, VI (2): Psychophysics, Applied Aspects and General Interpretations* (pp. 3–154). New York: Springer Verlag.

Helmholtz, H. (1867). *Handbuch der Physiologischen Optik.* Leipzig: Leopold Voss. [Translated in English (1924, 1962) *Treatise on Physiological Optics.* New York, Dover].

Henn, V. (1984). E. Mach on the analysis of motion sensation. *Hum. Neurobiol.* 3: 145–148.

Henn, V. (1987). Vestibular system. In G. Adelman (ed.) *Encyclopedia of Neuroscience.* Boston: Birkhäuser.

Henn, V., Cohen, B., & Young, L. R. (1980). Visual-vestibular interaction in motion perception and the generation of nystagmus. *Neurosci. Res. Program Bull.* 18: 457–651.

Henn, V., & Young, L. R. (1975). Ernst Mach on the vestibular organ 100 years ago. *ORL J. Otorhinolaryngol. Relat. Spec.* 37: 138–148.

Henson, O. W. (1974). Comparative anatomy of the middle ear. In W. D. Kiedel & W. D. Neff (eds.) *Handbook of Sensory Physiology: The Auditory System Vol. V/1* (pp. 39–110). New York: Springer-Verlag.

Hepp, K., Van Opstal, A. J., Suzuki, Y., Straumann, D., Hess, B. J. M., & Henn, V. (1997). Listing's Law: visual, motor, or visuo-motor. In M. Fetter, T. Haslwanter, H. Misslisch, & D. Tweed (eds.) *Three-Dimensional Kinematics of Eye, Head, and Limb Movements* (pp. 34–42). Amsterdam: Harvard Academic Publishers.

Hering, E. (1886). *Die Lehre vom binocularen Sehen.* Leipzig: W. Engelmann. [translated by: Bridgeman, B., Stark, L. (1977). *The theory of binocular vision.* New York: Plenum Press].

Hess, B. J. M., & Angelaki, D. E. (1999). Inertial processing of vestibular-ocular signals. In B. Cohen & B. J. M. Hess (eds.) *Otolith Function in Spatial Orientation and Movement* (pp. 148–161). New York: Annals of the New York Academy of Science.

Hirn, G. A. (1858). Recherches expérimentales sur la valeur de l'équivalent méchanique de la chaleur. [Experimental studies on the amount of the mechanical equivalent of heat]. *Revue d'Alcace.*

Hitzig, E. (1871). Ueber die beim Galvanisieren des Kopfes entstehenden Störungen der Muskelinnervation und der Vorstellungen vom Verhalten im Raume [About perturbations of muscle innervation and imaginations of behavior in space during head galvanisation]. *Reichert u. du Bois-Reymond's Arch.* 716–770.

Holton, G. (1993). *Science and Antiscience.* Cambridge, MA: Harvard University Press.

Horsley, V. A. H., & Clarke, R. H. (1908). The structure and functions of the cerebellum examined by a new method. *Brain* 31: 45–124.

Howard, I. P. (1982). *Human Visual Orientation.* Chichester: John Wiley & Sons.

Hudspeth, A. J., & Corey, D. P. (1977). Sensitivity, polarity and conduction change in the response of vertebrate hair cells to controlled mechanical stimuli. *Proc. Natl. Acad. Sci.* 746: 2407–2411.

Knight, T. A. (1806). On the direction of the Radicle and Germen during the vegetation of seeds. *Phil. Trans.* 99–108.

Koenig, E., Dichgans, J., & Dengler, W. (1987). Pursuit opposite to the vestibulo-ocular reflex (VOR) during sinusoidal stimulation in humans. *Acta Otolaryngol* 103: 24–31.

Kron, G. (1975). Advanced simulation in undergraduate pilot training: G-seat development. In A. F. B. Wright-Patterson (ed.) *Ohio, Air Force Human Resource Laboratory, Technical Report 75-59 (II).*

Land, E. H. (1959). Experiments in color vision. *Scientific American* 200: 84–99.

Land, E. H. (1964). The retinex. *American Scientist* 54: 247–264.

Lithgraw, J. D. (1920). *J. Laryngol*, 3: 81.

Mach, E. (1886). *Betraege zur Analyse der Empfindungen.* Jena: G. Fischer. [Translated by Williams, C.M. (1897, reprinted 1984) *Contributions to the Analysis of the Sensations.* Chicago: The Open Court Publishing Co.]

Mach, E. (1893). *The Science of Mechanics - a Critical and Historical Exposition of its Principles* [Translated by McCormack, T.J., from the second German edition: *Die Mechanik in ihrer Entwicklung, Historisch-Kritisch Dargestellt*]. Chicago: Open Court Publishing Co.

Mach, E. (1897). Über Orientierungsempfindungen. *Vorträge des Vereines zur Verbreitung naturwissenschaftlicher Kenntnisse in Wien*, XXXVII, Heft 12.

Malcolm, R, & Melvill-Jones, G. (1974). Erroneous perception of vertical motion by humans seated in the upright position. *Acta Otolaryngol. (Stockh.)* 77: 274–283.

Malcolm, R., & Melvill-Jones, G. (1970). A quantitative study of vestibular adaptation in humans. *Acta Oto-Laryngol.* 70: 126–135.

Mayne, R. (1974). A systems concept of the vestibular organs. In H. H. Kornhuber (ed.) *Handbook of Sensory Physiology: The Vestibular System* (pp. 493–580). New York: Springer Verlag.

Melvill-Jones, G., & Young, L. R. (1978). Subjective detection of vertical acceleration: A velocity-dependent response? *Acta Otolaryngol. (Stockh.)* 85: 45–53.

Melvill-Jones, G., Barry, W., & Kowalsky, N. (1964). Dynamics of the semicircular canals compared in yaw, pitch, and roll. *Aerospace Med.* 35: 984–989.

Ménière, P. (1861). Mémoire sur des lésions de lóreille interne donnant lieu à des symptômes de congestion cérébrale apoplectiforme. *Gaz. Méd. Paris* 16: 379–380.

Merfeld, D. M. (1966). When does the axis of the VOR shift? *Brain Research Bulletin* 40: 318–319.

Merfeld, D. M., Young, L. R., Tomko, D. L., & Paige, G. D. (1991). Spatial orientation of VOR to combined vestibular stimuli in squirrel monkey. *Acta Otolaryngo.l (Suppl.)* 481: 287–292.

Mittelstaedt, H. (1996). Somatic graviception. *Biological Psychology* 42: 53–74.

Money, K. E., & Scott, J. W. (1962). Functions of separate sensory receptors of non-auditory labyrinth of the cat. *Am. J. Physiol.* 202: 1211–1220.

Moore, D. M., Hoffman, L. F., Beykirch, K., Honrubia, V., & Baloh, R. W. (1991). The electrically evoked vestibulo-ocular reflex: I. Normal subjects. *Otolaryngol. Head Neck Surg.* 104: 219–224.

Müller, G. E. (1916). Ueber das Aubert'sche Phänomen [About Aubert's phenomenon]. *Z. Psychol. Physiol. Sinnesorg.* 49: 109–246.

Müller, J. (1826). *Zur vergleichenden Physiologie des Gesichtssinnes der Menschen und Thiere nebst einem Versuch über die Bewegungen der Augen und über den menschlichen Blick [The comparative physiology of the visual sense in men and animals and a treatise about the movements of the eyes and about the human gaze]*. Leipzig: C Cnobloch.

Müller, J. (1834–1840). *Lehrbuch der Physiologie des Menschen.* Wiesbaden: J Hölscher.

Necker, L. A. (1832). Observations on some remarkable phenonoma seen in Switzerland; and an optical phenomenon which occurs on viewing of a

crystal or geometrical solid. *Philosophical Magazine, 3rd series* 1: 329–337.

Oberndorf, C. P. (1953). Autobiography of Josef Breuer [translated and edited]. Int. J. Psychoanalysis 34: 64-67.

Oman, C. M. (1990). Motion sickness: A synthesis and evaluation of the conflict theory. *Canadian J. Physiol. Pharmacol.* 68: 294–303.

Oman, C. M., & Young, L. R. (1972). The physiological range of pressure difference and cupula deflections in the human semicircular canal. *Acta Otolaryngol.* 74: 324–331.

Paige, G. D. (1989). The influence of target distance on eye movement responses during vertical linear motion. *Exp. Brain Res.* 77: 585–583.

Paige, G. D., & Tomko, D. L. (1991). Eye movement responses to linear motion in the squirrel monkey. I. Basic characteristics. *J. Neurophysiol.* 65: 1170–1182.

Peng, G. C., Hain, T. C., & Peterson, B. W. (1996). A dynamic model for reflex activated head movements in the horizontal plane. *Biological Cybernetics* 75: 309–319.

Pfaltz, C. R. (1969). The diagnostic importance of the Galvanic test in otoneurology. *Pract. Oto-Rhino-Laryng.* 31: 193–203.

Plücker, J. (1853). Noch ein Wort ueber die Fessel'sche Rotationsmaschine. *Poggendorffs Annalen der Physik und Chemie* 90: 348–350.

Plücker, J. (1853). Ueber die Fessel'sche Rotationsmaschine. *Poggendorffs Annalen der Physik und Chemie* 90: 174–177.

Poinsot, L. (1834). *Théorie nouvelle de la rotation des corps présentée à l'institute le 19 mai 1834. [English translation by Whitley, C.: Outlines of a new theory of rotatory motion, Cambridge, England: Newby, 1834.]* Paris: Bachelier.

Popper, A. N., & Fay, R. R. (1973). Sound detection and processing by fish: A critical review. *J. Acoust. Soc. Am.* 53: 1515–1529.

Popper, A. N., & Platt, C. (1993). Inner ear and lateral line. In D. H. Evans (ed.) *The Physiology of Fishes.* Boca Raton: CRC Press.

Purkinje, J. E. (1820). Beyträge zur näheren Kenntniss des Schwindels aus heautognostischen Daten [Contributions towards a better understanding of vertigo from psychophysical data]. *Med Jahrb k u k österr Staates* 6: 79–125.

Raphan, T., Matsuo, V., & Cohen, B. (1979). Velocity storage in the vestibulo-ocular reflex arc (VOR). *Exp. Brain Res.* 35: 229-48.

Reason, J. T., & Brand, J. J. (1969). *Motion Sickness.* London: Academic Press.

Ritter, J. W. (1803). Ueber die Anwendung der Voltaischen Säule [About the use of the Voltaic column]. *Journal der practischen Heilkunde von CW Hufeland, Berlin* 17(part 3): 30–53.

Rollmann, W. (1868). Pseudoskopische Erscheinungen. *Poggendorffs Annalen der Physik* 134.

Ruben, R. J. (1989). The development and acceptance of the association of diseases of the ear and disorders of balance. In J. B. Nadol JR (ed.) *Second International Symposium on Ménière's Disease* (pp. 3–11). Amsterdam: Kugler & Ghedini Publishers.

Schofield, R. E. (1963). *The Lunar Society of Birmingham. A social history of provincial science and industry in eighteenth-century England.* Oxford: Clarendon Press.

Sedgwick, H. A. (1986). Space Perception. In K. R. Boff, L. Kaufman, & J. P. Thomas (eds.) *Handbook of Perception and Human Performance.* New York: Wiley.

Sherrington, C. S. (1918). *Brain* 41: 332.

Shiers, G. (1971). The induction coil. *Scientific American*, 224, 80-87.

Skinner, B. F. (1979). *The Shaping of a Behaviorist.* New York: Knopf.

Smee, A. (1840). On the galvanic properties of metallic elementary bodies, with a description of a new chemico-mechanical battery. *London, Philosophical Magazine. Series III*, 16.

Smee, A. (1849). *Elements of Electro-Biology.* London: Longman, Brown, Green and Longmans.

Stevens, J. K., Emerson, R. C., Gerstein, G. L., Kallos, T., Neufeld, G. R., Nichols, C. W., & Rosenquist, A. C. (1976). Paralysis of the awake human: Visual perceptions. *Vision Res.* 16: 93–98.

Stevens, S. S. (1975). *Psychophysics.* New York: Wiley.

Straumann, D. (1999). In memoriam Volker Henn. *J. Vest. Res.* 9: 155.

Suzuki, J. I., & Cohen, B. (1966). Integration of semicircular canal activity. *J. Neurophysiol.* 29: 981–995.

Suzuki, J. I., Goto, K., Tokumasu, K., & Cohen, B. (1969). Implantation of electrodes near individual vestibular nerve branches in mammals. *Ann. Otol. Rhinol. Laryngol.* 7: 815–827.

Thomson, H. P. J. J. (1882–88). *Thermochemische Untersuchungen [Translated to English by Burke, K.A.: Thermochemistry].* Leipzig.

Töpler, A., & Boltzmann, L. E. (1870). Ueber eine neue otische Methode, die Schwingungen tönender Luftsäulen zu analysiren. *Poggendorffs Annalen der Physik und Chemie* 141: 321–352.

Van Egmond, A. A. J., Groen, J. J., & Longkees, L. W. B. (1949). The mechanics of the semicircular canal. *J. Physiol. London* 110: 1–17.

Waespe, W., & Henn, V. (1977). Neuronal activity in the vestibular nuclei of the alert monkey during vestibular and optokinetic stimulation. *Exp. Brain Res.* 27: 523–538.

Wundt, W. (1873–74). *Grundzüge der Physiologischen Psychologie.* Leipzig: Engelmann [Translation of the 5th edition: Titchener E.: Principles of Physiological Psychology].

Yasui, S., & Young, L. R. (1975). Perceived visual motion as effective stimulus to pursuit eye movement system. *Science* 190: 906-908

Young, L. R. (1983). Perception of the body in space: Mechanisms. In *Handbook of Physiology – The Nervous System III, (pp. 1023–1066).* Bethesda: American Physiological Society.

Young, L. R., & Oman, C. M. (1969). Model for vestibular adaptation to horizontal rotation. *Aerosp. Med.* 39: 606–608.

Yue, Q., Straumann, D., & Henn, V. (1994). Three-dimensional characteristics of rhesus monkey vestibular nystagmus after velocity steps. *J. Vestib. Res.* 4: 313–323.

Figure A-9: Henn (left) and Young working on this book, using their laptop computers at a Roman ruin in Portugal, November 1997 *(photograph courtesy of Jody Young).*

Index